.

# Diffraction of Electromagnetic Waves by Small Apertures

## The ACES Series on Computational and Numerical Modelling in Electrical Engineering

Andrew F. Peterson, PhD – Series Editor

The volumes in this series will encompass the development and application of numerical techniques to electrical and electronic systems, including the modelling of electromagnetic phenomena over all frequency ranges and closely related techniques for acoustic and optical analysis. The scope includes the use of computation for engineering design and optimization, as well as the application of commercial modelling tools to practical problems. The series will include titles for senior undergraduate and postgraduate education, research monographs for reference, and practitioner guides and handbooks.

## Titles in the Series

K. Warnick, **"Numerical Methods for Engineering,"** 2010

W. Yu, X. Yang and W. Li, **"VALU, AVX and GPU Acceleration Techniques for Parallel FDTD Methods,"** 2014.

A.Z. Elsherbeni, P. Nayeri and C.J. Reddy, **"Antenna Analysis and Design Using FEKO Electromagnetic Simulation Software,"** 2014.

A.Z. Elsherbeni and V. Demir, **"The Finite-Difference Time-Domain Method in Electromagnetics with MATLAB Simulations 2nd Edition,"** 2015.

M. Bakr, A.Z. Elsherbeni and V. Demir, **"Adjoint Sensitivity Analysis of High Frequency Structures with MATLAB,"** 2017.

O. Ergul, **"New Trends in Computational Electromagnetics,"** 2019.

D. Werner, **"Nanoantennas and Plasmonics: Modelling, design and fabrication,"** 2020.

K. Kobayashi and P.D. Smith, **"Advances in Mathematical Methods for Electromagnetics,"** 2020.

V. Lancellotti, **"Advanced Theoretical and Numerical Electromagnetics, Volume 1: Static, stationary and time-varying fields,"** 2021.

V. Lancellotti, **"Advanced Theoretical and Numerical Electromagnetics, Volume 2: Field representations and the method of moments,"** 2021.

S. Roy, **"Uncertainty Quantification of Electromagnetic Devices, Circuits, and Systems,"** 2021.

# Diffraction of Electromagnetic Waves by Small Apertures

Applications to transmission, absorption, and scattering resonances

Young Ki Cho

**ACES**

The Institution of Engineering and Technology

Published by SciTech Publishing, an imprint of The Institution of Engineering and Technology, London, United Kingdom

The Institution of Engineering and Technology is registered as a Charity in England & Wales (no. 211014) and Scotland (no. SC038698).

The Institution of Engineering and Technology
Futures Place
Kings Way, Stevenage
Hertfordshire SG1 2UA, United Kingdom

www.theiet.org

**British Library Cataloguing in Publication Data**
A catalogue record for this product is available from the British Library



ISBN 978-1-78561-809-3 (hardback)
ISBN 978-1-78561-810-9 (PDF)



Typeset in India by MPS Limited


Cover Image: Ocean Waves/Saemilee/DigitalVision Vectors via Getty Images

# Contents

# About the author

**Young Ki Cho** is an emeritus professor at Kyungpook National University, where he worked in the Department of Electronics until his retirement in 2020. He taught applied electromagnetics including antenna and microwave engineering. In 2008 he served as the president of the Korean Institute of Electromagnetic Engineering and Science. From 1994 to 2012, he also served as the official member of commission B (fields and waves) and the president of the Korea URSI committee consecutively.

# Preface

This book deals with low-frequency diffraction characteristics of small aperture structures such as a narrow slit and a small hole and their periodic structures with a main emphasis on the transmission of maximum phenomena through those structures. As is well-known, a narrow slit structure in a conducting plane has been used as a simple model for a narrow slot planar antenna for example, whereas a small hole structure has been widely used as an aperture-coupling element in a transmission cavity filter or a directional coupler in the microwave regime.

In writing this book, this author has attempted to provide a guidebook that will be useful in understanding a wide variety of resonance-related device technologies in the microwave and optics areas. Therefore, the structure of this book is loosely divided into three parts: (1) transmission resonance (Chapters 3, 4, and 5), (2) absorption resonance (Chapter 6), and (3) scattering resonance (Chapter 7).

Two points are worthy of note. The first is that, for the sake of investigating various underlying physics of diffraction phenomena of a narrow slit structure, we deal with a narrow slit formed in a conducting plane over a grounded dielectric (Chapter 2), that is, a narrow slit loaded with guiding structure of a parallel-plate waveguide to include a feeding structure rather than simple narrow slit structure in a conducting plane. This is because the former is much more preferred in an equivalent circuit representation for various evaluations such as the transmission resonance via aperture coupling to nearby scatterers, if any, as well as impedance matching characteristics. Another point is that Chapter 7 dealing with the scattering resonance begins with an analysis of the reflection-type grating of periodic strip over a grounded dielectric with our main interest centered on the resonant Wood's anomaly, which corresponds to the Bragg blazing phenomena. Then we analyze the transmission-type grating of double periodic strips filled with a dielectric slab between them and discuss the essential similarity between the scattering resonance and guided mode resonance (GMR).

Chapter 8, the final chapter and as a supplement, discusses the transmission characteristics through a finite or an infinite array of apertures in both parallel and collinear configurations with the main concern centered on the relation between transmission cross section (TCS) and the directivity and add some discussions on enhanced transmission through single aperture and its periodic array for both perfect electric conductor (PEC) and real metal planes as well as comparison between extraordinary optical transmission (EOT) and GMR.

The key feature of this book is summarized as follows. First, throughout the book, the equivalent circuit representations are used for the explanation of transmission resonance (Chapters 3–5), absorption resonance (Chapter 6), and scattering resonance (Chapter 7) as well as diffraction by the narrow slit (Chapter 2).

The transmission resonance problem through a single aperture can be viewed and analyzed as an impedance matching problem whether it is a transmission resonant aperture (TRA) type or a transmission resonant cavity (TRC) type. In the case of an absorption resonance problem made of a single aperture coupled to a lossy cavity, assuming that we focus our concern on the maximum power transmitted through the input aperture into the lossy cavity structure, this absorption problem can be viewed as an impedance matching problem. This is because the maximum power absorbed into the lossy cavity is approximately equal to the maximum transmitted power through the transmission resonant cavity. It is important to note that the main concern should be focused on the high throughput in the case of a transmission resonance problem, whereas it should be focused on the high speed of operation in the case of an absorption resonance problem.

The concepts of impedance matching in both the transmission resonance in a single aperture (Chapters 3–5) and the absorption resonance through a single small hole coupled to a lossy cavity (Chapter 6) can be extended to their counterpart of periodic array type. In order words, composing the periodic structure of TRA or non-TRA leads to the general FSS structure whereas composing the periodic structure of the absorption resonant structure and then eliminating the four sidewalls of the lossy cavity structure leads to a kind of absorbing FSS (Chapter 6). In these periodic structure cases, the former impedance matching concept for the single TRA and single absorption resonant structures is employed to explain and present numerical results. The bandpass filter structure of GMR type associated with the scattering resonance condition in the periodic leaky waveguide can be viewed as another total transmission problem.

Due to the reasoning described above, the three different resonances (transmission, absorption, and scattering) under consideration can be viewed the same as an impedance matching problem, which concerns about efficient transmission of electromagnetic energy, regardless of its structure and configuration: the single aperture transmission problems (Chapters 3–5), the absorption resonance problem through a small hole (Chapter 6), periodic structures corresponding to the transmitting FSS (Chapter 8), the absorbing FSS (Chapter 6), or the GMR type of bandpass filter (Chapter 7). Most chapters are written in this context, which corresponds to the second key feature of this book.

In particular, the total transmission through the square periodic array of Bethe's small holes in an infinite thin conducting plane is discussed in comparison with the total transmission through Bethe's single hole to emphasize the essential similarity between these total transmission resonance phenomena and the extraordinary optical transmission (EOT) phenomena reported by Ebbesen *et al.*

From the study on the total transmission (corresponding to the transmission resonance) through Bethe's small hole in a waveguide, it has been found that this total transmission is associated with Rayleigh wavelength defined for the magnetic current source array along the E-plane. Along with the resonant type of Wood's anomaly (in Chapter 7), this type of Rayleigh wavelength is found to correspond to one of the two types of Wood's anomalies proposed by Hessel and Oliner.

It is worthwhile to mention that in the reflection grating made of a periodic strip over grounded dielectric (in Chapter 7), the resonant Wood's anomaly in which specularly reflected power is transformed into the first-order mode power under the assumption of propagating modes of $n = -1$ and 0 is shown to occur. The reason for the choice of periodic strip grating over the grounded dielectric is to deal with the physically realizable structure rather than the original theoretical model proposed by Hessel and Oliner. For both types of Wood's anomalies, Fano-like resonance, which shows a rapid change from the transmission power maximum to zero transmission, is observed. In addition, such transmission resonance phenomena are attributed to the coupling of an incident wave to the bound mode to the surface such as quasi bound state (QBS) mode type for the Rayleigh wavelength type or the leaky wave mode for the resonant type. Between the two types, the resonant type of Wood's anomaly is discussed in connection with the guided mode resonance (GMR).

It is hoped that this book may help the students and researchers in this applied electromagnetics area to understand the underlying physics of the various resonance phenomena in microwave and optics areas.

My sincere appreciation goes to the School of Electronics Engineering, College of Information Technology (IT), Kyungpook National University (KNU) for their support of my writing this book. Many valuable comments, critiques, and creative ideas received from former graduate students at the School of Electronics Engineering are very much appreciated.

My thanks go to Professors Jong Ig Lee of Dongseo University, Ji Whan Ko of Kumoh National University of Technology, Ki Chai Kim of Yeungnam University, Sang Wook Nam of Seoul National University, Hong Koo Kim of the University of Pittsburgh, and Hyo Joon Eom and Seong Ook Park of Korea Advanced Institute of Science and Technology (KAIST) for sharing valuable discussions during the preparation of this manuscript. Excellent assistance from Dr. Seung Soon Kwak in typewriting the manuscript and drawing figures is greatly acknowledged. I also wish to express my gratitude for the patience and support of my family during the writing of this book. Finally, any comments and critiques regarding the text would be gratefully received.

Daegu, Korea
Young Ki Cho

# Glossary

Aperture body resonance (ABR)

- ABR means a kind of resonance phenomenon in which the transmitted power through an electrically small aperture becomes much larger when the nearby scattering body near the aperture is present than when the body is absent.

Absorption cross section (ACS)

- ACS of the apertures is defined as the area for which the incident wave contains the same power as absorbed inside the lossy cavity coupled to the aperture.

Absorption and scattering resonances

- The meanings of absorption and scattering under consideration are different from those of the two same words for the classical light scattering by small particles for which the sum of the cross sections for the scattering and the absorption is related to that for the extinction. In particular, the absorption resonance here is focused on the maximum absorption phenomena via the small coupling hole into the lossy cavity whose inside wall is made of the lossy wall. What is more, the scattering resonance means that resonance phenomena like Wood's anomalies occur as the result of the interaction between the incident wave and the homogeneous leaky wave solution of the scatterers as the periodic reflection/transmission gratings. This is also the same in the case of the guided mode resonance (GMR). So the scattering resonance is also referred to as the GMR.

Bragg blazing

- This refers to the phenomenon that all the incident power is scattered back into the direction of incidence which equals the first-order Bragg backscattering angle, which corresponds to the retroreflection in terms of the language in the area of metasurface.

Off-Bragg blazing

- This means the case in which its scattered wave propagates in a direction different from the Bragg backscattering angle.

Extraordinary optical absorption (EOA)

- The EOA means the unusual absorption phenomena characterized by an absorption enhancement factor that will exceed or will be comparable to the enhancements seen for the extraordinary optical transmission (EOT) through slot in analogy with the EOT phenomena.

Extraordinary optical transmission (EOT)

- EOT means the phenomena in which the transmitted fraction of incident light through the small aperture array structure can significantly exceed the incident power upon the open fraction of the array for certain wavelengths and is also called enhanced or resonant transmission for frequencies ranging from the near-infrared to those of microwaves.

Guided mode resonance (GMR)

- GMR refers to rapid variations in the intensity of propagating waves that are diffracted by a waveguide grating occurring when the incident wave is coupled to a leaky waveguide mode by phase matching with a second-order grating, under which the guided mode is determined through the use of the transverse resonance condition.

Lorentz-like and Fano-like resonances

- The Lorentz-like resonance are general resonance curve that is characterized by its symmetric line shape. In contrast to the Lorentz-like resonance, the Fano-like resonance exhibits a distinctly asymmetric line shape whose microscopic origin arises from the interference of a narrow discrete resonance with a broad spectral line of continuum.

Metamaterial

- This is a general term for materials that are typically engineered with novel or artificial structures to produce electromagnetic properties that are unusual or difficult to obtain in nature.

Metasurface

- This refers to rationally designed two-dimensional arrangements of sub-wavelength scattering structures that can modify the phase, polarization, and amplitude of the propagating optical waves.

Nanoaperture

- This means the subwavelength aperture structure which can be used as an NSOM probe having the aperture size down to nanometer size for the sake of over-coming the diffraction limit while the transmission efficiency and spatial con-finement are maintained superior to those of conventional aperture.

Near-field scanning optical microscopy (NSOM)

- This is a scanning probe-based technology that employs a probe of a sub-wavelength size and an object that is mounted in the near field of the probe so that the spatial resolution is determined by the probe size rather than by the wavelength. The availability of this scanning probe microscopy makes it possible to overcome the diffraction limit and to open up fields of studies that were previously inaccessible.

Negative reflection and refraction

- The terminology "negative" means that the reflected and transmitted wave can be bent in arbitrary directions on the same side of the interface normal for the same complex root of the leaky wave chosen for both the reflection and the transmission gratings.

Optical antenna

- This concept is defined as a device designed to efficiently convert free-propagating optical radiation to localized energy, and vice versa like a planar bow-tie antenna structure with an open-circuited terminal for application to the near-field optical probe with transmission efficiency of order unity at microwave frequencies.

Parallel and collinear configuration

- Parallel configuration is the case where the equivalent magnetic current sources over the adjacent apertures for finite/infinite arrays are put side by side with the magnetic current source vectors parallel to each other. On the other hand, the collinear configuration is the case where the equivalent magnetic current sources over two adjacent apertures are put side by side with their magnetic current vectors lying along the array axis. The nomenclature holds also for the electric dipole array if the magnetic current vector is replaced by the electric current vector.

Photonic crystal

- This means general materials patterned with a periodicity in dielectric constant, which can create a range of forbidden frequencies called a photonic bandgap.

Transmission cavity resonance (TCR)

- TCR means a kind of resonance phenomenon in which the transmitted power through the transmission cavity becomes maximum depending on the resonant standing wave mode like the Fabry–Perot mode inside the cavities such as an aperture-coupled transmission cavity or the narrow slot type in the thick conducting screen etc.

Transmission cross section (TCS)

- TCS of the apertures is defined as the area for which the incident wave contains the same power as transmitted through the apertures.

Transmission resonance

- This means the maximum transmission phenomena through the small resonant apertures whose TCS as a measure of transmission efficiency amounts to $\frac{2G\lambda^2}{4\pi}$ [m$^2$] much larger than the physical area of the aperture. Here $G$ means the directivity of the aperture at the operating wavelength of $\lambda$.

Transmission resonant aperture (TRA)

- TRA means the aperture whose transmission resonance frequency is made to be decreased so that the TCS may be significantly increased by modifying the conventional non-TRA type such as a rectangular or circular aperture in shape.

Non-transmission resonant aperture (non-TRA)

- Non-TRA means the apertures such as rectangular or circular apertures whose TCS is much smaller than that of the TRA at the transmission resonance frequency of the TRA. Usually, the TCS of the non-TRA at the transmission resonance frequency of the non-TRA is comparable to the physical aperture area of the non-TRA.

Transmission resonant cavity (TRC)

- TRC is the general term for the structures in which the transmission cavity resonance phenomena occur.

Wood's anomalies

- These are effects of rapid variations in the intensity of the various diffracted spectral orders in certain narrow frequency bands which are observed in the scattering of electromagnetic waves from planar periodic structures such as reflection/transmission grating. There are two types of them. One is the Rayleigh type due to one of the spectral orders appearing at the grazing angle. The other one is the resonance type due to possible complex waves supportable by the grating.

*Chapter 1*

# Introduction

## Synopsis

A brief history of relevant subject matters under consideration is touched upon here. Then the scope of this book and an outline of each chapter are introduced. Among various diffraction problems relevant to small aperture structures, such as narrow slits and small holes, three different types of resonance problems are discussed with the main focus centered on transmission, absorption, and scattering resonances. In particular, as a typical structure designed for enhanced transmission, two types of aperture structures are considered: one is a single transmission resonant structure such as transmission resonant aperture (TRA) (Chapter 4) or transmission resonant cavity (TRC) (Chapter 5) and the other is an aperture array structure in the E-plane, which corresponds to a parallel configuration (Chapter 8).

## 1.1 Brief history of the low-frequency diffraction by narrow slit and small hole

In the long history of diffraction study of optics, the following aperture geometries are typically considered as the simplest nontrivial ones: a long slit and a circular aperture. As is well-known the diffraction phenomena by a long slit and a circular aperture in an infinite perfectly conducting screen are closely related to the so-called complementary problem involving long strips and circular disks of the same size and shape. The relation between the two sets of problems is established by the Babinet principle.

In the types of problems involving various disks including a circular disk, it is usually the total scattering cross section or the backscattering cross section that is of prime interest; in the related problems involving apertures such as long slit and circular aperture, the principal concern is with the transmission characteristics such as transmission cross section.

Chronologically the research progress on the diffraction by a disk and an aperture concerns the problem of finding scattering or transmission cross section at high frequencies first. On the other hand, the low-frequency diffraction characteristics of the above structures have been investigated afterward, after a long time. As an illustration, the diffraction by a freestanding narrow slit structure in an infinite conducting plane was dealt with first by R. F. Millar [1] in 1960. An application of a narrow slot at the open end of a wide microstrip line was proposed to the microstrip patch antenna concept [2] which is made of two narrow slots coupled through the lossy cavity of the finite length at the wide microstrip line like a patch type. Here the "loss" is due to radiation through the narrow slot formed in the open-end structure of the wide microstrip line. It is worthwhile to mention that the radiation resonance through the two narrow slots in which the radiation becomes maximum is essentially the same as the transmission resonance in which the transmission becomes maximum in the narrow thick slit problem as discussed later. However, when the concept of a microstrip patch antenna was put forward, they have received little attention.

Historically it was not until the early 1970s when there came a new demand for compact and lightweight antennas in many applications. Since then, various practical physics [3] on the radiation pattern, radiation impedance, and mutual coupling for the two narrow slot models for the microstrip patch antenna as well as a single narrow slot itself have been widely researched for the application to the planar radiator in microwave band. The microstrip antenna is a kind of small antenna, printed on a grounded dielectric substrate with a thickness commonly very small in wavelength. So the radiating edge slot in the microstrip patch antenna structure belongs to the typical narrow slot type. In addition, a narrow-slit problem [4] in the thick conducting screen case has been also investigated with a main emphasis on the transmission resonance under which the power transmitted through the slit becomes maximum. It should be noted that this problem corresponds to the slit model in a thick metal screen for which model calculations [5,6] have been widely undertaken while the original work on the extraordinary optical transmission (EOT) problem by Ebbesen *et al.* [7] was strictly limited to holes. Here, for the sake of clarification of terminology in this field, the slit refers to an infinitely long one whereas the slot is for a relatively short one.

It should also be mentioned that the two slots are mutually coupled in the maximum radiation problem of the microstrip patch structure whereas the two slots are not in the transmission resonance problem through the thick narrow slit. So except for this difference, the essential similarity between the two problems can be understood. It is well known that the Fabry–Perot resonant mode [5,6] in a single narrow thick slit under the transmission resonance is closely related to the extraordinary transmission phenomena observed in a periodic slit grating [8]. This is a brief historical sketch relevant to the contents to be dealt with in this book.

On the one hand, the diffraction problem by a small hole was first dealt with by H. A. Bethe [9]. Since then, this small hole structure has been widely used as a coupling aperture in filters and directional couplers, etc. in the microwave regime.

In this era of communication and information technologies, as the complexity level of their environment becomes rising, various aperture coupling problems of this low-frequency nature have emerged also in many practical situations in the areas of electromagnetic compatibility (EMC) and electromagnetic interference (EMI) [10,11]. Furthermore, with the advancement of subwavelength optics, the study of the diffraction problem by a small aperture has been naturally extended to the design technology of subwavelength aperture [12,13] offering both transmission efficiency and spatial confinement for high spatial resolution beyond the diffraction limit.

A periodic structure of small circular holes [14] whose diameter is much smaller than the wavelength deserves further comment. If we constitute the square periodic structure of the small hole whose period $d$ is equal to the wavelength $\lambda$, i.e., $\lambda = d$, such that the first propagating diffraction channel (first grating lobe) is on set, the Wood's anomaly [15] with total reflection occurs. At the same time, the total transmission phenomenon is also observed at wavelengths $\lambda$ immediately above the square lattice period $d$. This total transmission phenomenon corresponds to the periodicity resonance in terms of the terminology associated with the EOT. This type of Wood's anomaly belongs to the Rayleigh wavelength type and is different from the resonant type of Wood's anomaly associated with leaky wave mode. The topic of the resonant Wood's anomaly constitutes important content in connection with the scattering resonance [16] in chapter 7. In addition, the guided mode resonance (GMR) [17] phenomenon which has contributed a lot to the progress in the reflection filter design technology of the diffractive optics is applied to the bandpass filter operating near the stopband in the first Brillouin zone. The relation between the resonant type of Wood's anomaly and the scattering resonance is also briefly discussed.

It is worthwhile to add that in the above square periodic structure of the small holes in a thin perfect electrical conducting (PEC) plane, the rapid change from the total transmission to the total reflection within a very narrow frequency range is characterized by the Fano resonance physics [18], which results from the superposition [19] of non-resonant contribution (referred to also as Bethe's small hole contribution) related to the direct transmission of the field through the subwavelength holes and resonant contribution related to the periodic structure.

The total transmission through a periodic array of subwavelength apertures is necessarily accompanied by the Fano profile characteristics at the wavelengths close to the period $\lambda$ of the array corresponding to the Rayleigh wavelength as mentioned above for the thin PEC plane case. Such a complete transmission through an array of subwavelength apertures, accompanied by the Fano profile, has been documented across the entire electromagnetic spectrum [20]. This includes wavelengths ranging from the visible to terahertz and microwave frequencies. This phenomenon holds true not only for highly conductive metals but also for scenarios involving arrays of subwavelength apertures within a real metal slab of finite thickness. For the Fano profile to be observed as a transmission [17] (reflection [21]) curve, the resonant type of Wood's anomalies satisfying the Bragg condition should occur as a necessary condition. So it can be concluded that the EOT problem (Ebbesen's experiment) observed in the periodic array of the subwavelength hole in the real metal slab can be explained in terms of the resonant Wood's anomalies [22].

In general, it is well recognized that there are two different types of resonances [23] that contribute to the enhanced transmission through the real metal film of finite thickness with periodic arrays of subwavelength holes. One is the localized resonances in individual holes and the other is the lattice resonances originating in the array periodicity. Also, in the periodic arrays of rectangular aperture [24] in the PEC conducting screen of finite thickness, such two kinds of resonances have been found to contribute to the enhanced transmission, if each aperture is made of the transmission resonant cavity (TRC) through which the transmission resonance can occur such that the transmission cross section (TCS) is significantly larger than the physical area of the aperture cross section. As is well known, in the rectangular aperture in the thick PEC screen under consideration, there are two kinds of transmission resonances as discussed in Cho *et al.* [25]: one is the Fabry–Perot type where the resonant mode is set up along the longitudinal direction of the cavity; the other is the transverse resonant type (corresponding to the zeroth order Fabry–Perot type) where resonant mode is set up across the transverse direction of the cavity so that the transmission resonance of the TRC is mainly determined by the transverse dimension of the cavity. So, the former depends entirely on the longitudinal length of the cavity corresponding to the finite thickness of the PEC slab. Note that for the above TRC to be formed, guided mode along the cavity should be supportable for the Fabry–Perot type of longitudinal resonance and near-cutoff mode as the zeroth order Fabry–Perot mode should be formed for the transverse resonant type. Remind that the EOT phenomena [7] were observed first in the periodic arrays of sub-wavelength holes in the real metal slab of finite thickness.

Comparison between the periodic arrays of subwavelength slots in the PEC slab of the finite thickness and the periodic array of subwavelength holes in the real metal slab of the finite thickness reveals that for the above general EOT characteristics, i.e., simultaneous appearance of the localized resonance in individual holes and periodicity resonance in the real metal slab case to be observed, there must be a guided mode along the cylindrical cavity inside the subwavelength hole. This has been found to be the case. In more detail it has been found that however small the diameter of the subwavelength hole is in the real metal slab case, the guided mode is supportable [26,27]. This is thought to constitute a basis for the above two resonance peaks, which may lead to the Fano profile characteristics as a more general description of the various resonance phenomena [20].

The original extraordinary optical transmission (EOT) phenomenon was observed at optical frequencies through a periodic subwavelength hole array drilled in a real metal slab. In this subwavelength hole structure, despite the small diameter of the hole, i.e., even though the physical aperture area of the cylindrical hole structure approaches zero, a guided mode along the cylindrical hole structure is found to be supported.

As a counterpart of the aperture perforated in the perfect conducting slab, the narrow rectangular slot type can be viewed, in the sense that in this case too, if the slot length is chosen to be approximately half wavelength, the guided mode of $TE_{10}$ mode is supported even if the physical area is made to approach zero by reducing the slot width. As a result, a transmission resonance occurs through the slot structure whose physical area approaches zero. This has been a brief sketch of the research works on the narrow slit and small aperture according to the historical developments of the research topics such as small aperture coupling and EOT and GMR phenomena in application areas of microwaves and optics.

## 1.2   Scope of this book

The readers of this book are assumed to be equipped with basic knowledge of the electromagnetism microwave circuit theory, antenna theory, and numerical methods such as the method of moments (MoM).

In Chapter 1, we provide an introduction to low-frequency diffraction by narrow slit and small hole structures, covering their brief histories and relevant technologies and then present an outline of this book.

In Chapter 2, the diffraction properties of narrow slits are discussed from the viewpoint of equivalent circuit representation. To understand the reflection, transmission, and radiation properties of the slits, the slits are assumed to reside in the conducting plane over the grounded dielectric of an arbitrary thickness to simulate both microstrip and waveguide structure cases. The magnetic wall property of the narrow slit is discussed in comparison with a flanged narrow slit structure which has been widely investigated in connection with the EOT (extraordinary optical transmission) phenomena. As a representative aperture-body resonance problem, the two coupling mechanisms of microstrip antenna structure were discussed for the simplified structures of the aperture-coupled and proximity-coupled microstrip antennas to investigate the transmission resonance corresponding to the maximum radiation in comparison with the EOT problem from the viewpoint of the impedance matching.

In Chapter 3, diffraction problems for both infinite long slit and rectangular slots in the thick conducting slab are considered with the main concern centering on the transmission resonance phenomena in which the transmitted power through slit/slot becomes maximum. For the rectangular slot case, the two types of transmission resonance, the near cutoff mode (having the near zero group velocity) as well as the above Fabry–Perot type are observed, unlike the former infinite long slit case where only the Fabry–Perot type exists due to the non-existence of the cutoff frequency. This is mainly due to the existence of the cutoff frequency of the dominant $TE_{10}$ mode inside the slot region, for which experimental verification is given. The background for the shortening factor for the magnetic dipole corresponding to that of the electric dipole antenna is also discussed.

In Chapter 4, the transmission resonant aperture (TRA) structures which provide much larger transmission cross section (TCS) than that of conventional small circular or rectangular apertures corresponding to the non-TRA type are dealt with along with their dual resonant scatterer structures. The transmission resonance condition is described from the viewpoint of the equivalent circuit representation. Various transmission resonant aperture of the TRA is discussed. The general impedance relation between the two dual problems is also discussed. Finally, the use of the small resonant scatterer for field measurement is considered along with other possible applications to the filter, frequency selective surface (FSS), near field scanning optical microscope (NSOM) probe design, etc.

Chapter 5 deals with transmission resonant cavity (TRC) structures which can give the transmission cross section comparable to the TCS of the transmission-resonant aperture (TRA). Various TRC structures are considered. The common feature between transmission resonance phenomena by TRA and TRC structures is discussed from the viewpoint of the TCS characteristics. The scattering phenomena unique to the TRC structure as a difference between scattering by TRA and TRC structures are also discussed. An application example is considered for the NSOM probe design for microwave and millimeter range. The linkage of the near-field optical probe using a bow-tie optical antenna to the TRC type is also discussed.

In Chapter 6, the absorption resonance condition for maximum absorption into the aperture-coupled cavity which is made of a lossy cavity wall is investigated under the normal incidence of the vertically polarized plane wave. The relation between the absorption resonance condition into the lossy cavity and the transmission resonance condition through the transmission resonant cavity under the same incident wave condition is considered. It is important to note that both the resonance phenomena are attributed to the resonant excitation of a mode of the structure and so the transmission and absorption peaks coincide though the main concern is different, i.e., in the former transmission resonance problem main concern is centered on the efficient confinement of high efficient flow through the small aperture whereas in the latter absorption resonance problem the main concern is centered on the high speed operation.

Then the absorption characteristics of the absorption frequency selective surface which is made by omitting all the side walls of the periodic structure composed of the absorption resonant cavity is investigated with the main concern centering on the absorption resonance (i.e., maximum absorption) condition.

In Chapter 7, we first consider the scattering resonance that is attributed to the coupling between the incident wave and the leaky wave sustained by the reflection grating made of the periodic strip grating over the grounded dielectric.

The Bragg and off-Bragg blazing phenomena as a kind of Wood anomalies that belong to the scattering resonance have been discussed. The Bragg blazing phenomena which correspond to the retroreflection phenomena in the metasurface area show the two types of Lorentz- and Fano-profile characteristics curve. Between these, the Fano-profile type which belongs to the genuine scattering resonance is observed to occur under the resonant Wood anomaly condition satisfying the Bragg condition. Then the essential nature between this condition for reflection grating and the guided mode resonance (GMR) condition in the transmission filter design in the diffractive optics area has been also discussed. How to implement the structure of the transmission grating which can show the refraction working in the same manner as the above retroreflection (negative reflection) is briefly discussed. The passband filter of the GMR type working for the normal incidence is also discussed in connection with the Fano-like characteristics. In addition, various engineering problems associated with the scattering resonance based upon the leaky wave excitation such as the scan blindness phenomenon in the phased array problem, Smith Purcell radiation, and a common feature of the working principle between dichroic surface and grating coupler are briefly touched upon.

In Chapter 8, as a supplementary discussion on the transmission resonance phenomena, we investigate the transmission of electromagnetic energy through finite and infinite periodic arrays of the resonant and non-resonant apertures. For this purpose, starting with a two-element array of collinear and parallel configurations, we investigate the transmission characteristics of TCS for an arbitrary number of aperture arrays. From this, the TCS of the finite hole array of the parallel configuration is seen to be significantly larger than that of the collinear configuration, which is compatible with the previous result on the basic unit of EOT as the minimal system that shows the EOT.

Equivalence between infinite long resonant slit source which has been widely investigated in connection with EOT phenomena and discrete TRA source is also considered from the viewpoint of the TCS. The transmission characteristics of the FSS composed of the TRA are compared with that of the FSS composed of the conventional circular aperture of non-TRA type. Interrelation among the Wood anomaly, periodicity resonance, and Fano profile characteristics in the conventional FSS structure is discussed. Enhanced transmission through a single hole and its periodic hole array are investigated in the PEC medium case and compared with that for the real metal case.

A scenario for the EOT phenomena for the real metal case is considered based upon the PEC slab model with a square periodic circular hole array where the physical area of the single hole approaches zero and the period corresponds to the wavelength. In this case, only the transmission peak associated with the periodicity is observed. By increasing the TCS of the single hole while maintaining the physical aperture area to the convergent value to zero, the transmission peak due to the single aperture resonance is made to occur along with the transmission peak due to the periodicity resonance. That is, by employing the perforated narrow slot type of TRC in a PEC slab of a finite thickness as a single aperture element in the periodic aperture array, the simultaneous appearance of two different transmission peaks associated with the single aperture resonance and periodicity resonance, respectively, are made to be observed, which is well known to be a general feature of the EOT phenomena found for the real metal slab with a periodic array of the subwavelength holes.

It is important to point out that for the above two types of transmission peaks to be observed, the required physical situation for the PEC slab case with periodic hole array is different from that for the real metal slab case with the subwavelength hole array as an original problem. In the PEC slab case, even in the limiting case where the physical area of the cross section of the single perforated aperture type of TRC approaches zero like a thick narrow slot, the guided mode is supportable and so the TCS of the TRC formed along the guided direction of the cavity should be significantly increased. On the other hand, in the real metal slab case, even in the limiting case where the diameter of the single subwavelength cylindrical hole approaches zero, the guided mode should be supportable along the cylindrical hole for the same reason as the above PEC case.

Finally, a common feature between the EOT resonance and the GMR is briefly discussed.

# References

[1]   Millar R. F. 'A note on diffraction by an infinite slit.' *Can. J. Phys.* 1960, vol. 38, pp. 38–47

[2]   Deschamps G. A. 'Microstrip microwave antennas.' *3rd USAF Symp. Antennas* 1953

[3]   James G. A., Hall P. S. (eds.). *Handbook of Microstrip Antennas*. London: Peter Peregrimus Ltd; 1989. pp. 527–578

[4]   Harrington R. F., Auckland D. T. 'Electromagnetic transmission through narrow slots in thick conducting screen.' *IEEE Trans. Antennas Propagat.* 1980. vol. 28(5), pp. 616–622

[5]   Takakura Y. 'Optical resonance in a narrow slit in a thick metallic screen.' *Phys. Rev. Lett.* 2001, vol. 86(24), pp. 5601–5603

[6]   Yang F., Sambles J. R. 'Resonant transmission of microwave through a narrow metallic slit.' *Phys. Rev. Lett.* 2002, vol. 89(6), pp. 063901-1–063901-3

[7]   Ebbesen T. W., Lezec H. J., Ghaemi H. F., Thio T., Wolff P. A. 'Extraordinary optical transmission through subwavelength hole arrays.' *Nature.* 1998, vol. 391, 667–669

[8]   Porto J. A., Garcia-Vidal F. J., Pendry J. B. 'Transmission resonance on metallic gratings with very narrow slits.' *Phys. Rev. Lett.* 1999, vol. 83(14), pp. 2845–2848

[9]   Bethe H. A. 'Theory of diffraction by small holes.' *Phys. Rev. Lett.* 1944, vol. 66, pp. 163–182

[10]  Leviatan Y. 'Study of near-zone fields of a small aperture.' *J. Appl. Phys.* 1986, vol. 60(5), pp. 1577–1583

[11]  Leviatan Y., Harrington R. F. 'A low frequency moment solution for electromagnetic coupling through an aperture of arbitrary shape.' *Arch. Elekr. Ubertr.* 1984, vol. 38(4), pp. 231–238

[12]  Shi X. Hesselink L., Thomton R. L. 'Ultrahigh light transmission through a C-shaped nanoaperture.' *Opt. Lett.* 2003, vol. 28(15), pp. 1320–1322

[13]  Shi X. Hesselink L. 'Design of a C-aperture to achieve $\lambda/10$ resolution and resonant transmission.' *J. Opt. Soc. Am. B* 2004, vol. 21(7), pp. 1305–1307

[14]  Garcia de Abajo F. J., Gomez-Medina R., Sàenz J. J. 'Full transmission through perfect conductor subwavelength hole arrays.' *Phys. Rev. E* 2005, vol. 72, pp. 016608-1–016608-5

[15]  Wood R. W. 'Anomalous diffraction gratings.' *Phys. Rev.* 1935, vol. 48, pp. 928–936

[16]  Brundrett D. L., Glytsis E. N., Gaylord T. K. Bendickson J. M. 'Effects of modulation strength in guided-mode resonant subwavelength gratings at normal incidence.' *J. Opt. Soc. Am. A* 2000, vol. 17(7), pp. 1221–1230

[17]  Ding Y., Magnusson R. 'Bandgaps and leaky-wave effects in resonant photonic-crystal waveguides.' *Opt. Express* 2007, vol. 15(2), pp. 680–694

[18]    Fano U. 'Effects of configuration interaction on intensities and phase shifts.' *Phys. Rev.* 1961, vol. 124(6), pp. 1866–1878

[19]    Genet C., van Exter M. P., Woerdman J. P. 'Fano-type interpretation of red shifts and red tails in hole array transmission spectra.' *Opt. Comm.* 2003, vol. 225, pp. 331–336

[20]    Masson J.-B., Podzorov A., Gallot G. 'Extended Fano model of extra-ordinary electromagnetic transmission through subwavelength hole arrays in the terahertz domain.' *Opt. Express* 2009, vol. 17(17), pp. 15280–15291

[21]    Lee C. H., Cho U. H., Cho Y. K. 'Revisited generalized Wood anomalies.' *IEEE Antennas and Propag. Soc. International Symp.* 1999, vol. 3, pp. 1754–1757

[22]    Sarrazin M., Vigneron J.-P., Vigoureux J.-M. 'Role of Wood anomalies in optical properties array of subwavelength holes.' *Phys. Rev. B*, 2003, vol. 67 (8), pp. 085415-1–085415-8

[23]    Ruan Z., Qiu M. 'Enhanced transmission through periodic arrays of sub-wavelength holes; The role of localized waveguide resonance.' *Phys. Rev. Lett.* 2006, vol. 96(23), pp. 233901-1-4

[24]    Medina F. Mesa F., Marqués R. 'Extraordinary transmission through arrays of electrically small holes from a circuit theory perspective.' *IEEE Trans. Microw. Theory Tech.* 2008, vol. 56(12), pp. 3108–3120

[25]    Cho Y. K., Kim K. W., Ko J. H., Lee J. I. 'Transmission through a narrow slot in a thick conducting screen.' *IEEE Trans. Antennas Propag.* 2009, vol. 57(3), pp. 813–816

[26]    Catrysse P. B., Fan S. 'Propagating plasmonic mode in nanoscale apertures and its applications for extraordinary transmission.' *J. Nanophotonics* 2008, vol. 2, pp. 021790-1–021790-20

[27]    Kim K. Y., Cho Y.K., Tae H. S., Lee J. H. 'Optical guided dispersions and subwavelength transmissions in dispersive plasmonic circular holes.' *Opto-Electroncis Rev.* 2006, vol. 14, pp. 233–241

*Chapter 2*
# Diffraction by a sub-wavelength slit in a waveguide

## Synopsis

Here we are going to investigate the diffraction property of the transverse narrow slit in the upper wall of the parallel-plate waveguide (PPW) in terms of the equivalent circuit representations. The reason for this choice of equivalent circuit approach is that it is useful to understand the various resonance phenomena such as the maximum radiation condition in the transmission line model of microstrip patch antenna and transmission resonance of the Fabry–Perot type in the slit of the thick conductor for explanation of extraordinary optical transmission and the interrelated physics between these two phenomena in microwave engineering and optics areas. The diffraction properties of the narrow slit are discussed in connection with the working physics of the various transmission resonance phenomena such as aperture body resonance (ABR).

## 2.1 Diffraction properties of the narrow slit

As is well known, the slit problem is an essential component in radiators [1] and transmission resonant structures [2,3], such as microstrip antennas and coupling slots in thick conductors, among others. As a representative example of prior research in this direction, some authors have examined the two-dimensional diffraction caused by a transverse slit in the upper wall of a parallel-plate waveguide. Millar [4] developed a pair of simultaneous integral equations over semi-infinite ranges, which he solved only for the wide slit case. Keshavamurthy and Butler [5] analyzed the slit problem with the additional complexity created by a discontinuity in the interior medium in the region of the slit using the moment method.

In this chapter, attention is given to the equivalent circuit description of the slit in a parallel plate guide, whose equivalent circuit approach is extended to the more complex coupling problem of the slit to the nearby conducting scatterer appearing later.

Some authors [6] obtained equivalent circuit parameters in centerline representation but the results of their work are valid only for sufficiently narrow slits.

Here both the centerline and edge representations of the equivalent circuit parameters are presented for narrow and arbitrary wide slits analyzed, respectively, by the method used in [6] and the method of moments. The notations of the centerline and edge representations in the equivalent circuit description follow those of the work of Oliner [7] and will be made clear as the discussion proceeds.

## 2.1.1  Analysis method

Figure 2.1 shows the geometry under consideration. The edges of the slit of width 2a are parallel to the y-axis, with the origin at the center of the slit. For region I the constants of the medium are as follows: permittivity $\varepsilon_o\varepsilon_{rI}$, permeability $\mu_o$ and propagation constant $k_I = \omega\sqrt{\mu_o\varepsilon_o\varepsilon_{rI}} = k_0\sqrt{\varepsilon_{rI}}$ where $\omega$, $\varepsilon_o, \mu_o, k_0$ denote angular frequency, permittivity, permeability, and propagation constant of the free space, respectively, and $\varepsilon_{rI}$ is the relative dielectric constant of region I. For region II the corresponding constants of the medium are $\varepsilon_o\varepsilon_{rII}$, $\mu_o$ and $k_{II} = \omega\sqrt{\mu_o\varepsilon_o\varepsilon_{rII}}$. A time harmonic transverse electromagnetic (TEM) wave, whose time dependence factor $e^{j\omega t}$ is suppressed throughout, is incident upon the slit. The electric field amplitude of the TEM wave is assumed to be unity and $k_I\,b$ is less than $\pi$ so that only the TEM mode can propagate.

Choosing the appropriate Neumann Green functions [8] in each region I, II and imposing the continuity condition of tangential electromagnetic fields in the slit, one sets up the following integral equation [9] for the y-component equivalent magnetic current $M_y$:

$$-\int_{-a}^{a} M_y(z')\left\{G(z,z') - \frac{k_{II}}{2\eta_{II}}H_0^{(2)}(k_{II}|z - z'|)\right\}dz' = \frac{1}{\eta_I}e^{-jk_Iz} \tag{2.1}$$

Here

$$\eta_{I(II)} = \sqrt{\frac{\mu_0}{\varepsilon_o\varepsilon_{rI(II)}}}$$

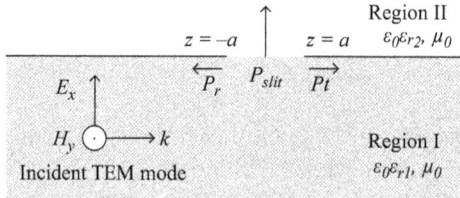

*Figure 2.1*   *Original boundary value problem and associated symbols (reprinted from Ref. [9] with permission from IEEE)*

And $H_0^{(2)}$ means the Hankel function of the second kind of order zero and $G(z, z')$ is represented by

$$G(z, z') = -\frac{1}{2\eta_I b} e^{-jk_1|z-z'|}$$

$$-\frac{jk_1}{\eta_I b} \sum_{n=1}^{\infty} \frac{\exp\left[-\sqrt{\left(\frac{n\pi}{b}\right)^2 - k_1^2}|z-z'|\right]}{\sqrt{\left(\frac{n\pi}{b}\right)^2 - k_I^2}} \tag{2.2}$$

where $G(z, z') = G(x, z; x', z')|_{x,x'=0}$ and $G(x, z; x', z')$ is the Green function satisfying

$$\left(\frac{\partial^2}{\partial x^2} + \frac{\partial^2}{\partial z^2} + k_I^2\right) G(x, z; x', z') = j\omega\varepsilon_o\varepsilon_{rI}\delta(x - x')\delta(z - z')$$

with boundary condition

$$\left.\frac{\partial G}{\partial x}\right|_{x=0,-b} = 0$$

Here we divide this integral equation problem into the narrow slit case and arbitrary slit-width case. For narrow slits an approximate analytic solution is obtained and for arbitrary width slits a numerical solution is obtained for the unknown magnetic current over the slit by use of the method of moments. From the knowledge of the magnetic current density over the slit, we obtain the equivalent circuit parameters for both cases.

### 2.1.1.1 Narrow slit case

If the slit width is narrow relative to the wavelength in region II, i.e., $k_{II}|z - z'| \ll 1$, then the Hankal function in (2.1) can be replaced by the small argument approximation [8]:

$$H_0^{(2)}(k_{II}|z - z'|) \cong -j\frac{2}{\pi}\left\{\ln(\frac{k_{II}|z - z'|}{2}) + \gamma + j\frac{\pi}{2}\right\} \tag{2.3}$$

where $\gamma$ is Euler's constant.

When the slit width is narrow compared to the waveguide height $b$ and wavelength, i.e., $2\pi k_I a \ll k_I b < \pi$, the Green function in (2.2) in region I can also be replaced by its small argument approximation [9].

$$G(z, z') = -\frac{1}{2\eta_I b} + \frac{jk_I}{\eta_I\pi} \cdot \left\{\ln\left(\frac{\pi|z - z'|}{b}\right) - \sum_0 \left(\frac{k_I b}{\pi}\right)\right\} \tag{2.4}$$

When the slit width is narrow, the incident magnetic field $H_{y.inc}\left(= (1/\eta_I)e^{-jk_I z}\right)$ can be approximated by a two-term Taylor series:

$$H_{y.inc} \cong \frac{1}{\eta_I}(1 - jk_I z) \tag{2.5}$$

By substituting (2.3), (2.4), and (2.5) into (2.1), after some algebraic manipulations, one obtains the approximate integral equation:

$$C_1 \int_{-a}^{a} M_y(z')\ln|z - z'|dz' + C_2 \int_{-a}^{a} M_y(z')dz' = \frac{1}{\eta_I}(1 - jk_I z) \tag{2.6}$$

where

$$C_1 = -\frac{j}{\pi}\left(\frac{k_I}{\eta_I} + \frac{k_{II}}{\eta_{II}}\right)$$

$$C_2 = \frac{1}{2\eta_I b} - \frac{j}{\pi}\left\{\frac{k_{II}}{\eta_{II}}\ln\left(\frac{k_{II}}{2}\right) + \frac{k_I}{\eta_I}\ln\left(\frac{\pi}{b}\right) + \frac{k_{II}}{\eta_{II}}\gamma + j\frac{k_{II}\pi}{2\eta_{II}} - \frac{k_I}{\eta_I}\sum_0\left(\frac{k_I b}{\pi}\right)\right\}$$

Making use of the method [10], one can obtain the following solution $M_y(z')$:

$$M_y(z') = \frac{\alpha + \beta z'}{\sqrt{a^2 - z'^2}} \tag{2.7}$$

where

$$\alpha = \frac{1}{\eta_I \pi\left\{-j\frac{k_{II}}{\eta_{II}\pi}\ln\left(\frac{k_{II}a}{4}\right) - j\frac{k_I}{\eta_I\pi}\ln\left(\frac{\pi a}{2b}\right) + \frac{1}{2\eta_I b} - j\frac{k_{II}}{\eta_{II}\pi}\gamma + \frac{k_{II}}{2\eta_{II}} + \frac{jk_I}{\eta_I\pi}\sum_0\left(\frac{k_I b}{\pi}\right)\right\}}$$

$$\beta = -\frac{k_I \eta_{II}}{k_{II}\eta_I + k_I\eta_{II}}$$

The magnetic field of the reflected TEM wave, far to the left ($z < -a$), where all higher order modes except TEM mode become negligible, can be obtained from knowledge of the induced magnetic currents and is readily found to be

$$H_{y.ref}^{TEM} = -\frac{\pi}{2\eta_I b}\{\alpha J_0(k_I a) - j\beta J_1(k_I a)a\}e^{jk_I z} \tag{2.8}$$

where $J_0$ and $J_1$ denote the Bessel functions of the first kind of order zero and one, respectively. If the magnetic field reflection coefficient $\Gamma_H$ which is the ratio of the reflected field in (2.8) to the incident field $(=(1/\eta_I)\,e^{-jk_I z})$ is determined at the reference plane $z = 0$ and if the relationship given in [9] between $\Gamma_H$ and the normalized series admittance $\overline{Y}\left(= \overline{G} + \overline{B}\right)$ in the transmission line with its characteristic admittance $Y_c$ unity is used, then the approximate functional form of the normalized conductance $G$ and the normalized susceptance $B$ in the centerline representation are as follows:

$$\overline{G} \cong \frac{k_0 b \varepsilon_{rII}}{2\sqrt{\varepsilon_{rI}}} \tag{2.9}$$

$$\overline{B} \cong \frac{k_0 b}{\pi\sqrt{\varepsilon_{rI}}}\left\{\varepsilon_{rII}\ln\left(\frac{4}{k_0 a\sqrt{\varepsilon_{rII}}}\right) + \varepsilon_{rI}\ln\left(\frac{2b}{\pi a}\right) - \gamma\varepsilon_{rII} + \varepsilon_{rI}\sum_0\left(\frac{k_I b}{\pi}\right)\right\} \tag{2.10}$$

Here it has been assumed that the equivalent $\pi$ circuit in Figure 2.2 in the centerline representation be approximated to a series element $\overline{Y}\left(= \overline{G} + j\overline{B}\right)$ for the narrow slit width case as shown in Figure 2.2. In other words, for the narrow slit width, the value of the shunt element in the original equivalent $\pi$ circuit is much smaller than that of the series element and so the shunt element is negligible.

For the special case $\varepsilon_{rI} = \varepsilon_{rII} = 1$, (2.10) reduces to

$$\overline{B} \cong \frac{4b}{\lambda_0}\ln\left(\frac{\lambda_0}{\pi a}\right) + \frac{2b}{\lambda_0}\left\{\ln\left(\frac{\lambda_0}{b}\right) - \gamma + \sum_0\left(\frac{k_0 b}{\pi}\right)\right\} \tag{2.11}$$

which is comparable to Marcuritz's expression [11] for the corresponding para-meter in his study of E-plane slit coupling of rectangular guides. Here $\lambda_0$ represents the free space wavelength. It is worthwhile to note also that if the normalized conductance $\overline{G}$ in (2.9) is transformed into the unnormalized conductance $G$ $(G = Y_c \times \overline{G}$; here $Y_c = \sqrt{\frac{\varepsilon_{rI}}{\eta_0 b}}$, where $\eta_0$ denotes the intrinsic impedance of the free space), one obtains

$$G = \frac{1}{120\lambda_0}\left(\frac{S}{m}\right) \tag{2.12}$$

which is identical to the expression [12,13] for the radiation conductance of the rectangular microstrip antenna of the wide width and is the approximate form of Harrington's radiation conductance of a flanged parallel-plate waveguide [14]. Note that this expression is used in the study of transmission resonance through narrow slots in thick conducting screens [15].

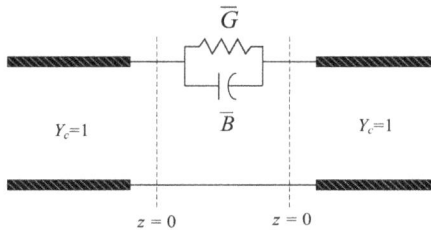

*Figure 2.2  Equivalent circuit representation at center line z = 0 for the narrow slit*

## 2.1.1.2    Wide slit width case

To obtain the equivalent pi circuit for an arbitrary slit width case (Figure 2.3), it is convenient to separate this problem into an even mode excitation case and an odd mode excitation case [16].

With even (odd) mode excitation in which the waveguide is excited by two TEM waves whose electric field amplitudes are symmetrical (anti-symmetrical) about the slit center, one obtains the following integral equation:

$$-\int_{-a}^{a} M_y^q(z')\left\{ G(z,z') - \frac{k_{II}}{2\eta_{II}} H_0^{(2)}(k_{II}|z - z'|) \right\} dz'$$

$$= \begin{cases} -j\dfrac{2}{\eta_I} \sin k_I z, & \text{for } q = v \\[2mm] \dfrac{2}{\eta_I} \cos k_I z, & \text{for } q = d \end{cases} \tag{2.13}$$

where the superscripts $q = v$, $d$ will be used to designate the even and odd excitation cases, respectively. $M_y^q$ is the y-component equivalent magnetic current induced in the slit due to the excitation. The even (odd) mode excitation case corresponds to the presence of a magnetic (electric) wall which closes off the guide in the $z = 0$ plane and so the equivalent circuit in Figure 2.3 is reduced to the equivalent circuits which are terminated by $\overline{Y_2}$ and $\overline{Y_2} + 2\overline{Y_1}$, respectively, according to the even and odd excitation modes.

Once $M_y^q$ is calculated, one obtains the reflected TEM magnetic field ($H_{\text{TEM}}^q$) traveling to the negative z-direction due to the excitation as follows:

$$H_{\text{TEM}}^q = \left( -\frac{1}{2b}\int_{-a}^{a} M_y^q(z')e^{-jk_I z'}\,dz' \pm 1 \right)\frac{e^{jk_I z}}{\eta_I} \tag{2.14}$$

where the minus (plus) sign should be chosen for $q = v(d)$ and this sign notation holds also for (2.15) which will be seen later. From (2.14), the two TEM magnetic field reflection coefficients which are the ratios of the reflected magnetic field in (2.14) to the incident field ($e^{-jk_I z}/n_I$) can be found in the even (odd) mode excitation case—the magnetic field reflection coefficients $\Gamma_{Hv(d)}^c$ in the centerline

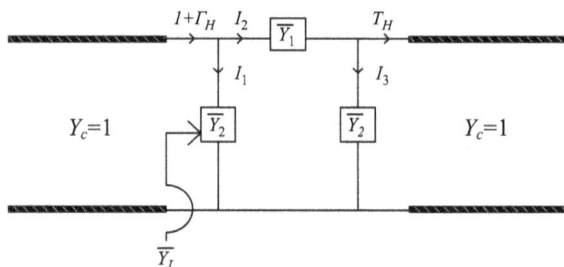

*Figure 2.3    General equivalent circuit representation for both wide and narrow slit cases*

representation and $\Gamma^e_{Hv(d)}$ in the edge representation—depending on the reference planes $z = 0$ and $z = -a$, respectively, as follows:

$$\Gamma^c_{Hq} = -\frac{1}{2b} \int_{-a}^{a} M^q_y(z') e^{-jk_l z'} \, \mathrm{d}z' \pm 1, \quad q = v, d \tag{2.15}$$

$$\Gamma^e_{Hq} = \Gamma^c_{Hq} \cdot e^{-2jk_l a} \tag{2.16}$$

Here the superscript $c(e)$ denotes the centerline (edge) representation. The relation between the magnetic field reflection coefficient and normalized input admittance according to the excitation mode yields

$$\Gamma^p_{Hq} = \frac{\overline{Y^p_q} - 1}{\overline{Y^p_q} + 1}, \quad p = c, e \quad q = v, d \tag{2.17}$$

where $p$ and $q$ follow the notations used above, $\overline{Y^p_v}\left(= \overline{Y^p_2}\right)$ and $\overline{Y^p_d}\left(= \overline{Y^p_2} + 2\overline{Y^p_1}\right)$ are the normalized input admittances in the even and odd mode excitation [17] respectively, as mentioned above. For brevity, the superscript of the admittance parameters has been omitted in Figures 2.2 and 2.3. From (2.15), (2.16), and (2.17) the equivalent circuit parameters in Figure 2.3 are determined to be

$$\overline{Y^p_2} = \overline{G^p_2} + j\overline{B^p_2} = \frac{1 + \Gamma^p_{H\cdot v}}{1 - \Gamma^p_{H\cdot v}}, \quad p = c, e \tag{2.18}$$

$$\overline{Y^p_1} = \overline{G^p_1} + j\overline{B^p_1} = \frac{1}{2}\left[\frac{1 + \Gamma^p_{H\cdot d}}{1 - \Gamma^p_{H\cdot d}} - \overline{G^p_2} - j\overline{B^p_2}\right], \quad p = c, e \tag{2.19}$$

By the scheme employing pulse expansion and point matching in the conventional moment method [18], the linearization of the integral equation for both even and odd mode excitations is established. Detailed calculation procedure for moment method solution and equivalent circuit parameters by use of (2.18) and (2.19) is performed by standard technique [9,18] and omitted here.

### 2.1.2  Numerical results and their physical interpretation

In Figure 2.4(a) it is seen that, as the slit width $2a$ increases, $|T_H|$ goes to zero while $|\Gamma_H|$ is constant for $k_0 a \geq 0 \cdot 1$ and when $k_0 a \geq 1 \cdot 1$, the ratio of the normalized transmitted power $|T_H|^2$ to the normalized radiated power $\left(1 - |\Gamma_H|^2 - |T_H|^2\right)$ is less than 1%. The coupling to the guide beyond the slit is therefore negligible for $k_0 a \geq 1 \cdot 1$. Hence the value of $|\Gamma_H|$ for wide slit ($k_0 a \geq 1 \cdot 1$) approaches that for $2a$ (slit width) $\rightarrow \infty$, which means that, effectively, the upper plate is removed beyond $z = a$.

When $k_0 a \cong 1 \cdot 1$, the value of the slit width $2a$ is equivalent to about 56% of the wavelength in the dielectric of $\varepsilon_{rl} = 2.55$. Therefore the normalized admittance $\overline{Y^e_L}\left(= \overline{G^e_L} + j\overline{B^e_L}\right)$ in edge representation for $k_0 a \geq 1 \cdot 1$ in Figure 2.4(a) can be interpreted as the normalized radiation admittance (of the radiating edge at $z = -a$)

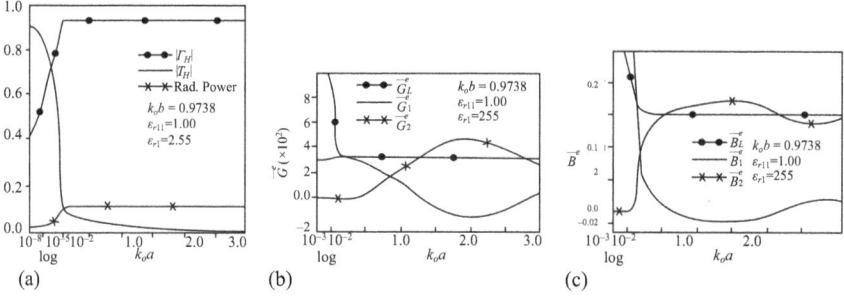

*Figure 2.4    Numerical data for diffraction property of the slit and equivalent circuit parameter. (a) Reflection, transmission, and radiation characteristics of the slit, (b) normalized conductances $(\overline{G_1^e}, \overline{G_2^e},$ and $\overline{G_L^e})$ in edge representation, and (c) normalized susceptance $(\overline{B_1^e}, \overline{B_2^e},$ and $\overline{B_L^e})$ in edge representation.*

of the geometry with the upper plate removed beyond $z = a$ in Figure 1.1 for which Bahl's expression in (2.12) of the conductance is derived.

Also one observes in Figure 2.4(b) that the values of $\overline{G_1^e}$ for the range of $k_0 a$ for which $e^{-j2k_1 a} \cong 1$ are the same as those of $\overline{G_L^e}$ for wide slit ($k_0 a \geq 1 \cdot 1$).

This observation, together with the above considerations, confirms the comment that multiplying $\overline{G}$ in (2.9) by characteristic admittance gives Bahl's expression for the radiation conductance of the rectangular microstrip antenna of the wide width. This means that the Bahl's expression of $1/(120 \lambda_0)$ [s/m] holds also for the unnormalized radiation conductance for the narrow slit which corresponds to $\overline{G_1^e} \times Y_c$.

Since $|\Gamma_H|$ at $z = -a$ in edge representation is constant for $k_0 a \geq 0 \cdot 1$, $\overline{Y_L^e} (= \overline{G_L^e} + j\overline{B_L^e})$ is also seen to be constant for $k_0 a \geq 0 \cdot 1$ from the relation $\Gamma_H = \frac{\overline{Y_L^e} - 1}{\overline{Y_L^e} + 1}$ as shown in Figures 2.4(b) and (c). This implies that as the slit width increases, the value of $\overline{Y_L^e}$ converges rapidly to that of $\overline{Y_L^e}$ for $k_0 a \geq 0 \cdot 1$. So the value of $\overline{Y_L^e}$ for $k_0 a \geq 0 \cdot 1$ can be used as the radiation admittance of the radiating end of the rectangular microstrip patch antenna whose electromagnetic field distribution for the fundamental radiating mode [19] under the patch is almost uniform along the radiating edge. Some more discussion on the equivalent circuit parameters in Figures 2.4(b) and (c) is given in the prior work [9].

The diffraction properties of the slit in the guide—including the reflection from the slit, transmission to the guide beyond the slit, and the radiation through the slit—drastically change depending on the guide height [20]. To demonstrate these diffraction properties of the slit, the variations in $P_r$, $P_t$, and $P_{slit}$ in Figure 2.1 as a function of $kb$ (where $k = \omega \sqrt{\mu_0 \varepsilon_0 \varepsilon_r}$) are illustrated in Figure 2.5. Results were obtained for $\varepsilon_r = 2.5$ and $ka = 0.001$. The reason for selecting $ka = 0.001$ was to

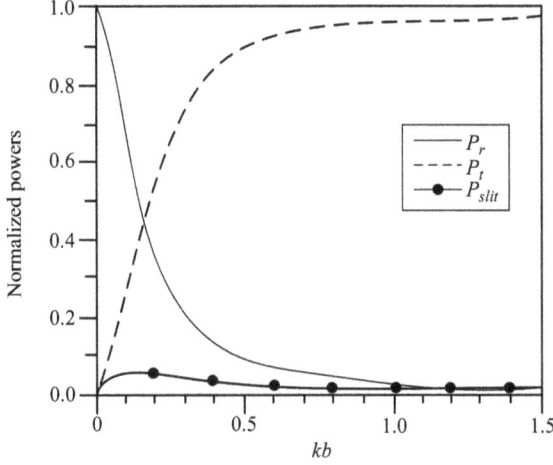

*Figure 2.5   Variations in reflected power $P_r$ from the slit, the transmitted power $P_t$ to guide beyond slit region and radiated through $P_{slit}$ due to variation in kb where $k = \omega\sqrt{\mu_0\varepsilon_0\varepsilon_r}$ and b = height of PPW. Here $\varepsilon_r = 2.5$ and ka = 0.001*

emphasize that even a very narrow slit cut behaves almost as an open circuit, particularly with a small value of *kb* as shown in Figure 2.5.

For a small value of *kb* comparable to a typical microstrip structure, a considerable amount of the incident power $P_{inc}$ is reflected from the slit. In contrast, $P_t$ and $P_{slit}$ are very small.

In contrast to the above, where a larger value of *kb* corresponds to the waveguide case, most of $P_{inc}$ is transmitted to the guide beyond the slit, whereas $P_r$ and $P_{slit}$ become very small. Two extreme cases are thus identified, the former of which will hereafter be referred to as the microstrip structure case and the latter as the waveguide structure case.

## 2.2   Microstrip structure case

As a representative example of a microstrip structure case, the variations in $P_r, P_t$ and $P_{slit}$ as a function of *2ka* (note that the slit width = *2a*) are investigated for typical values of $\varepsilon_r = 2.5$ and $K_b = 0.09425$ in a commercial microstrip substrate, as illustrated in Figure 2.6. Two observations can be made from the figure. The first of these shows that as the slit width 2a approaches zero, $P_r$ and $P_{slit}$ slowly decrease to zero, while $P_t$ slowly increases to unity. In contrast, the second observation establishes that once the slit width 2a exceeds a certain small critical value (e.g., *2ka* = 0.1 in Figure 2.6), $P_r$, $P_{slit}$, and $P_t$ rapidly approach their own respective values: 0.93, 0.07, and 0. These observations are compatible with the earlier comment that even a very narrow slit behaves like an open circuit.

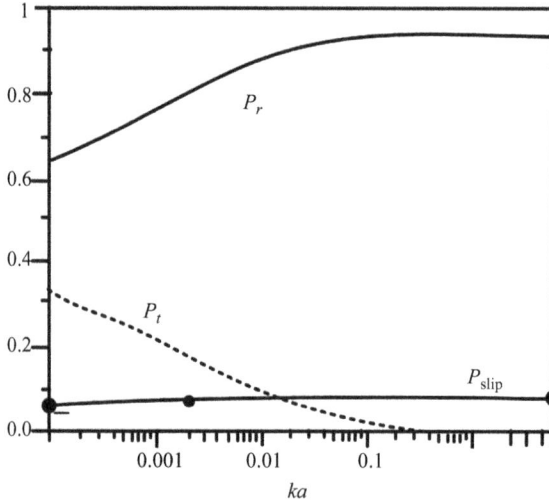

*Figure 2.6    Variations in $P_r$, $P_t$, and $P_{slit}$ as a function of ka for typical microstrip structure cases where $\varepsilon_r = 2.5$, $b = 0.015\lambda$, wavelength inside the PPW $\lambda = 2\pi/k$, and kb = 0.09425*

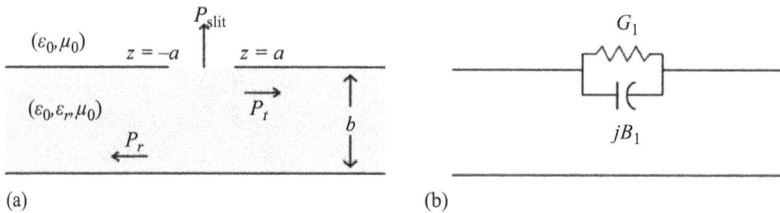

*Figure 2.7    Equivalent circuit representation for a narrow slit. (a) Narrow slit in the microstrip structure. (b) The equivalent circuit where the shunt admittance element $Y_2$ is neglected and $Y_1 = \overline{Y_1} \times Y_c$, $Y_c = \frac{\sqrt{\varepsilon_r}}{\eta_0 b}$.*

## 2.2.1    Equivalent circuit representation for a narrow slit

For a narrow slit where $2ka$ is comparable to or smaller than a quarter of $K_0b$ (roughly speaking), $\overline{Y_2}$ (shunt element) is negligible compared to $\overline{Y_1}$ (series element). In this case, the slit can be represented by a single series admittance element $\overline{Y_1}$ as shown in Figure 2.7.

The behavior of this series element $\overline{Y_1}(=\overline{G_1}+j\overline{B_1})$ for such a narrow slit range is illustrated as a function of $2ka$ in Figure 2.8. In this figure—within the limit of $2ka$ approaching zero— $\overline{B_1}$ approaches infinity, whereas $\overline{G_1}$ remains constant regardless of the slit width [9]. More specifically, within the limit of $2ka \to 0$, $\overline{B_1}$ logarithmically increases to infinity ([9] as seen from (2.10) and (2.11)).

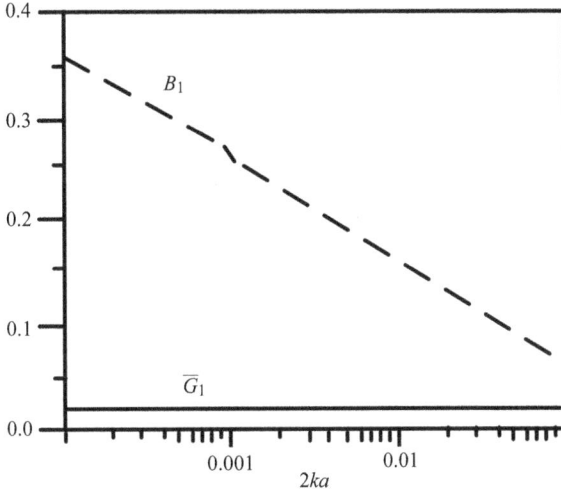

*Figure 2.8*    *Admittance parameter $Y_1 (= G_1 + jB_1)$ versus 2ka, where $\varepsilon_r = 2.5$ $_r =$*
       *2.5, b = 0.015λ, and kb = 0.09425*

To demonstrate such slow divergent behavior, Figure 2.8 illustrates the admittance curve versus $2ka$ on a semi-logscale.

     Furthermore, the non-normalized conductance $G_1 (= \overline{G}_1 \times Y_c)$ is identified as $(120\lambda_0)^{-1}$ [s/m]. Accordingly, this clearly explains how the series admittance $\overline{Y}_1$ for the narrow slit case diverges within the limit of $ka \rightarrow 0$. As a result, $\overline{Z}_1 = (\overline{Y}_1)^{-1}$ approaches zero, as expected.

     It is also interesting to note the above expression $(120\lambda_0)^{-1}$ for $G_1 (\overline{G}_1 \times Y_c)$ is the same as that for the aperture conductance of both a flanged parallel-plate waveguide [14] and a simplified microstrip open-end structure for a two-dimensional problem [12] as discussed above.

### 2.2.2   Equivalent circuit representation for a wide slit

As mentioned above, Figure 2.6 shows that the power $P_t$ transmitted to the guide beyond the slit becomes zero (whereas $P_r$ and $P_{slit}$ remain constant at 0.93 and 0.07, respectively), regardless of the slit width 2a, once the slit width exceeds a small critical value. A wide slit in this instance signifies $P_t \cong 0$. It should, however, be noted that the discussion in this section is restricted to a thin substrate (i.e., small height $b$ of the guide) where the surface wave can be neglected. Some discussion associated with the surface wave is presented later.

     An equivalent pi ($\pi$) circuit representation is not as useful in those wide slit examples in which the slit has an arbitrary width of $P_t \cong 0$ [9].

     A more useful indication is the load admittance $\overline{Y}_L (= \overline{G}_L + j\overline{B}_L)$, which signifies input admittance looking rightward as shown in Figure 2.3. Because, under the condition that $P_t \cong 0$, even if the upper plate beyond $z = a$ is removed, the

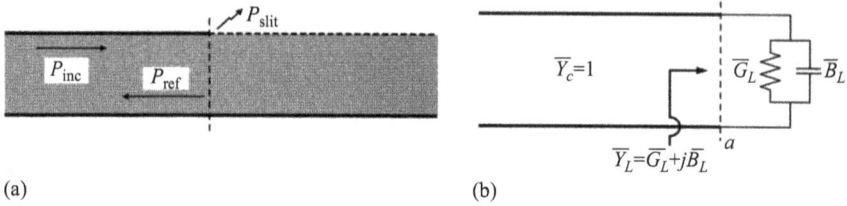

Figure 2.9    Reduced structure for $Y_L(=\overline{Y_L} \times Y_C)$. (a) Reduced structure, (b) its
equivalent circuit

original boundary condition does not change significantly, the original problem
structure can be reduced to the structure in Figure 2.9. In this case, the load
admittance $Y_L$ in Figure 2.3 is appropriate as an equivalent circuit representation as
shown in Figure 2.9(b). As discussed above, $\overline{G}_L$ for the wide slit case is the same as
$\overline{G}_1$ for the narrow slit case. Similarly, the non-normalized $G_L(\overline{G}_L \times Y_c)$ becomes
$(120\lambda_0)^{-1}$ [s/m].

Within the limit of $kb \rightarrow 0$, $G_L$ and $B_L$ both approach zero, while $B_L$ is larger
than $G_L$. Accordingly, this is the basis for using the magnetic wall concept in the
cavity model of microstrip antenna theory.

Next, the waveguide structure case will be considered where the height $b$ of the
guide is much larger than that of the microstrip structure case.

## 2.3    Waveguide structure case

It has already been shown that the reflection and transmission characteristics of a
narrow slit are quite different in the waveguide structure case, compared to the
above microstrip structure case where $kh$ is much smaller than that in the waveguide structure case, as illustrated in Figure 2.5. To investigate more thoroughly the
diffraction properties—including reflection, transmission, and radiation—due to a
slit in the waveguide structure case, the variations in $P_r$, $P_t$, and $P_{slit}$ as a function of $2k_0a$ (where $k_0 = \omega\sqrt{\mu_0\varepsilon_0}$) are calculated for the values of $b = 0.35\lambda_0$
$(k_0b = 0.21999)$ and $\varepsilon_r = 1.0$ as illustrated in Figure 2.10.

This figure reveals that while $P_{slit}$ increases as the slit width increases, $P_t$
decreases (it should be noted that $P_r$ is very small, regardless of the slit width), thus
the variations of $P_r$, $P_t$, and $P_{slit}$ relative to $2ka$ are quite different from those of the
microstrip structure case shown in Figure 2.6.

### 2.3.1    Equivalent circuit representation for a narrow slit

To examine the behavior of the series admittance $\overline{Y}_1(=\overline{G}_1 + jB_1)$ curve for the
narrow slit case, the variations in the admittance versus $2ka$ are calculated, as
illustrated in Figure 2.11. For this waveguide structure case, as the slit width $2a$
becomes smaller, the susceptance $\overline{B}_1$ increases, whereas the conductance $\overline{G}_1$
becomes constant.

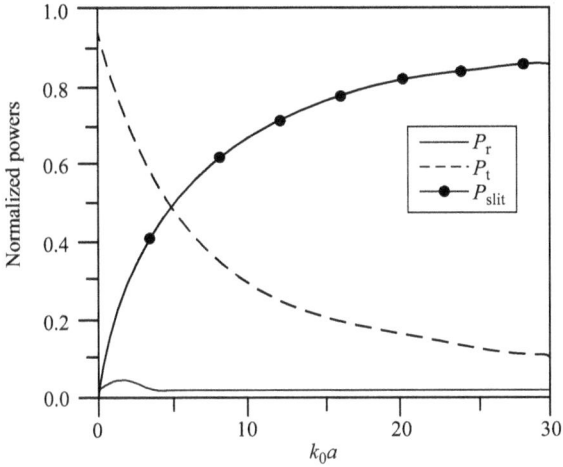

*Figure 2.10    Variations in $P_r$, $P_t$, and $P_{slit}$ as a function of $k_0a$ for hollow waveguide structure, where $\varepsilon_r = 1.0$ and $k_0b = 2.1999$ (b = 0.35$\lambda_0$)*

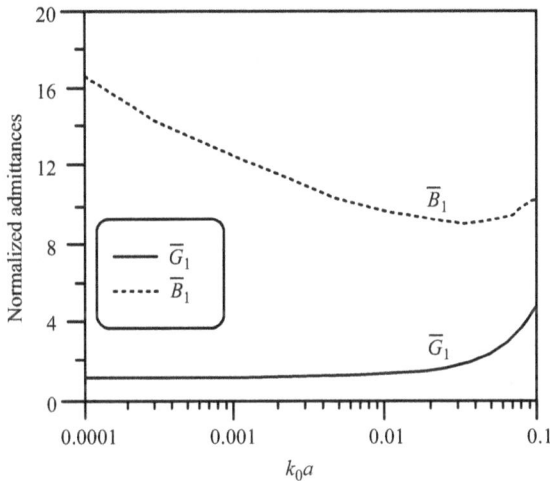

*Figure 2.11    Admittance parameter $Y_1(=G_1 + jB_1)$ versus $k_0a$ where $\varepsilon_r = 1.0$ and $k_0b = 2.1999$ (b = 0.35$\lambda_0$)*

Therefore, within the limit where $2ka \rightarrow 0$, the non-normalized conductance $\overline{G}_L(\overline{G}_L \times Y_c)$ converges to a constant value of $(120\,\lambda_0)^{-1}$ [s/m], which is the same as in the microstrip structure case. However, the salient difference is that as the slit width decreases to zero, the susceptance $\overline{B}_1$ diverges much more rapidly in comparison with the microstrip structure case. The variation of $\overline{B}_1$ as a function of $2ka$ or $k_0a$ for microstrip and waveguide cases are compared in Figures 2.8 and 2.11.

These figures show that even a narrow slit can cause significant perturbation to the guiding structure in the microstrip structure case, yet not in the waveguide structure case.

### 2.3.2    Diffraction properties of wide slit

We deal with diffraction properties of the wide slit for both hollow waveguide cases and dielectrically filled waveguide cases.

To investigate the radiation characteristics from a wide slit in the case of a hollow waveguide, the aperture source distribution and radiation pattern are obtained when $\varepsilon_r = 1.0, b = 0.35\lambda_0$, and $2k_0a = 30$. For reference, $P_r$, $P_t$, and $P_{slit}$ are found to be 1.24, 10.15, and 88.61%, respectively. The results for the aperture source distribution are illustrated in Figure 2.12(a), where the solid(dashed) line represents the real(imaginary) part of the normalized source distribution $E_Z(0, z)$ with respect to the magnitude $|E_{inc}|$ of the incident electric field. The source distribution is reminiscent of one typical of a leaky wave-type traveling wave antenna in that the amplitude of the distribution appears to decay exponentially along the z-direction. In Figure 2.12(b), the pattern indicated by the dashed line shows the corresponding radiation pattern for the case of a hollow waveguide, where the gain is 8.4[ $dB_i$ ] and the maximum beam direction 21°.

For the wide slit problem in a hollow waveguide, the antenna efficiency $(= P_{slit}/P_{inc})$ can be improved by increasing the aperture area. However, this is not the case with a dielectrically filled waveguide, as described below.

Figure 2.13 shows the variations of $P_r$, $P_t$, and $P_{slit}$ as a function of $2ka$ where $\varepsilon_r = 2.5$, $kb = 2.1999$ (corresponding to $b = 0.35\lambda$). The variations of $P_r$ and $P_{slit}$ are quite different from those for the hollow waveguide in Figure 2.10.

As the slit width 2a increases, $P_t$ and $P_{slit}$ are observed to approach their convergent values of approximately 69.7% and 27.6%, respectively (although some fluctuations are observed). This signifies that most radiation occurs near the two discontinuities at both ends of the slit, i.e., near the conducting edges at $z = \pm\frac{a}{2}$, whereas the radiation from the grounded slab region is negligible.

The aperture source distribution for $ka = 30.0$ is given in Figure 2.14 where $P_r = 2.72\%$, $P_t = 69.71\%$, and $P_{slit} = 27.57\%$. Figure 2.14 also shows the traveling wave feature of the tangential ($z$) component electric field over most of the slit width 2a. It is observed that the slit width 2a corresponds to a length of approximately four times the wavelength $\lambda_{sf}$ of the surface wave.

With a grounded dielectric slab waveguide corresponding to the slit region, the propagation constant $\beta_{sf}$ for the fundamental $TM_0$ surface wave [21] is found to be $\beta_{sf} = 1.327 \ k_0$. As a result, $2\beta_{sf}a$ is identified as $4.0075 \times 2\pi$. Thus it can be determined that the slit width is $2a = 4.0075\lambda_{sf}$ in terms of the wavelength $\lambda_{sf}$ of the fundamental surface wave mode. For a comparison between the cases of a hollow and dielectrically filled waveguide, the respective radiation patterns are provided in Figure 2.12(b). The comparison shows that the radiation leakage from the slit region is significantly reduced when a surface-waveguiding structure is formed [22]. This means that in an antenna structure, such as a metal-strip-loaded

(a)

(b)

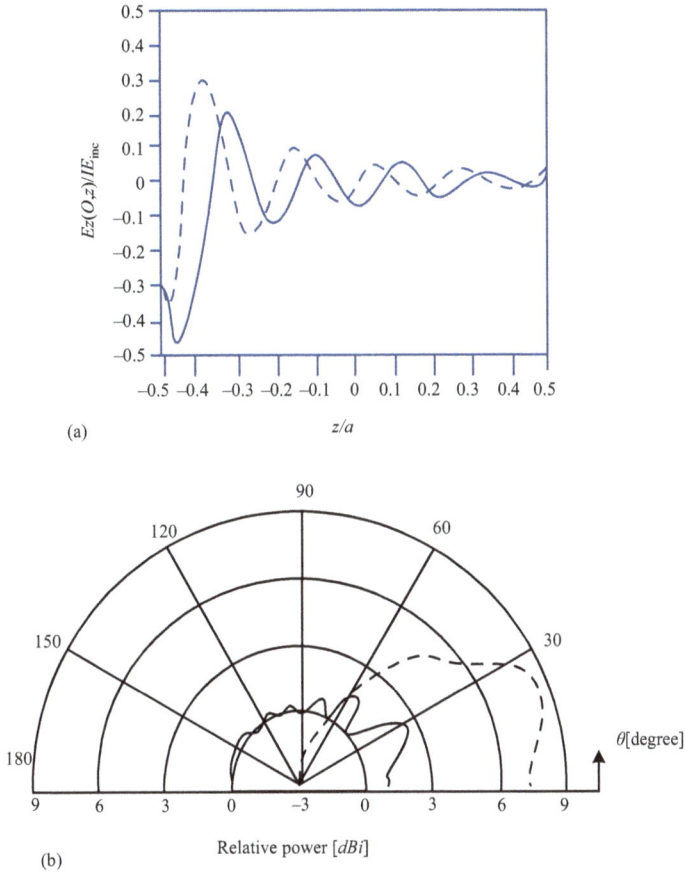

*Figure 2.12* *Distribution of tangential electric field over a wide slit and radiation pattern for hollow waveguide cases where $\varepsilon_r = 1.0$, $k_0 b = 2.1999 (h = 0.35\lambda_0)$, and $k_0 a = 30$ (for reference, $P_r = 1.24\%$, $P_r = 10.15\%$, $P_{slit} = 88.61\%$, Gain = 8.42 [dBi] and maximum beam angle $\theta_m = 21.0°$). (a) Distribution of a tangential electric field, normalized to the magnitude of the incident TEM wave electric field $E_{inc}$, over a wide slit. (b) Radiation pattern (dashed line). For comparison, the radiation pattern corresponding to the source distribution in Figure 2.14 is also drawn using a solid line.*

dielectric antenna [23,24], or two-dimensional microstrip patch element phased array [12], as the dielectric slab thickness $b$ increases, the attenuation constant $\alpha_0$ of the leaky wave decreases and the propagation constant $(\beta_0 - j\alpha_0)$ increases. This may be helpful in understanding the physical grounds for determining the complex propagation constant in such leaky wave-related antenna structures. The reason for six grating lobes being observed in the dielectrically filled case in Figure 2.12(b)

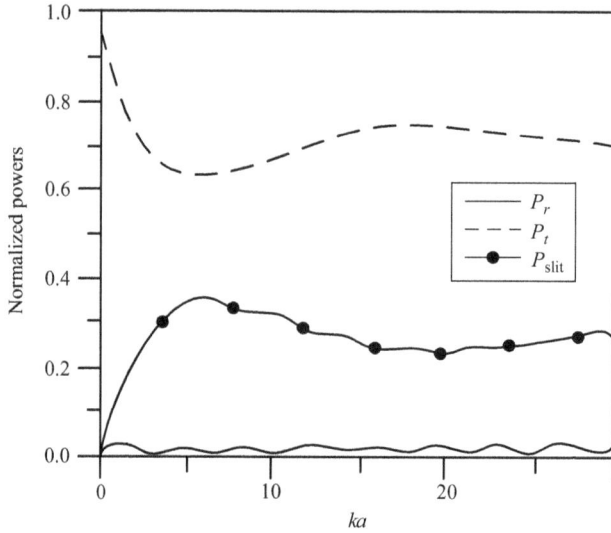

*Figure 2.13    Variations in $P_r$, $P_t$, and $P_{slit}$ as a function of ka for dielectrically filled waveguide cases where $\varepsilon_r = 2.5$ and $kb = 2.1999$ (b = 0.35$\lambda$)*

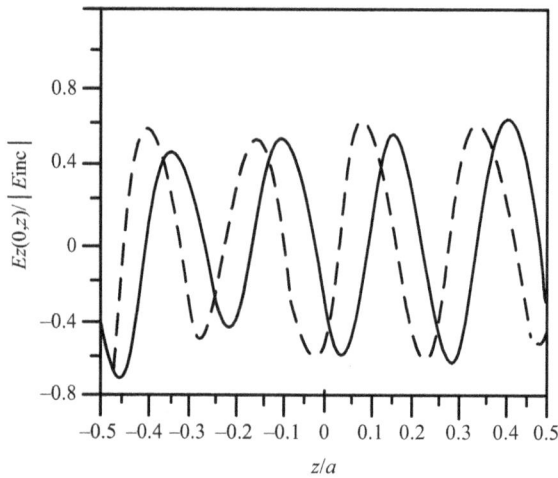

*Figure 2.14    Distribution of a normalized tangential electric field over a wide slit of dielectrically filled waveguide case, where $\varepsilon_r = 2.5$, $kb = 2.1999$ (b = 0.035$\lambda$), and ka=30 (for reference, $P_r = 2.72\%$, $P_t = 69.71\%$, and $P_{slit} = 27.57\%$, Gain = 3.62 [dBi], and maximum beam angle $\theta_m = 25.7°$)*

can be explained by finding that the slit width $2a$ corresponds to $3.02 \, \lambda_0$ in terms of free space wavelength $\lambda_0$. As such, two magnetic current line sources (infinitely long along the y-axis), separated from each other by this distance, produce six grating lobes as shown in Figure 2.12(b). This underlines the previous observation that most radiation occurs near the edge discontinuities at $z = \pm \frac{a}{2}$ between the PPW region and the grounded dielectric slab region.

### 2.3.3 Magnetic wall properties of a narrow slit

When the guide height $b$ is much smaller than the wavelength in Figure 2.9(a), the load admittance $Y_L$ is much smaller than the characteristic admittance $Y_c$ of the guide region. Note that the reflection coefficient $\Gamma$ at the aperture plane of $z = -a$ is given by $\Gamma = (Y_c - Y_L)/(Y_c + Y_L)$. From this, it is seen that as $b$ decreases to zero, $Y_c$ becomes larger than the load admittance $Y_L$ and so $\Gamma$ approach +1.

This corresponds to the magnetic wall. This concept has been widely used in the formulation of the cavity model [25], in the area of microstrip antenna theory, and the Fabry–Perot-like transmission resonance model in the research area of extraordinary optical transmission [26]. Note that the magnetic wall model for the extraordinary optical transmission (EOT) problem has been frequently chosen to be the open end of the flanged parallel-plate waveguide, whose structure and its equivalent circuit are given in Figure 2.15. As mentioned above, the values of $G$ in Figure 2.15 are almost the same as those of $G$ in Figure 2.7. It is worthwhile to mention that when tightly bound propagating mode in the guiding structure is incident upon the open end discontinuity, the incident wave undergoes a strong reflection whose reflection coefficient is close to that from the open end of the magnetic wall type. The reflection at the open end of the cylindrical small hole in the plasmonic medium is thought to belong to this type of reflection under the

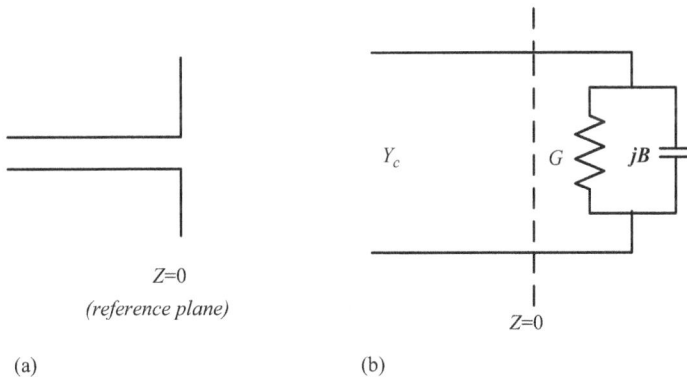

(a)                                        (b)

*Figure 2.15* *The open-end of the narrow slit type in the flanged parallel-plate waveguide and its equivalent circuit representation. (a) Open-end of the flanged parallel-plate waveguide, (b) its equivalent circuit representation.*

assumption that the guided mode be supportable however small the diameter of the hole is, as discussed later.

## 2.4   Narrow slit coupling to nearby conductor

So far the diffraction and equivalent admittance properties of a narrow slit in the guide have been investigated for two contrasting cases corresponding to typical microstrip and waveguide structures.

Here we focus our attention on discussing the narrow slit coupling problem to a nearby conductor with the main interest centering on two types of radiative coupling mechanisms: cavity and parasitic types.

The first accurate treatment for the coupling problem was developed by Mannikko *et al.* [27]. They dealt with the general case where the nearby conducting scatterer to the slit is of an arbitrary shape. However, they did not examine the radiative coupling mechanisms, which are the focus of this study. The analysis method used in this paper is borrowed from Mannikko *et al.* with some modifications for an embedded conducting strip in a dielectric slab, as shown in Figure 2.16, which was selected as representing the simplified geometry of an aperture–coupled microstrip antenna [28].

The similarity between the cavity-type radiative coupling problem and a transmission resonance problem through a narrow slit in thick screens [15] is also described in connection with the aperture-body resonance phenomenon [29].

Furthermore, the main difference in the slit admittance behaviors of the cavity and parasitic types of radiation is discussed from the viewpoint of equivalent circuit description [9].

### *2.4.1   Analysis method*

The problem under consideration is shown in Figure 2.16, where the center of the slit is determined as the origin of the coordinate. The material of the upper half-free

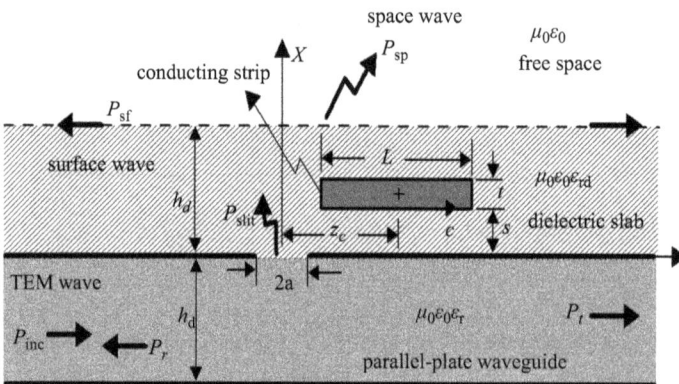

*Figure 2.16   Geometry under consideration*

space, dielectric slab, and inside of the parallel-plate waveguide (PPW) is characterized by $(\varepsilon_0, \mu_0)$, $(\varepsilon_0 \varepsilon_{rd}, \mu_0)$, and $(\varepsilon_0 \varepsilon_r, \mu_0)$, respectively. It is assumed that only a transverse electromagnetic (TEM) wave can propagate along PPW and that the incident TEM magnetic field can be given by $H_y^{inc} = \left(\frac{V}{\eta h}\right) e^{-jkz}$, in which $V$ is the potential difference across the plates, $h$ is the height of the PPW, $k = \omega \sqrt{\mu_0 \varepsilon_0 \varepsilon_r}$, and $\eta = \sqrt{\mu_0 / (\varepsilon_0 \varepsilon_r)}$. The time variation $\exp(j\omega t)$ is suppressed throughout.

Following the magnetic field integral equation (MFIE) approach of Mannikko et al. [27] and Butler and Nevels [30], a pair of coupled integral equations are obtained whose unknowns are the induced electric current density $J_l$ on the surface of the conducting strip and the equivalent magnetic current density $M_y$ over the shorted slit as

$$H_y^{Jd}(\rho) + H_y^{Md}(\rho) = -J_l(\rho)$$
$$\rho \downarrow C \quad \text{(on the surface of the strip)} \tag{2.20}$$

and

$$\left(H_y^{Jd} + H_y^{Md}\right)_{x=0+} = \left(H_y^M + H_y^{inc}\right)_{x=0-} |z| < a, x = 0 \text{(over the slit)} \tag{2.21}$$

where

Subscript $y$ is the y-component,

$\ell$ is the tangential component counterclockwise to the strip surface contour C,

$\rho(= x\widehat{x}_0 + z\widehat{z}_0)$ is the observation point,

$H_y^{Jd}(\rho)$ and $H_y^{Md}(\rho)$ are the magnetic fields in the dielectric slab region due to $J_l$ and $M_y$, respectively, and

$H_y^M$ the magnetic field inside the PPW due to $-M_y$.

Notations in (2.20) and (2.21) follow those of Butler et al. [31].

The expressions for $H_y^M$ and $H_y^{Md}(p)$ are given in the work of Butler et al. [30]. The expression for $H_y^{Jd}(y)$ is found to be as follows using the methods shown in [27,30].

$$\widehat{y}_0 \mu_0 H_y^{Jd}(\rho) = \nabla \times \int_c J_l(\rho') \left[\widehat{x}_0 l_x G_x^d + \widehat{z}_0 l_z G_z^d\right] d\ell \tag{2.22}$$

The expression in the bracket indicates a vector potential Green's function in the grounded dielectric slab region due to the uniform current distribution along the y-direction whose vector includes x- and z-components transverse to the y-axis. $G_x^d$ and $G_z^d$ are given by

$$G_u^d(x, z; x', z') = -j\frac{\mu_0}{4} \left\{ H_0^{(2)} \left[ k_d \sqrt{(x - x')^2 + (z - z')^2} \right] + q H_0^{(2)} \left[ k_d \sqrt{(x - x')^2 + (z - z')^2} \right] \right\}$$
$$+ j\frac{\mu_0}{2\pi} \int_{-\infty}^{\infty} \frac{\Gamma e^{-j2\beta_d h_d}}{1 + \Gamma e^{-j2\beta_d h_d}} \frac{e^{-jk_z(z-z')}}{\beta_d} \cdot [\cos\beta_d(x - x') + q\cos\beta_d(x + x')] dk_z$$

$$\tag{2.23}$$

where

$\rho'(x'\widehat{x}_0 + z'\widehat{z}_0)$ is the source point on $C$,

$l_x$ and $l_z$ are $x$- and $z$-components of a unit tangential vector $\ell' = \ell'_x\widehat{x}_0 + \ell'_z\widehat{z}_0$, respectively,

$G_u^d$ in (2.23): $G_x^d$ or $G_z^d$ appearing in (2.22) and when $G_u^d$ corresponds to $G_x^d(G_z^d)$,

$f$: $1(-1)$

$k_d$: $\omega\sqrt{\mu_0\epsilon_0\varepsilon_{rd}}, \beta_d = \sqrt{k_d^2 - k_z^2}$,

$H_0^{(2)}$ is the Hankel function of the second kind of order zero, and

$\Gamma$: $\frac{\varepsilon_{rd}\beta_0 - \beta_d}{\varepsilon_{rd}\beta_0 + \beta_d}$ with $\beta_0 = \sqrt{k_0^2 - k_z^2}$ (here $k_0 = \omega\sqrt{\mu_0\varepsilon_0}$).

Once the coupled integral equations (2.20) and (2.21) are solved numerically as in [32], the quantities of interest can be calculated, such as the reflected power $P_r$ from the slit, transmitted power $P_t$ beyond the slit in the PPW, coupled power $P_{slit}(= P_{sf} + P_{sp})$ through the slit, surface-wave power $P_{sf}$ launched along the dielectric slab, radiated power $P_{sp}$ into the upper half space, and far-zone field pattern $\left|H_y^0(\rho, \phi)\right|$. All the power quantities are normalized with respect to the incident power $P_{inc}$. The normalized equivalent circuit parameters $(\overline{Y}_1 = \overline{G}_1 + j\overline{B}_1)$ as a series element for the narrow coupling slit are also calculated by use of the method [9].

## 2.4.2   Electromagnetic coupling mechanisms

To understand the underlying physics of the electromagnetic coupling mechanism through the slit to a nearby conductor, we investigate the effect of the nearby scatterer on the radiated power through the slit and reflection and transmission properties inside the guide along with the equivalent circuit representation of $\overline{Y}_1(= \overline{G}_1 + j\overline{B}_1)$ for narrow slit case.

As an example of the case where the conducting strip is just above and very close to the slit, $P_r$, $P_t$ in PPW, $P_{slit}$, and $\overline{Y}_1(= \overline{G}_1 + j\overline{B}_1)$ were calculated as a function of the strip offset position $\frac{z_c}{\lambda_d}$ (here $\lambda_d = \lambda_0/\sqrt{\varepsilon_{rd}}, \lambda_0 =$ free space wavelength) with the separation $S = 0.0135\lambda_d$ between the strip and the slitted plane at $x = 0$ fixed when $\varepsilon_r = \varepsilon_{rd} = 2.2, h = 0.015\lambda\left(\lambda = \frac{\lambda_0}{\sqrt{\varepsilon_r}}\right), h_d = 0.0155\lambda_d, 2a = 0.003\lambda$ and the strip thickness $t = 0.002\lambda_d$. Here the strip length $L = 0.475\lambda_d$ was chosen to achieve maximum coupling through the slit as the strip offset $z_c/\lambda_d$ was varied, with the other parameters fixed as given above.

The results are illustrated in Figures 2.17 and 2.18, where $P_r$, $P_t$, $P_{slit}$, and $\overline{Y}_1(= \overline{G}_1 + j\overline{B}_1)$ in the case of no strip are also given using dashed lines for the sake of comparison. All the numerical data presented were calculated for the fixed case of $h_d = S + t$ with $t = 0.002\lambda_d$.

In Figure 2.18, the normalized equivalent circuit parameter $\overline{Y}_1$ for the narrow slit with respect to the characteristic admittance $Y_c(= 1/\eta b)$ of the PPW means the series

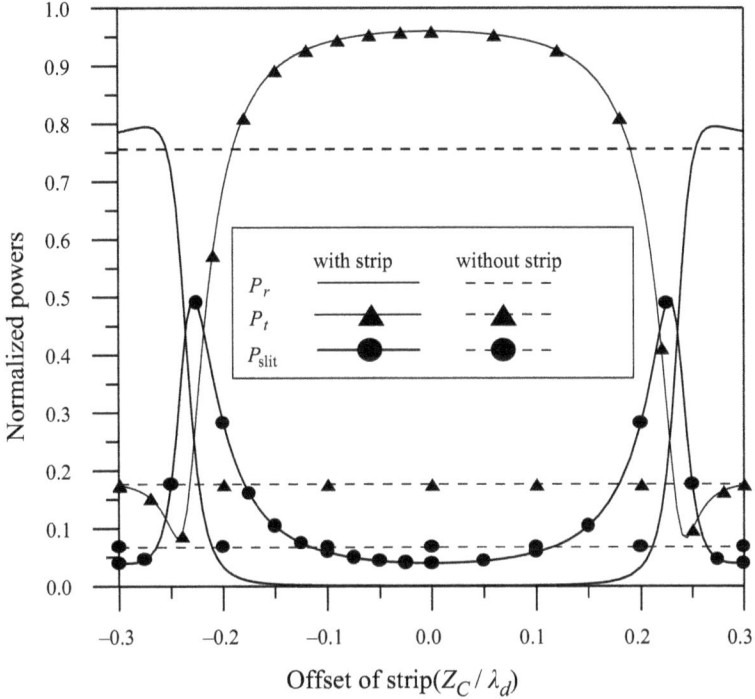

*Figure 2.17*   *Plots of $P_r$, $P_t$, and $P_{slit}$ versus $z_c/\lambda_d$ for $\varepsilon_r = \varepsilon_{rd} = 2.2$, $h = 0.015\lambda$*
*$\left(\lambda = \lambda_0/\sqrt{\varepsilon_r}\right)$, $h_d = 0.0155\lambda_d$, $2a = 0.003\lambda$, $S = 0.0135\lambda_d$,*
*$L = 0.475\lambda_d$, and $t = 0.002\lambda_d$*

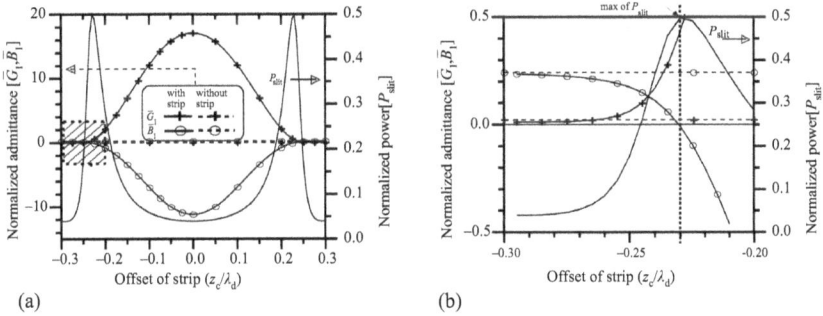

(a)                                              (b)

*Figure 2.18*   *Radiation from the slit and its equivalent circuit representation.*
*(a) Plots of normalized slit conductance $\overline{G_1}$ and susceptance $\overline{B_1}$*
*(along with $P_{slit}$) versus $Z_c/\lambda_d$ for the same parameters as those in*
*Figure 2.17. (b) Enlarged view of curves (in shaded region) for*
*$\overline{G_1}, \overline{B_1}$ and $P_{slit}$ near the offset position for maximum coupling.*

element representation, which is only valid for a narrow slit case where the shunt element $\overline{Y}_2$ for a slit of arbitrary width ($2a$) can be neglected as discussed above.

From Figure 2.17, with the offset position $Z_c/\lambda_d = \pm 0.23$, two main peaks of $P_{slit}$ were observed, indicating where the maximum coupling through the slit occurred and amounting to about 50% of the incident power $P_{inc}$ in the PPW. This quantity of 50% is the maximum radiative coupled power available from the slit, as shown in the equivalent circuit for the slit in Figure 2.19. In detail, at the two positions, the conductance, $\overline{G}_1$ was observed to be 0.43, which was comparable to the value of 0.5 corresponding to the series combination of the two-characteristic admittance (as the input admittance at the slit downward into the lower feeding PPW), as shown in Figure 2.19. The susceptance became almost zero ($\overline{B}_1 = -0.01$, though slightly inductive), as seen from the enlarged view in Figure 2.18(b). For reference, the ratio of the radiated power $P_{sp}$ to $P_{slit}$ amounted to 96% or more. As such, the surface wave power $P_{sf}$ was negligible.

In the present case of $S = 0.0135\ \lambda_d$, the strip embedded in the dielectric slab and the slitted upper plane of the PPW is considered as a transmission line terminated by open ends (or magnetic walls) on both ends of the strip, which forms an electromagnetic cavity. Therefore, when the cavity was fed through the narrow slit by an incident TEM wave in the PPW whose frequency corresponded to the cavity resonance, a fundamental $TM_{01}$ (to the x-direction) mode is set up inside the cavity independent of the offset position of the strip. For the x-component electric field of the $TM_{01}$ (to the x-direction) mode inside the cavity, one half-cycle variation along the z-direction is used. As a result the strip offset, which produces the maximum coupling through the slit, is observed at two offset positions symmetrical about the zero offset position. This is the same as in the case of a normal rectangular microstrip patch antenna where the impedance matching points are found at two points symmetrical about the center of the patch length. The cavity resonance that gives the maximum radiative coupling of $P_{slit}\left(\cong P_{sp}\right)$ (see Figure 2.17) is found to occur when the length $L$ of the strip was slightly less than half-wavelength in the dielectric slab region. This resonance length $L$ ($\approx 0.475\lambda_d$) is approximately the same as that ($\approx 0.476\lambda_d$) obtained by the conventional transmission-line model in microstrip antenna theory. In this model [1], the resonance length $L$ is approximately given by $L \cong 1 - 2Im\left(\frac{\overline{Y}_L^e}{\pi}\right)\lambda_d/2$ under the assumption that $\left|\overline{Y}_L^e\right| \ll 1$, where $\overline{Y}_L^e$ means the normalized radiation admittance for the open end (radiating edge) of

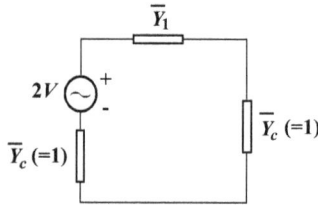

*Figure 2.19    Equivalent circuit for coupling slit*

the strip with respect to the characteristic admittance $(Y_c)$ of the PPW structure formed between the strip and the upper plane of the feeding guide as discussed in [9] and $Im\{\cdot\}$ denotes the imaginary part of $\{\cdot\}$. It should be noted that $Im\{\overline{Y}_L^e\}$ here corresponds to $B/Y_c$ in Figure 5 in [1]. In the calculation of $\overline{Y}_L^e$, the method in [9] was used because the strip was very thin.

As discussed, the coupling mechanism described above is the same as the radiation mechanism in the conventional transmission-line model of a rectangular microstrip antenna. In this sense, the aforementioned coupling used in this paper will hereafter be referred to as "cavity-type radiation" to distinguish it from parasitic-type radiation, which will be discussed later.

It is worth noting that the above expression for the resonance length for maximum radiation is the same as that for the resonant thickness for maximum transmission in the previously mentioned narrow slit transmission problem in thick conducting screens [15, p. 619, equation (43)] except that the radiation susceptance for the former case is replaced by that of a flanged PPW in the latter case.

Accordingly, the transmission problem through the narrow slit in thick conducting screens [15] and the problem related to the present cavity-type radiation are, in essence, the same, in that the two geometries form a cavity and maximum transmission and coupling occur when the same resonance condition is met. The only difference is in the location of the feeding point.

As seen in Figure 2.18, when no strip was present, the conductance of the narrow slit itself was very small and the susceptance was capacitive. In contrast, in the case of maximum coupling, the presence of a nearby strip increased the conductance to about 0.43, thereby providing the required inductive susceptance for resonance, as mentioned above. As a result, the maximum coupling (radiation) occurred when the susceptance was very near the zero crossing point, as seen in the enlarged view in Figure 2.18(b), which shows the usual resonant feature. It is interesting to note that once a cavity structure between the strip and the upper plate of the PPW was formed, the maximum value and two main peaks for the coupled power $P_{slit}$ through the slit were observed to be almost independent of the separation $S$ between the strip and the upper plate of the PPW for a small range of values for $S$ (as partly shown in Figures 2.17, 2.18, and 2.20). Note that $S = 0.0135\lambda_d$ in Figures 2.17 and 2.18, whereas $S = 0.015\lambda_d$ in Figure 2.20. This is analogous to the previous observation that the transmission-slot width product is independent of the actual slot width for a narrow slot at a resonance in the aforementioned transmission resonance problem [15]. Therefore, the present maximum coupling problem is an example of the aperture-body resonance problem presented in [29].

As the separation S between the strip and the upper plate of the PPW increases, the two sharp peaks of the maximum coupling $P_{slit}$ becomes weaker and the strip length, which gives the maximum coupling of $P_{slit}$, decreased gradually, as shown in Figure 2.20, that is, as the separation S increases, the two sharp peaks approach each other plus; simultaneously, the region of the small coupling level of $P_{slit}$ between the two peaks gradually swells up. Consequently, a different coupling feature occurs, as shown by the dashed line marked with dots in Figure 2.20.

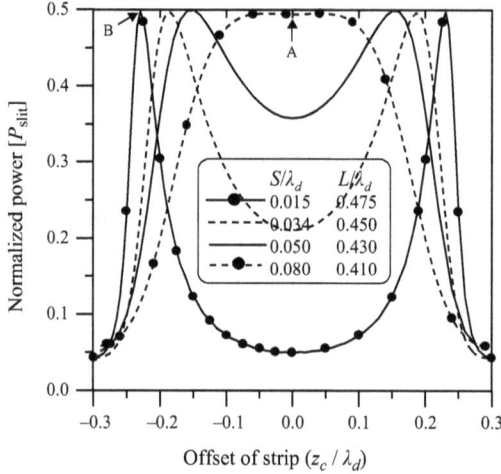

*Figure 2.20    Plots of $P_{slit}$ versus $z_c/\lambda_d$ with the separation S, dielectric slab thickness $h_d(= s + t)$, and the strip length L as parameters for $\varepsilon_r = \varepsilon_{rd} = 2.2, h = 0.015\lambda, 2a = 0.003\lambda$, and $t = 0.002\lambda_d$*

Therefore, $P_{slit}$ becomes maximum when the offset is zero. In addition, $P_{slit}$ is insensitive to the offset near the zero offset position and thus remains almost constant over the range of $|z_c/\lambda_d| < 0.1$. This is in contrast to the case of cavity-type radiation since the role of the strip is similar to that of the parasitic director in a three-element Yagi–Uda array in that main beam enhancement is in the broadside direction from the slit toward the strip for a larger separation S and shorter strip length L in comparison with the cavity-type radiation case.

Therefore, this type of radiation is called "parasitic-type radiation" as opposed to cavity-type radiation. It would appear that the recently reported chip-sized antenna [33] employs this parasitic type of radiation.

In the case of parasitic-type radiation, an undesired surface-wave problem can result from an increased dielectric slab thickness. For example, the surface-wave power $P_{sf}$ amounted to 13% of $P_{slit}$ in the case of point "A" in Figure 2.20. In the low dielectric case without such a surface-wave problem, where $P_{slit}$ becomes $P_{sp}$, the radiated power through the slit when $\varepsilon_{rd} = 1$ is given in Figure 2.21.

In this calculation, the separation values S and the strip length $L$ for maximum radiation are determined with the other parameters being the same as those in Figure 2.20. In Figure 2.21, the admittance behavior of the slit in the parasitic case is observed to be quite different from that in the cavity-type case. That is, the maximum radiative coupling $(P_{slit} = P_{sp})$ through the slit only occurs when the normalized conductance $(\approx 1/2)$ and susceptance of the slit become stationary values, i.e., their own maximum and minimum, respectively, which is in contrast to the admittance behavior in the cavity-type case. Furthermore, in the parasitic-type case, the maximum radiated power from the slit is observed to amount to about

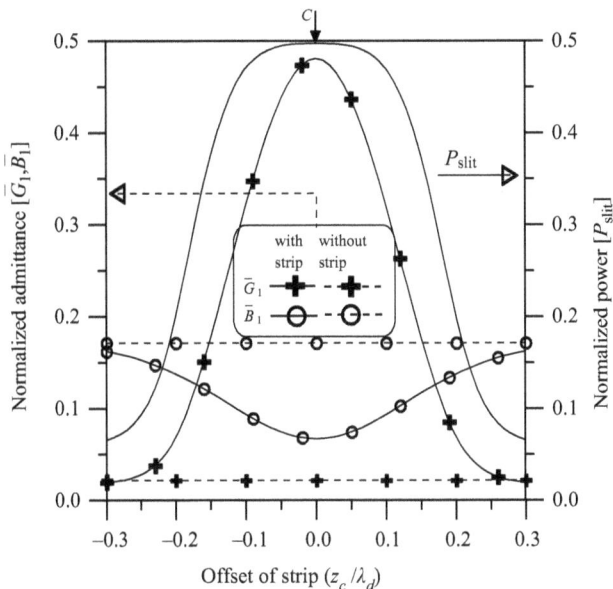

*Figure 2.21* *Plots of the normalized slit conductance $\overline{G}_1$ and susceptance $\overline{B}_1$ along with $P_{slit}$ versus $\frac{z_c}{\lambda_d}$ for $\varepsilon_r = 2.2, \varepsilon_{rd} = 1, h = 0.015\lambda$, $2a = 0.03\lambda, s = 0.054\lambda_0, L = 0.4\lambda_0$, and $t = 0.002\lambda_0$*

50% of $P_{inc}$, which is the maximum available radiated power from the slit, as in the aforementioned cavity-type case. For reference, the far-zone field patterns $\left|H_y^0(p, \phi)\right|$ for both the cavity and parasitic types of radiation are given for comparison in Figure 2.22, along with the pattern for the case where no strip and no dielectric slab are present. There the radiation beam for the parasitic type is observed to be more directive than that for the cavity type. The validity of the numerical results is confirmed by checking the power conservation relations, $P_r + P_t + P_{slit} = P_{inc}$ inside the PPW and $P_{sp} + P_{sf} = P_{slit}$ outside the PPW.

So far we have discussed two electromagnetic coupling mechanisms, cavity-type and parasitic-type radiation, observed in the two-dimensional coupling problem related to a conducting strip through a narrow slit in a parallel-plate waveguide.

## 2.4.3 Impedance matching structures

In the previous section, we have dealt with electromagnetic coupling structures whose maximum radiation efficiency amounts to roughly 50%.

The 50% of the radiation efficiency is found to be the maximum bound of the coupling structure in Figure 2.16 irrespective of whether cavity coupling or proximity coupling mechanism is employed. Here we consider coupling structures whose radiation efficiency can be made to be almost 100%.

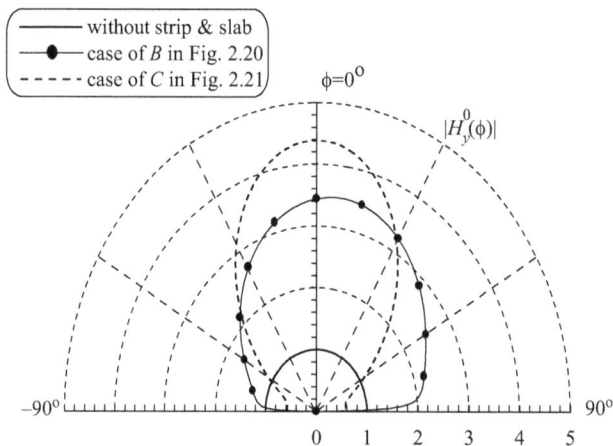

*Figure 2.22   Far-zone field patterns $\left| H_y^0(p, \varnothing) \right|$ for both the cavity type (case of B in Figure 2.20) and parasitic type (case of C in Figure 2.21) in comparison to cases with no strip and no dielectric slab*

### 2.4.3.1   Aperture-coupled and proximity-coupled structures

As perfect matching circuit structures, we deal with simplified models of aperture coupled [28] and proximity-coupled microstrip antenna structures [34], which use narrow and wide slits as feeding apertures respectively.

Here a narrow slit means a slit whose width is so narrow that the shunt admittance element $\overline{Y_2}$ in the equivalent $\pi$-circuit description for the slit are neglected and so the narrow slit discontinuity only appears as a simple series admittance $\overline{Y_1}$ as shown in Figure 2.7. In contrast, a wide slit means a slit that is so wide that the power transmitted to the guide beyond the slit region becomes negligible. As a result, the semi-infinite upper conducting plane beyond the wide slit region can be removed without affecting the boundary value problem as shown in Figure 2.9. The equivalent circuit for this case is represented in terms of the load admittance $\overline{Y_L}$ (input admittance looking toward the right side from the open end at $z = -a$, as shown in Figure 2.9). When used as radiators, narrow and wide slit structures both have capacitively stored energy and very poor radiation efficiency. However, when a parasitic conductor, such as a strip, is placed near the slit along with either a shorting stub in the case of a narrow slit or just a nearby conductor in the case of a wide slit, most of the incident power can be radiated into the upper-half free space.

Accordingly, here the emphasis is on describing two coupling mechanisms that can produce such perfect matching in the cases of narrow and wide slits as a closely related discussion to small flush antenna structures [35] in small antenna technology.

The wide slit means the open-end structure and is so suitable for feeding the proximity-coupled type structure.

As a slit coupling structure to the nearby scatterer in which perfect matching (or 100% radiation efficiency) can be achieved, we consider two structures. One is the simplified model of the aperture-coupled microstrip antenna structure which is fed by use of the narrow slit as shown in Figure 2.23(a). The other one is the simplified model of the proximity-coupled microstrip antenna structure which is fed by use of the wide slit, i.e., open-end structure as shown in Figure 2.23(b). The analysis method used here is borrowed from the methods [27,30,31] with certain modifications to the conducting strip on the dielectric slab [36] as shown in Figure 2.23 so omitted here. Two types of coupling mechanisms can facilitate perfect matching, i.e., cavity type and parasitic type coupling as discussed in the previous Section 2.4.2, with a narrow slit to the strip conductor coupling problem and no short stub introduced in the feeding PPW where the maximum radiative coupling efficiency through the slit is about 50%.

These two types of coupling mechanisms are observed also for both cases of aperture-coupled and proximity-coupled types under consideration. Figures 2.24(a) and (b) show typical frequency responses of the radiated power $P_{slit}$ through the slit and the equivalent circuit parameters for cavity type and parasitic type respectively, in the case of the aperture-coupled and the proximity-coupled structure as shown in Figure 2.23(a).

A closer look at the two curves of Figure 2.24(a) and (b) reveals that the $Q$ of the cavity type is significantly larger than that of the parasitic type. In other words, in cavity-type coupling, the matching performance is very sensitive to variations in the slit location, as expected from the resonance property of the cavity, whereas the matching performance in parasitic-type coupling is insensitive to an offset near $z_c = 0$, thereby remaining almost constant over the range of $|z_c/\lambda_d| < 0.1$ as in [36]. It is to be noted also that the parasitic type of coupling occurs with larger values of $h_d$ and a smaller strip length L than in cavity-type coupling.

Under perfect matching conditions, cavity-type coupling can be thought of as the extreme case where the separation $h_d$ is so small that a strong resonant mode

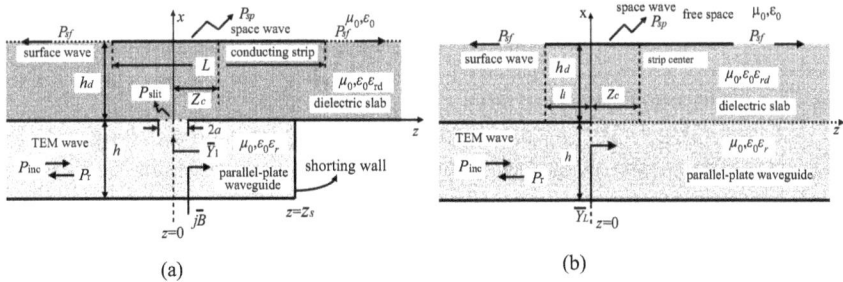

(a)                                                          (b)

*Figure 2.23*    *Simplified structures of the two representative microstrip antennas of aperture coupled- and proximity-coupled types. (a) Geometry for narrow slit problem as a simplified structure of aperture-coupled microstrip antenna. (b) Geometry for wide slit problem as a simplified structure of proximity-coupled microstrip antenna.*

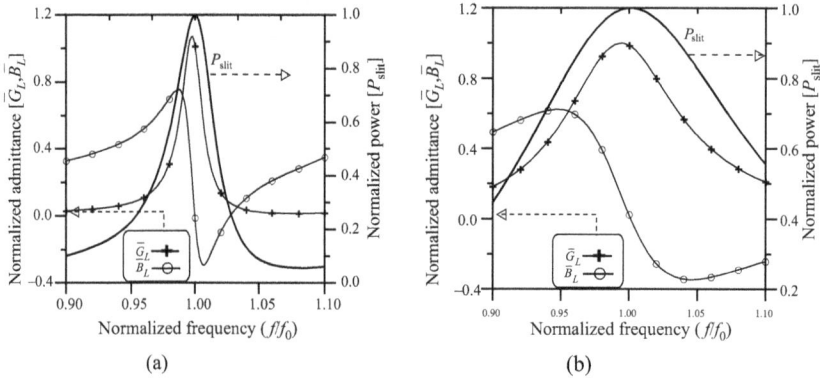

*Figure 2.24*    *Radiation and input impedance characteristics of the cavity- and*
*parasitic types. (a) Normalized coupled power $P_{slit}$ through slit and*
*load admittance $\overline{Y_L} \left( = \overline{G_L} + j\overline{B_L} \right)$ relative to frequency $(f/f_0)$,*
*$\varepsilon_r = \varepsilon_{rd} = 2.2, h = 0.015\lambda, 2a = 0.002\lambda, hd = 0.01\lambda d,$*
*$L = 0.482\lambda d, z_c = 0.226\lambda d,$ and $z_c = 0.5\lambda atf/f_0 = 1$ (cavity-type).*
*(b) $P_{slit}$ and $\overline{Y_L} \left( = \overline{G_L} + j\overline{B_L} \right)$ relative to frequency $(f/f_0)$,*
*$\varepsilon_r = \varepsilon_{rd} = 2.2, h = 0.015\lambda, 2a = 0.002\lambda, hd = 0.066\lambda d,$*
*$L = 0.417\lambda d, z_c = 0,$ and $z_s = 0.5\lambda$ at $f/f_0 = 1$ (parasitic type).*

field is set up underneath the strip. Conversely, parasitic-type coupling corresponds
to the other extreme and only occurs when the values of $h_d$ are considerably larger
than those in the cavity-type coupling case.

Also in the case of the proximity-coupled type, two types of coupling
mechanisms are observed in the same way as the aperture-coupled type from the
viewpoint of frequency characteristics of radiated power through the slit and
equivalent circuit.

So to avoid redundancy, we do not go into the discussion on the coupling
mechanism for the proximity-coupled type. For more details on this, see [36].

### 2.4.3.2   Field configuration and current distribution

Here we discuss the field configuration underneath the nearby conducting strip and
the current distribution on the conducting strip for the two cases in the aperture-
coupled type of structure.

The cavity-type coupling is defined as the type observed when the fundamental
cavity mode (electric) field is strongly excited underneath the strip, as shown in
Figure 2.25. This mode field is exactly the same as that for the $TM_{01}$ mode
(in the x direction) in a normal rectangular microstrip antenna structure. This type
of coupling only occurs when the separation $h_d$ between the strip and the upper
plate of the PPW is so small that an electrical cavity is formed underneath the
strip. As such, for perfect matching, the strip length L should be equal to the strip
resonance, as discussed in Lee *et al.*'s study [32].

strip

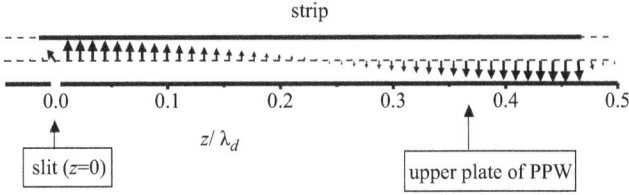

*Figure 2.25 Electric field configuration for cavity type.*
$\varepsilon_r = \varepsilon_{rd} = 2.2, h = 0.015\lambda, 2a = 0.002\lambda,$
$hd = 0.01\lambda d, L = 0.482\lambda d, z_c = 0.226\lambda d, and\ z_s = 0.5\lambda$

strip

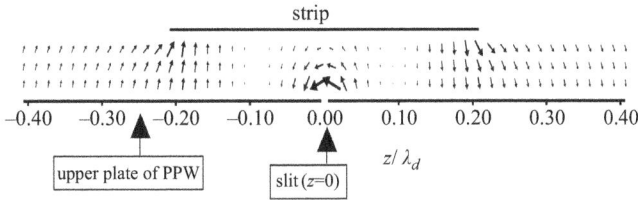

*Figure 2.26 Electric field configuration for parasitic type. $\varepsilon_r = \varepsilon_{rd} = 2.2$,*
$h = 0.015\lambda, 2a = 0.002\lambda, hd = 0.01\lambda d, L = 0.417\lambda d, z_c = 0,$
*and $z_s = 0.5\lambda$*

In contrast, parasitic-type coupling is defined as near-field coupling where the coupling field configuration from the slit to the strip (over the region $|z_c/\lambda_d| \leq 0.1$ underneath the strip) is essentially the same as that of the electrostatic field when a DC voltage source is applied across the slit and a parasitic conductor is present near the slit, as in the structure shown in Figure 2.26. The difference in the field configuration is the fundamental distinction between the two types of coupling.

As mentioned above, cavity-type coupling uses a smaller separation $h_d$ than parasitic-type coupling. As such, since a strong fundamental mode field ($TM_{01}$ mode) is excited underneath the strip, the magnitude of the induced electric current density on the strip (as well as the fields underneath the strip) is considerably larger than that in parasitic-type coupling, as shown in Figure 2.27.

In the case of cavity-type coupling, due to the half-cycle variation of the $x$-component electric field of the fundamental $TM_{01}$ mode along the $z$-direction underneath the strip, the strip offset $Z_c$, which produces perfect matching, was observed at two offset positions (near the strip edges) symmetrical about the zero offset position, similar to the feeding locations of normal rectangular patch antennas [37]. Conversely, in the case of parasitic-type coupling, the matching point location of the slit was near the strip center, similar to the discussion in [32].

In the case of parasitic-type coupling, an undesired surface wave problem can result from an increase in the dielectric slab thickness $h_d$. Yet when a dielectric substrate is employed with a low dielectric constant, such as $\varepsilon_{rd} = 1.2$, the surface

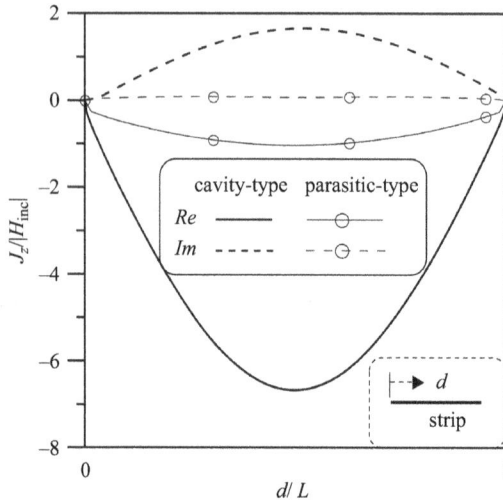

*Figure 2.27*   *Current distributions over conducting strip for both cavity-type and parasitic-type. Cavity-type: case in Figure 2.25. Parasitic-type: case in Figure 2.26.*

wave power level became negligible, as discussed in [32]. In the case of a low dielectric constant (e.g., as $\varepsilon_{rd} = 1.2$) without such a surface wave problem, the main beam radiation became more directed in comparison with that in the cavity-type coupling, as described in [32]. As such, a gain enhancement of almost 1.8 dB was achieved in the case of parasitic-type coupling. This is also one of the main differences between the two types of coupling.

Due to the gain enhancement along with the coupling field characteristics shown in Figure 2.26, the strip behaves more as a director of the two-element Yagi–Uda array instead of forming a cavity along with the upper plate of the PPW, thereby explaining the use of the term "parasitic-type" in contrast to "cavity-type", as described in [32].

Although the two types of coupling are quite distinct from each other, in both cases, the role of the strip is the same in that its presence increases a very small load conductance, which the load admittance would have without the presence of the strip, to almost the same value as the characteristic admittance of the feeding PPW, plus it provides the required inductive susceptance for resonance (i.e., cancellation of reactive component).

## 2.5   Aperture-body resonance problem

So far we have dealt with the electromagnetic coupling problem through a narrow slit to a nearby conducting body, in which case a fundamental guided mode wave is assumed to be incident upon the slit in the guiding structure. In this case, we describe the coupling problem from the viewpoint of impedance matching.

On the other hand, when we deal with an electromagnetic coupling problem between two half-space regions separated by a slot-perforated conducting screen, it is more convenient to describe the problem from the viewpoint of transmission cross section [38] or transmission width [15]. The concept of transmission cross section (TCS) is useful to grasp the meaning of funneling and high-density energy stream through a narrow slit or small aperture at resonance. The reason for dealing with this aperture-body resonance (ABR) problem here is to contrast with the transmission-cavity resonance (TCR) problem in Chapter 5 as two representative transmission resonance problems.

### 2.5.1 Transmission resonance of the aperture-body resonance problem

In a broad sense, aperture-body resonance (ABR) problem means the problem which is composed of a small aperture backed by a nearby conducting body with the main interest focused on investigating the condition under which the transmitted power through the small aperture is increased significantly over that of the case in which the scatterer near to the small aperture is not present.

As a simplified model of the ABR problem, the two-dimensional problem composed of a narrowly slitted conducting screen and a nearby conducting strip, as shown in Figure 2.28(a) is considered. In Figure 2.28(a), $\mu_0$ and $\varepsilon_0$ are the permeability and permittivity of the free space, respectively. $E_{inc}(H_{inc})$ is the electric (magnetic) field incident on the slit with incidence angle $\theta_0$ with respect to the z-axis, $H_0$ is the amplitude of the incident magnetic field, and $k_0\left(=\frac{2\pi}{\lambda_0}, \lambda_0;\ \text{free space wavelength}\right)$ is the propagation constant in the free space. $a$ is the width of the slit and $d$ is the distance between the slitted conducting screen and the nearby conducting strip. The equivalent circuit representation for this structure with a narrow slit can be given as in Figure 2.28(b). $I_s$ is the equivalent

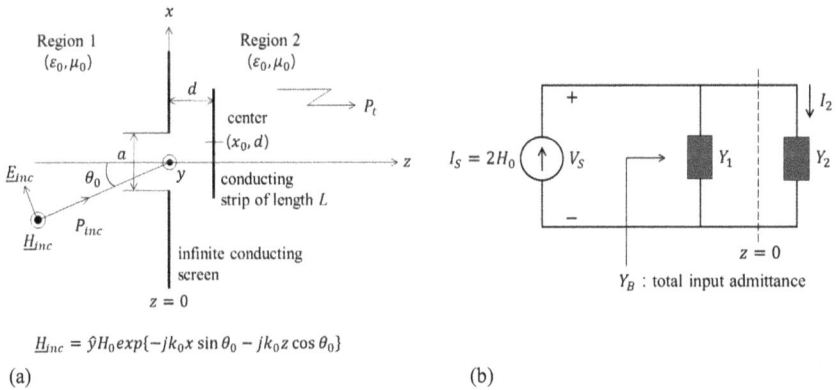

$$H_{inc} = \hat{y}H_0 exp\{-jk_0 x \sin\theta_0 - jk_0 z \cos\theta_0\}$$

(a)                                                                (b)

*Figure 2.28   (a) The ABR structure under consideration, (b) its equivalent circuit representation*

source current corresponding to a sum of the incident and reflected magnetic fields in Figure 2.28(a).

For this analysis, the equivalence principle is employed as in the analysis method in Section 2.1.1 in order to divide the original problem into two equivalent situations by placing appropriate equivalent magnetic currents over both sides of the shorted slit. Pair of coupled integral equations, whose unknowns are the magnetic current density on the shorted slit and the induced electric current density on the conducting strip, are set up. Next, the pulse basis and Galerkin scheme for the unknown magnetic current densities over the slit, together with the piecewise Galerkin scheme for the unknown induced current densities on the strip, are employed in order to reduce the aforementioned coupled integral equation to a linear equation system. Accordingly, based on knowledge of the unknown current densities, the quantities of interest are obtained, including the coupled (transmitted) power $P_t$ through the slit into the region 2, transmission width [15] or apparent width $T$ of the slit, and the equivalent circuit parameters of Figure 2.28(b) for the input admittances, $Y_1$ and $Y_2$, as seen by the slit looking into the region 1 and 2 respectively, at the slit. For further details, see [27,37].

When the problem is specialized to the case of a narrow slit, i.e., $k_0 a \ll 1$, with no nearby conducting strip present, the admittance of Figure 2.28(b), $Y_1$ and $Y_2$, looking into the left and right half spaces, respectively, are the same as

$$Y_1 = Y_2 = G + jB = k_0/2\eta_0 \left[ 1 - j\frac{2}{\pi}\log(K_0 aC) \right] \tag{2.24}$$

In which $C = 0.1987$ [15]. Note that the conductance $G$ ($= k_0/2\eta_0 = 1/120\lambda_0$[S/m]) is the typical value of conductance appearing in the various types of narrow slit problems, such as narrow and wide transverse slit [9,20] in a parallel-plate waveguide (PPW) of small guide height, and a narrow slit in a flanged PPW [39], where $\eta_0$ is the intrinsic impedance of the free space and is given by $120\pi$. It is also worth mentioning that the susceptance $B$ becomes approximately equal to that of a narrow slit in a flanged PPW of a small guide height. This is because, in the latter case, most of the reactive power near the slit is confined to the exterior region, and so the reactive power is very small in the interior region near the flanged slit [39]. In the case of a narrow slit with no nearby scatterer present, the conductance $G$ of the slit is much smaller than the susceptance, i.e., $G \ll B$. From the equivalent circuit in Figure 2.28(b), the transferred power $P_t$ to the load $Y_2$, corresponding to the transmitted power through the slit into the region 2, is given as

$$P_t = \frac{1}{2}\text{Re}\{V_s I_2^*\} = \frac{1}{2}\frac{H_0^2 G}{G^2 + B^2} \text{[W/m]} \tag{2.25}$$

which is seen as being very small in the case of a narrow slit.

The power transmitted by an electrically narrow slit backed by a nearby conducting strip, however, can be much larger than that when the strip is absent. In the equivalent circuit for a narrow slit, when the strip is present in region 2, the admittance $Y_2$ would be changed, whereas the admittance $Y_1$ remains the same, irrespective of the presence of the strip in region 2. Maximum power transmission and its condition have been searched by inspecting the transmitted power $P_t$ for various combinations of the strip length $L$ and the location $(X_0, d)$ of the nearby strip center with a narrow slit width kept unchanged.

When the maximum power transmission occurs, the input admittance $Y_2$ looking into the region 2 is found to be a complex conjugate of $Y_1$, i.e., $Y_2 = Y_1^* = G - jB_1$. As a result, the total admittance $Y_s(= Y_1 + Y_2 = 2G = 1/60\lambda_0 \,[\text{s/m}])$ becomes real value. This corresponds to the ABR condition, under which the maximum transmission power $P_t$ through the slit is obtained as $H_0^2/2G = 60\lambda_0 H_0^2 \,[\text{W/m}]$ by use of the simple circuit theory in Figure 2.28(b). Note that $P_t$ has the physical dimension of [W/m], i.e., transmitted power per unit length along the y-axis, whereas the projected component of the incident power density $P_{inc}$ on the screen for an arbitrary incidence angle $\theta_0$ is given by

$$P_{inc} = \frac{1}{2}\eta_0 H_0^2 \cos\theta_0 \,\left[\text{W/m}^2\right] \tag{2.26}$$

Therefore, if we define the ratio of $P_t/P_{inc}$ so that the ratio has a dimension of length, this ratio $T(= P_t/P_{inc})$ [m] can have a physical meaning of transmission width or apparent width of the slit along the x-axis transverse to the slit axis. This ratio will be interchangeably used as the transmission cross section $T$ in Chapter 3. In particular, for normal incidence (i.e., $\theta_0 = 0$), this ratio becomes a maximum of $\lambda_0/\pi \,[\text{m}]$, which, interestingly enough, corresponds to the effective height of a half wavelength dipole antenna [40] as well as the transmission cross section T for the narrow slit in the thick conducting slab which is dealt with in Chapter 3.

This can be explained in a physical sense as follows: though the plane wave is incident upon the slit whose actual width is denoted as "a", the incident power upon the apparent slit width of $\lambda_0/\pi$ [m], greatly exceeding the actual width "a" is effectively concentrated into the slit and then radiated into the right half-space, all within the framework of the ABR condition [41]. As a result of this, the power transmission through the slit is remarkably enhanced, regardless of the actual slit width.

We have investigated the variation of the transmission width $T$ of the slit against the lateral strip offset $X_0$ for sets of $\{L, X_0, d\}$ found for the maximum transmission under the ABR condition. Figure 2.29 shows two contrastive types of the ABR phenomena. The curve of a solid line shows two sharp peaks of the power transmission at the two offset positions, $X_0/\lambda_0 \cong \pm 0.24$. This means that the variation of $T$ is very sensitive to the offset $X_0$.

On the other hand, the curve of the dashed line reaches its maximum when the offset is zero. In addition, this curve at maximum is insensitive to the offset near the zero offset position $X_0 = 0$. Note that the maximum values of $T(= \lambda_0/\pi)$ for both the former and the latter cases are the same. For the former case, the cavity mode field is strongly excited underneath the strip as in a normal rectangular microstrip antenna structure. This type of maximum power transmission only occurs when the separation d between the strip and the conducting screen is so small that an

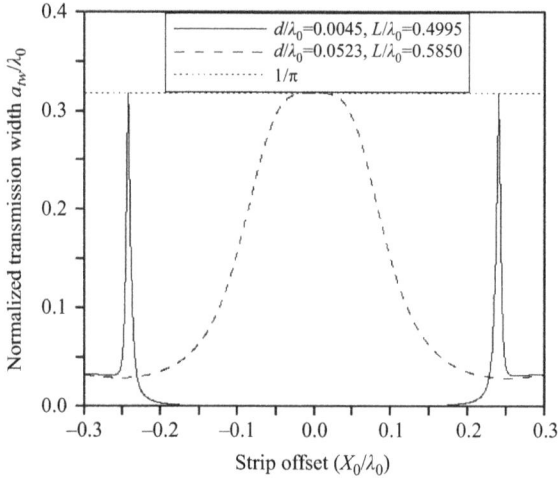

*Figure 2.29    Normalized transmission width $a_{tw}/\lambda_0$ of the slit versus the normalized strip offset $(X_0/\lambda_0)$*

electrical cavity is formed underneath the strip. Accordingly, under the condition of maximum power transmission, the former type can be regarded as an extreme case in which the separation $d$ is so small that a strong resonant mode field is set up underneath the strip, as mentioned above. Conversely, the latter type of maximum power transmission corresponds to the other extreme case and only occurs when the values of d are considerably larger than those in the former case. Although the two cases of maximum transmission are different from each other, the role of the strip is the same in that its presence provides the required inductive susceptance for transmission resonance; in other words, $Y_2$ is made to be $Y_1^*$, i.e., $Y_2 = Y_1^*$. These two kinds of coupling are similar to those in the prior work on the slit-to-strip coupling in the PPW structure [36].

The ABR problem is essentially the same as the transmission cavity resonance (TCR) problem from the viewpoint of power funneling and TCS. The similarity between the two problems is to be discussed in Chapter 3.

## 2.5.2    Resonance condition of the rectangular microstrip patch antenna

As is well-known, a microstrip patch antenna of a halfwave rectangular type radiates from the open ends. So the microstrip radiating element can be represented by two slots separated by a transmission line of low characteristic impedance. Because of the reflection coefficient approaching +1 at the slot, a magnetic wall is set up along the slot. As a result lossy cavity is formed where the loss means radiation from the two slots. Rectangular microstrip patch [1] can be represented by an equivalent circuit as shown in Figure 2.30. Note that the reflection coefficient $\Gamma$ is near +1 at the slit and so a magnetic wall is set up.

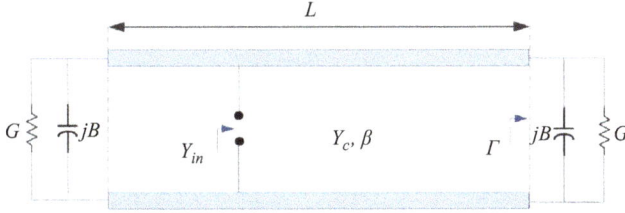

*Figure 2.30    Equivalent circuit representation of the rectangular microstrip patch antenna*

In Figure 2.30, L means the rectangular patch length, $G + jB$ means the radiation admittance, $B$ and $Y_c$ mean the propagation constant and characteristic admittance, respectively. For the equivalent circuit to be resonant, the input admittance $Y_{in}$ at an arbitrary point should be real, i.e., an imaginary part of the input admittance should be canceled. From this condition, the desired resonant condition for maximum radiation is found to be

$$\tan\beta L = \frac{2Y_c B}{G^2 + B^2 - Y_c^2}.$$    (2.27)

It is interesting to note that this expression is essentially the same as that for the transmission resonance in the array structure of slits [42], which has been widely investigated in connection with the extraordinary optical transmission (EOT) phenomena [3]. It is worthwhile to mention that the above condition corresponds to the lowest order mode of the Fabry–Perot transmission resonance condition [42] observed in the thick slit structures relevant to the EOT phenomena.

## 2.5.3    Other transmission resonant structures

Transmission resonant cavity structure is formed also by overlapping two parallel conducting plates while keeping a narrow gap between them as shown in Figure 2.31(a).

The equivalent circuit is obtained through the use of the equivalent circuit model for the open end of the one-dimensional microstripline [9,20]. The partially overlapped part of two parallel plates can therefore be viewed as a resonator composed of a finite length ($L$) of transmission line with both ends terminated by a lossy magnetic wall as shown in Figure 2.31(b). In the figure, $Y_c$ and $\beta$ signify, respectively, the characteristic admittance and propagation constant of the TEM mode in the PPW region, while $G + jB$ signifies the radiation admittance, which can be obtained through the use of the method shown in [9]. In the present case, the conductance $G(= 1/120\lambda_0)$ is much smaller than the susceptance $B$, i.e., $G \ll B$ as discussed already.

It is observed that the transmission cross section for normal incidence amounts to $\lambda_0/\pi$[m], which is identical to the transmission width or the apparent width of the

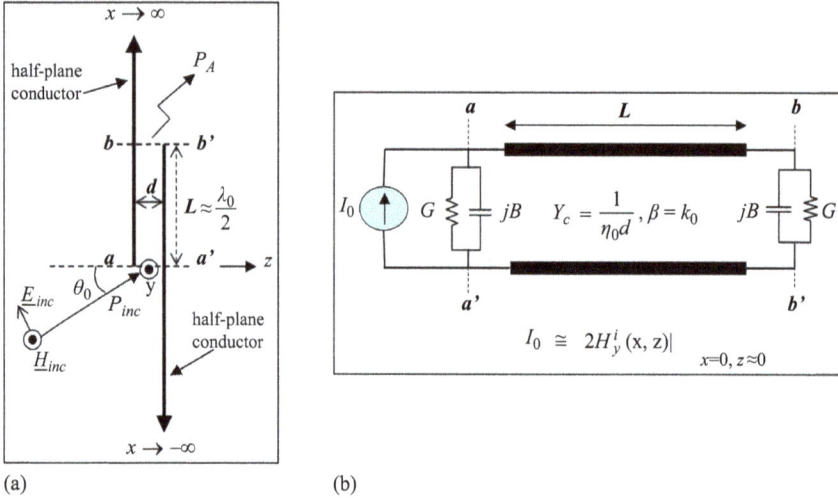

(a)                                    (b)

*Figure 2.31     Another transmission resonant structure and its equivalent circuit. (a)
Another transmission resonant structure, (b) its equivalent circuit.*

slit in [15], which is dealt with as a terminology of the transmission cross section in Chapter 3.

In addition when the transmitted power $P_t$ reaches its maximum which corresponds to the TCS value of $\lambda_0/\pi$, the total admittance at the reference plane a–a′ is found to be purely real value, i.e., 2G, after transforming the radiation admittance of the other end at the plane b–b′ along the transmission line of the electrical length $\beta L$ to the reference plane a–a′. Under this transmission resonance condition, we have

$$\tan\beta L = \frac{2Y_c B}{G^2 + B^2 - Y_c^2} \cong -\frac{2B}{Y_c}$$

since $G^2 + B^2 \ll Y_c^2$. This resonance condition is essentially the same as that of the radiative resonance condition in the transmission line model for rectangular microstrip antenna structure [1], as well as that of the maximum power transmission through the narrow slit in a thick conducting screen [15]. These three phenomena presuppose the formation of the resonant cavity in which a strong resonant field is established—and through which the maximum transmission or radiation results. In this sense, the present problem and the narrow slit coupling problem through a thick conducting screen can be described as a transmission cavity resonance problem as discussed in Chapter 3.

The maximum value $(\lambda_0/\pi[m])$ of the ratio $(P_t/P_{inc})$ can be physically interpreted as this. The normal incident power (for $\theta_0 = 0$) that corresponds to the width of $\lambda_0/\pi[m]$, which is much larger than that on the actual slit width a, is funneled [40] into the incident side slit $S_a$ and radiated through the output side slit $S_A$ into the region of the right half-space, irrespective of the actual slit width

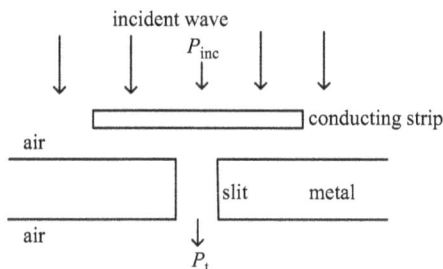

*Figure 2.32* *Transmission resonance structure formed by a nearby conducting strip just above the slit in a thin metallic film*

corresponding to the distance d between the planes under the transmission resonance condition as shown in [15]. It is interesting to note that the value of $\lambda_0/\pi$[m] is identical to the effective height of the half wavelength $(\lambda_0/2)$ dipole antenna as mentioned above.

The prior works on the transmission resonance [43] and EOT phenomena [44] have a direct bearing on the transmission resonance phenomena under consideration.

Leviatan [43] dealt with the electromagnetic coupling problem between two half-space regions separated by two slot-perforated parallel conducting planes. In his work, the resonance peaks of transmission coefficient were observed to occur at spacings between two slot centers along a transverse direction which approaches multiples of half wavelengths. Such transmission resonance is thought to belong to the transmission resonance type under consideration. Some authors [44] dealt with enhancing extraordinary transmission of light through a metallic nano slit with a nanocavity antenna as shown in Figure 2.32.

In this work, the transmission resonance structure is composed by introducing a combination of cavity and narrow slit in metal film with finite thickness. In this case too, the TCS is expected to be $\lambda_0/\pi$ by reciprocal consideration (by reversing the direction of the incident wave) of the present ABR problem. In this sense, this structure is thought to be categorized into the slit to nearby coupling problem under consideration.

# References

[1] Derneryd A.G. 'Linearly polarized microstrip antenna'. *IEEE Trans. Antennas Propagat.* 1976, vol. 24(6), pp. 846–851

[2] Takakura Y. 'Optical resonance in a narrow slit in a thick metallic screen'. *Phys. Rev. Lett.* 2000, vol. 86(24), pp. 5601–5604

[3] Ebbesen, T.W., Lezec, H.J., Ghaemi, H.F., Thio, T., and Wollf, P.A. 'Extraordinary optical transmission through subwavelength hole array'. *Nature(London)*. 1998, vol. 391(6668), pp. 667–669

[4]   Millar, R.F. 'Radiation and reception properties of a wide slot in a parallel-plate transmission line, part I and II'. *Can. J. Phys.* 1959, vol. 37(2), pp. 144–169

[5]   Keshavamurthy, T.L., Butler, C. M. 'Characteristics of a slotted parallel plate waveguide filled with a truncated dielectric'. *IEEE Trans. Antennas Propagat.* 1981, vol. 29(1), pp. 112–117

[6]   Cho, Y.K., Son, H. 'Characteristics of a parallel-plate waveguide with a narrow slit in its upper plate'. *Electron. Lett.* 1986, vol. 22(22), pp. 1166–1167

[7]   Oliner, A. A. 'Equivalent circuits discontinuities in balanced strip transmission line'. *IRE Trans. Microwave Theory Tech.* 1955, vol. 3(2), pp. 134–143

[8]   Abramowitz, M., Stegun I. A. (1968): *Handbook of Mathematical Functions.* Dover; 1968.

[9]   Cho, Y.K. 'On the equivalent circuit representation of the slitted parallel-plate waveguide filled with a dielectric'. *IEEE Trans. Antennas Propagat.* 1989, vol. 37(9), pp. 1193–1200

[10]  Butler, C.M., Wilton, D.R. 'General analysis of narrow strips and slots'. *IEEE Trans. Antennas Propagat.* 1980. vol. 28(1), pp. 42–48

[11]  Marcuvitz N. *Waveguide Handbook.* McGraw-Hill; 1951. no. 21

[12]  Bahl I. J. 'Build microstrip antennas with paper-thin dimensions'. *Microwaves.* 1979, pp. 50–63

[13]  James J.R., Hall P.S. *Handbook of Microstrip Antennas.* IEE Press; 1989

[14]  Harrington R. F. *Time-Harmonic Electromagnetic Fields.* McGraw-Hill; 1961. pp. 180–186

[15]  Harrington R. F., Auckland O. 'Electromagnetic transmission through narrow slots in thick conducting screen'. *IEEE Trans. Antennas Propagat.* 1980, vol. 28(5), pp. 616–622

[16]  Schwinger J., Saxon D.S. *Discontinuities in Waveguide.* Gordon and Breach; 1968. Ch. 1

[17]  Katehi, P.B., Alexopoulos, N.G. 'Frequency-dependent characteristics of microstrip discontinuities in millimeter-wave integrated circuits'. *IEEE Trans. Microwave Theory Tech.* 1985, vol. 33(10), pp. 1029–1035

[18]  Harrington R. F. *Field Computation by Moment Methods.* Macmillan; 1968, Ch. 3

[19]  Chang D.C. 'Analytical theory of an unloaded rectangular microstrip patch; *IEEE Trans. Antennas Propagat.* 1981, vol. 29, pp. 54–60

[20]  Lee J. I., Ko J.H., Cho Y.K. 'A note on diffraction and equivalent admittance properties of a transverse slit in a parallel plate waveguide filled with a homogeneous dielectric'. *IEICE Trans. Commun.* 2003, vol. E86-B(12), pp. 3600–3605

[21]  Collin R. E. *Foundation for Microwave Engineering.* McGraw-Hill; 1966

[22]  Liu C. C., Shmoy J., Hessel A. 'E-plane performance trade-offs in two-dimensional microstrip-patch element phased array'. *IEEE Trans. Antennas Propagat.* 1982, vol. 30(6), pp. 1201–1206

[23] Encinar J.A. 'Mode-matching and point-matching techniques applied to the analysis of metal-strip loaded dielectric antennas'. *IEEE Trans. Antennas Propagat.* 1990, vol. 38(9), pp. 405–412

[24] Lee C.W., Lee J.I., Cho Y.K. 'Analysis of leaky waves from a periodically slotted parallel-plate waveguide for finite number of slots'. *Electron. Lett.* 1994, vol. 30(20), pp. 1633–1634

[25] Garg R., Bhartia P., Bahl I., Ittipiboon A. *Microstrip Antenna Design Handbook.* Artech-House; 2001, pp. 90–102

[26] Jiso X., Wang P., Tang L, Lu Y. Li Q., Zhang D. *et al.* 'Fabry–Perot-like phenomenon in the surface plasmons resonant transmission of metallic gratings with very narrow slits', 2005, *Appl. Phys. B* 80, pp. 301–305

[27] Mannikko P.D., Courtney C.C., Butler C.M. 'Slotted parallel-plate waveguide coupled to a conducting cylinder'. *Proc. Inst. Elect. Eng. Pt. H.* 1992, vol. 139(2) pp. 193–201

[28] Sullivan P.L., Schaubert D.H. 'Analysis of an aperture coupled microstrip antenna'. *IEEE Trans. Antennas Propagat.* 1986, vol. 34(8), pp. 978–984

[29] Harrington R. F. 'Resonant behavior of a small aperture backed by a conducting body'. *IEEE Trans. Antennas Propagat.* 1982 vol. 30(2), pp. 205–212

[30] Butler C.M., Nevels R.D. 'Coupling through a slot in a parallel-plate waveguide covered by a dielectric slab'. *AEU.*, 1988, vol. 42, pp. 46–53

[31] Butler C.M., Courtrey C.C., Mannikko P.D., Silvestro J.W. 'Flanged parallel-plate waveguide coupled to a conducting cylinder'. *Proc. Inst. Elect. Eng. Pt. H.* 1991, vol. 138(6), pp. 549–559

[32] Lee J.I., Lee C.H., Cho Y.K. 'Electromagnetic coupling mechanism to a conducting strip through a narrow slit in a parallel-plate waveguide'. *IEEE Trans. Antennas Propagat.* 2001, vol. 49(4), pp. 592–596

[33] Sakai S., Arai H. 'Gain enhancement of chip-sized antenna using notched parasite element.' In *Proc. 1998 Asia-Pacific Microwave Conference*, vol. 3, Yokohama, Japan, Sept. 1998, pp. 1307–1310

[34] Herd J.S. 'Full wave analysis of proximity coupled rectangular microstrip antenna arrays'. *Electromagnetics* 1991, vol. 11, pp. 21–46

[35] Wheeler H.A. 'Small antenna'. *IEEE Trans. Antennas Propagat.* 1975, vol. 23, pp. 462–469

[36] Lee J.I., Cho Y.K. 'Maximum electromagnetic coupling to a nearby conducting strip through narrow and wide slits in a parallel plate waveguide'. *IEEE Trans. Antennas Propagat.* 2003, vol. 51(4), pp. 693–699

[37] Bahl J.I., Bhartia P. *Microstrip Antennas.* Artech House; 1980, pp. 48–55

[38] Park J.E., Cho Y.K. 'Comparison of transmission resonance phenomena through small coupling apertures between two kinds of transmission resonance structures'. *Proc. URSI International Symposium on Electromagnetic Theory*; 2010, pp. 898–902

[39] Lee J.I., Ko J.H., Cho Y.K. 'Coupling through a slit in a flanged parallel-plate waveguide with a conducting strip'. *Proc. International Symposium Antennas Propagation (ISAP).* 2004; vol. 1, pp. 589–592

[40]  Kraus J.D., Marhefka R.J. *Antennas for All Applications.* McGraw-Hill; 2003, pp. 30–36

[41]  Astilean S., Lalanne Ph., Palamaru M. 'Light transmission through metallic channels much smaller than the wavelength'. *Opt. Commun.* 2000, vol. 175, pp. 265–273

[42]  Francisco Medina, Francisco Mesa, Diana C.S. 'Extraordinary transmission though arrays of slits: a circuit theory model'. *IEEE Trans. Microwave Theory Tech.* 2010, vol. 58(1), pp. 105–115

[43]  Leviatan Y. 'Electromagnetic coupling between two half-space regions separated by two slot-perforated parallel conducting screen'. *IEEE Trans. Microwave Theory Tech.* 1988, vol. 36(1), pp. 44–52

[44]  Cui Y., He S. 'Enhancing extraordinary transmission of light through a metallic nano slit with a nano cavity antenna'. *Opt. Lett.* 2009, vol. 34(1), pp. 16–18

*Chapter 3*

# Transmission resonances in infinitely long and rectangular slot in thick conducting screen

## Synopsis

We examine the transmission resonance phenomena in two different scenarios: first, as a 2-dimensional problem involving an infinitely long narrow slit within a thick conducting screen, and then as a 3-dimensional problem, focusing on the transmission resonance phenomena through a small rectangular slot within the same thick conducting screen. It is important to note that, depending on whether there is cutoff frequency or not, the transmission resonance characteristics of the Fabry–Perot type become significantly changed. The 2-dimensional narrow slit structure is a representative one in which the cutoff frequency does not exist. It is worth noting that extensive research has been conducted on this 2-dimensional narrow slit structure in relation to the extraordinary optical transmission phenomenon.

While only one kind of Fabry–Perot transmission resonance phenomenon is observed in the 2-dimensional slit problem, there are two kinds of transmission resonance—Fabry–Perot type and near cutoff mode (having near zero group velocity), in the 3-dimensional slot problem as discussed later.

The Fabry–Perot type is related to the resonance along the guided direction of the transmission cavity, whereas the near cutoff mode corresponding to the zeroth order Fabry–Perot type is related to the transverse resonance which is observed at near cutoff.

The transmission cross section (TCS) for the 2-dimensional long slit problem is expressed as $\lambda/\pi$[m] per unit length. On the other hand, the TCS for the 3-dimensional small aperture problem is expressed as $\frac{2G\lambda^2}{4\pi}$ [m$^2$]. where $G$ and $\lambda$ mean, respectively, the directivity of the aperture and wavelength. The experimental verification for the presence of two kinds of transmission resonance phenomena is described for a 3-dimensional problem where the cutoff frequency exists for the lowest dominant mode. The background for the shortening factor of the magnetic dipole, i.e., slot dipole is discussed based on the observation of transmission resonance in the limit where the conducting slab thickness approaches zero.

In parallel, the physical origin of the nanoaperture shape such as C-shaped or H-shaped aperture, which is applicable to the design of near-field probes in the area of compact data storage or nanolithography, etc., is also discussed based upon the observation of the transmission resonance in the ridged aperture in thick conducting screen.

## 3.1    Transmission resonance through a long slit in a thick conducting screen

In this section, we consider the transmission resonance problem through an infinitely long slit in a thick conducting screen as a representative 2-dimensional problem [1].

Figure 3.1 illustrates a coupling issue occurring with a long slit when a plane wave is incident at a normal angle, and its electric field is oriented perpendicular to the axis of the slit. Note that the physical dimensions of $P_{inc}$ and $P_t$ are chosen to be those of the incident power density [powers/m$^2$] and of the transmitted power through the slit respectively such that the ratio of $P_t/P_{inc}$ may be the dimension [m$^2$] of the area meaning the transmission cross section (TCS) [2].

### 3.1.1    Analysis method

For the general case of dealing with more complex slit coupling problems including arbitrary dielectric loading and arbitrary slit width, a more rigorous approach of integral equation formulation and method of moments (MOM) solution should be employed. Here we consider only an approximation method that deals with solutions for narrow slits following the method in [1].

To this end, we go into the transmission problem through a narrow slit in a thick conducting screen with the main concern centering on the transmission resonance where the transmitted power through the slit becomes maximum.

That is, the transmitted power $P_t$ through the slit is calculated to be dissipated power in the load admittance $Y_L$ by the use of the equivalent circuit representation [1] as shown in Figure 3.2.

From the equivalent circuit, the transmitted power through the slit $P_t$ is found to be

$$P_t = |V_2|^2 \, \text{Re}(Y_L), \tag{3.1}$$

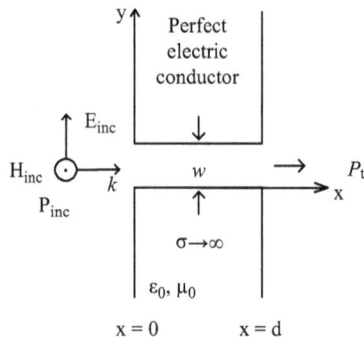

*Figure 3.1    2-dimensional slit of width w in thick conducting screen*

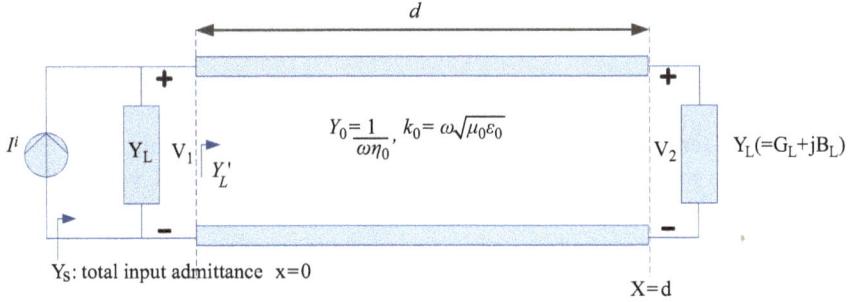

*Figure 3.2   Equivalent circuit representation for the transmission resonance problem in Figure 3.1*

where $Y_L$ means the radiation admittance $G_L + jB_L$ of the slit. Expressing it in terms of the transfer admittance

$$Y_{12} = \frac{I^i}{V_2}, \tag{3.2}$$

$P_t$ is given to be

$$P_t = \left|\frac{I^i}{Y_{12}}\right|^2 \text{Re}(Y_L), \tag{3.3}$$

where

$$Y_{12} = 2Y_L \cos k_0 d + j\left(Y_0 + \frac{Y_L^2}{Y_0}\right) \sin k_0 d \tag{3.4}$$

Here the load admittance, i.e., radiation admittance of the slit $Y_L$ is given by

$$Y_L = \frac{1}{\eta_0 \lambda_0}[\pi - 2j\ln(ck_0 w)], \tag{3.5}$$

where the constant c represents $\frac{\gamma}{\pi e} \approx 0.2086$.

Here, $\gamma = 1.781 \ldots$ and $e = 2.718 \ldots$ [1].

In the case of a normal incident plane wave whose magnetic field is

$$H_z^{\text{inc}} = H_0 e^{-jk_0 x}, \tag{3.6}$$

the incident power density is

$$P_{\text{inc}} = \eta_0 |H_0|^2.$$

The short circuit magnetic field $H_z^{\text{sc}}$ is twice the incident magnetic field on the conductor, i.e., $H_z^{\text{sc}} = 2H_0$. from which the impressed current source $I^i$ in the equivalent circuit representation in Figure 3.2 is

$$I^i = \frac{1}{w}\int_0^w H_z^{\text{sc}}\mathrm{d}_y = 2H_0 \tag{3.7}$$

which is the average electric current over the short-circuited aperture.

Here, as a measure of transmission efficiency through the slit, we can define the transmission coefficient as

$$T = \frac{P_t}{P_i}$$

If $P_i$ is defined to be the power incident on the slit per unit length in the z direction, the transmission coefficient is given as

$$T = \frac{4\text{Re}(Y_L)}{w\eta_0|Y_{12}|^2} \tag{3.8}$$

Note here that $T$ will be the maximum when $|Y_{12}|$ is minimum. Inserting the expression for $Y_L(= G_L + jB_L)$ into (3.4) and separating into real and imaginary parts leads to the following expressions:

$$\text{Re}(Y_{12}) = 2G_L\cos k_0 d - \frac{2}{Y_0}G_L Y_L \sin k_0 d \tag{3.9}$$

$$Im(Y_{12}) = 2B_L\cos k_0 d + \left(Y_0 + \frac{G_L^2 + B_L^2}{Y_0}\right)\sin k_0 d \tag{3.10}$$

For narrow slits, it is seen that $B_L \gg G_L$ from $Y_L = G_L + jB_L = \frac{1}{\eta_0\lambda_0}[\pi - 2j\ln(ck_0 w)]$ and that $Im(Y_{12}) \gg \text{Re}(Y_{12})$.

So we can minimize $|Y_{12}|$ by setting $Im(Y_{12}) = 0$. As $w$ approaches zero, since $Y_0 \gg \frac{G_L^2 + B_L^2}{Y_0}$, $Im(Y_{12})$ is approximated to be

$$Im(Y_{12}) \cong 2B_L\cos k_0 d + Y_0\sin k_0 d. \tag{3.11}$$

Then $Im(Y_{12}) = 0$ when $\tan k_0 d = -\frac{2B_L}{Y_0}$. which is the condition for the transmission resonance through the slit in a thick conducting screen.

Since $2B_L \ll Y_0$, a first approximation to the transmission resonance is

$$k_0 d \cong n\pi \text{ or } d \cong n\frac{\lambda_0}{2}, n = 1, 2, 3 \ldots \tag{3.12}$$

Assuming that $\tan k_0 d$ varies linearly in the vicinity of each zero, we have, as a better approximation to the resonant thickness than is equation (3.12)

$$d \cong \left(n - \frac{2B_L}{\pi Y_0}\right)\frac{\lambda_0}{2}, n = 1, 2, 3 \ldots \tag{3.13}$$

At resonance, $\cos k_0 d \cong \cos n\pi = (-1)^n$ and $\sin k_0 d \cong 0$. From (3.9), $\text{Re}(Y_{12})$ approaches $(-1)^n 2G_L$ as $w$ approaches zero. Since $\text{Re}(Y_L) = G_L = \frac{1}{120\lambda_0}$ [s]

and, at resonance, $|Y_{12}| = |\text{Re}(Y_{12})|$, and in the limit where $w \to 0$, (3.8) reduces to

$$T_r = \frac{\lambda_0}{w\pi} \tag{3.14}$$

where subscript $r$ means resonance. As a word of caution, do not confuse the above transmission coefficient with the transmission cross section (TCS) concept [2] which will be used in the definition of the transmission resonant aperture (TRA) and transmission resonant cavity (TRC) in the forthcoming chapters. Note that the transmission coefficient is dimensionless whereas the dimension of the TCS is given as that of the length corresponding to the slit width. To get the expression for TCS, if we define the transmission width $T_w$ to be the above transmission coefficient multiplied by the slot width $w$, it is given as

$$T_w = T \cdot w = \frac{4\text{Re}(Y_L)}{\eta_0 |Y_{12}|^2} \tag{3.15}$$

In this case, in the limit where the slit width $w$ approaches zero, $T_w$ approaches $\lambda_0/\pi$ [m].

Multiplying the incident power density by the transmission width $T_w$ gives power transmitted by the slit. In this sense, the $T_w$ is called the transmission cross section (TCS) [2].

Note that this quantity of TCS, called also transmission width in [1], approaches $1/\pi$ wavelength as $w \to 0$ regardless of its actual slit width.

As a numerical result [1] for transmission resonance phenomena, the transmission coefficient is shown in Figure 3.3, where the equivalent circuit results were obtained from (3.15). As the slit width $w$ approaches zero, $T$ becomes maximum at a given cavity length which approaches multiples of $\lambda_0/2$ as given by (3.12).

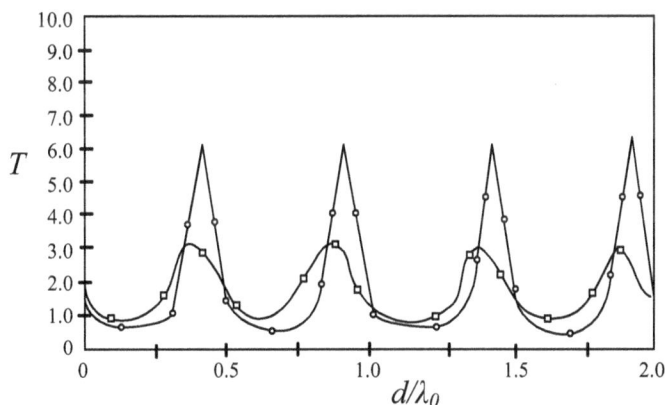

*Figure 3.3   Transmission coefficients* T *versus* $d/\lambda_0$ *for cases of*
*$w = 0.05\lambda_0$ (circles), $w = 0.1\lambda_0$ (squares)—extracted from*
*Harrington and Auckland [1].*

The peak values of $T$ for $w = 0.01\lambda_0$, are $T = 30.5$ occurring at $\frac{d}{\lambda_0} = 0.47, 0.97, 1.47 \ldots$. This is in good agreement with the prediction that the TCS approaches $\lambda_0/\pi$ at values of $d$ slightly less than integral multiples of $\lambda_0/2$ when the slit width $w$ approaches zero. It is also seen that as the slit width $w$ increases the $Q$ and the peak values become smaller based on the data for $w = 0.05\lambda_0$ (circles), and $w = 0.1\lambda_0$ (squares).

This repetitive occurrence of the transmission resonance corresponds to the typical Fabry–Perot resonance [1,3–11] which has been widely investigated in connection with the extraordinary optical transmission (EOT) [11] phenomena in the optics area.

It is interesting to note that the TCS for a narrow slit at resonance is independent of the actual slit width. This phenomenon is analogous to that of scattering by resonant scatterers or transmission by small transmission resonant apertures. For example, a loss-free short dipole resonated by a resonant scatterer has a scattering cross section of $\frac{3}{2\pi}$ square wavelengths [12] regardless of the actual size of the resonant scatterer as discussed in Chapter 4. The small transmission resonant aperture has a TCS [2,13] of $\frac{3}{4\pi}$ square wavelengths regardless of the actual size of the transmission resonant aperture.

### 3.1.2 Similarity between ABR and TCR problems

The narrow slit coupling problem in a thick conducting screen under consideration exemplifies the TCR problem, which presupposes a transmission cavity. In the case of the structure in Figure 3.1, the gap region $\left(0 \leq x \leq d, |y| \leq \frac{w}{2}\right)$ constitutes a transmission resonant cavity, as seen from the equivalent circuit represented in Figure 3.2. There, the inside of the gap region is represented by the transmission line of PPW and the flanged aperture at both ends of the gap region is represented by the radiation admittance $Y_L (= G_L + jB_L, G_L = 1/120\lambda_0)$ [s/m]. For this problem, the slit width $w$ is assumed to be much smaller than the free space wavelength $\lambda_0$ so that only the transverse electromagnetic (TEM) mode is guidable. Therefore, the characteristic admittance $Y_c$ is expressed by $Y_c = 1/\eta_o w$. Note that $Y_c \gg |Y_L|$, $B_L \gg G_L$, and $G_L = 1/120\lambda_0$ [s/m] for $w \ll \lambda_0$, as discussed in [14,15]. In this case, the reflection coefficient at the aperture plane is roughly close to unity. This is essentially the same as the magnetic wall concept in the cavity model of the microstrip antenna theory [16].

Therefore, the present structure in itself constitutes the lossy resonant cavity. It should, however, be remembered that, in the ABR problem, there are two coupling mechanisms that give the same maximum TCS-cavity type and parasitic types—as discussed in Section 2.4.2. Between these two, only the cavity type has a similarity to the TCR phenomenon in the sense that the strong resonant field is established underneath the nearby conducting strip.

In addition to the TCR problem under consideration, the transmitted power $P_t$ through the slit into the right half-space reaches its maximum when the total admittance $Y_s$ becomes a real value, i.e., when the admittance $Y_L'$ (obtained after transforming $Y_L$ at $z = d$ along the transmission line of the PPW region to the left input port at $x = 0$) becomes the complex conjugate of $Y_L$, i.e., $Y_L' = Y_L^* = G_L - jB_L$. and so $Y_L$ at $z = 0$ and $Y_L'$ are summed to be purely real, i.e., $Y_s = 2\text{Re}\{Y_L\} = 2G_L$. Under this condition, the TCS of the slit becomes $\frac{\lambda_0}{\pi}$ [m] [17,18] for normal incidence cases ($\theta_0 = 0$) regardless of the actual slit width $w$. Therefore, the transmitted power through the slit is considerably increased by way of the funneling mechanism, which is the same as that for the ABR problem.

The difference between the ABR problem and the TCR problem is that the mutual coupling between the radiation from the two slits should be taken into account in the ABR problem in contrast to the TCR problem where such a mutual coupling does not exist.

Even if the mutual coupling is taken into account in the ABR problem, the mutual coupling effect becomes minimum [19] where the transmission resonance occurs, i.e., the transmission cavity length corresponds roughly to integral multiples of the half waveguide wavelength. For convenience, consider the case where all the dielectrics constant is assumed to be that of the free space ($\varepsilon_0$). For this reason, the repetitive occurrence of the transmission resonance peaks is clearly observed for both ABR and TCR problems in almost the same manner as seen in Figure 3.3.

### 3.1.3 Other examples of Fabry–Perot transmission resonance

Similar phenomena to the repetitive occurrence of maximum peaks of transmission resonance at $d \cong n\frac{\lambda_0}{2}$ ($n$: integer) under consideration in Figure 3.3 is observed also in the impedance matching problem in the case of aperture-coupled microstrip type [20] in Figure 2.23(a) in Chapter 2. In Figure 2.23(a), once the cavity is formed underneath the nearby conducting strip whose length is $L$ such that almost the perfect matching is achieved if the overall length $L'(= L + \Delta L)$ is increased by an integer multiple of $\Delta L = \lambda_d/2$ ($\lambda_d$: wavelength in the dielectric slab).

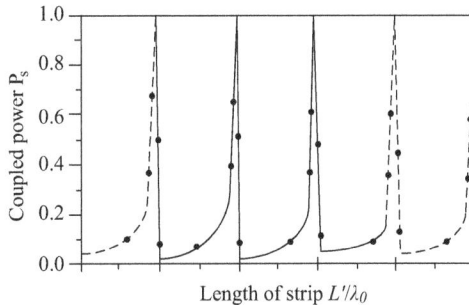

*Figure 3.4*    *Coupled power $P_s$ through the slit versus $L'(= L + \Delta L)/\lambda_d$ for the case that $\varepsilon_r = 2.2, \varepsilon_{rd} = 2.2, h = 0.015\lambda, L' = L + \Delta L. L = 0.482\lambda_d.$ $x_0 = 0.01\lambda_d.$ and $z_0 = 0.226\lambda_d + 0.5\Delta L. (x_0, z_0;$ conducting strip center)*

Figure 3.4 shows such repetitive occurrence of maximum coupling, i.e., impedance matching at $L' \cong n\lambda_d/2$.

Also, in the case where the flanged parallel-plate waveguide (PPW) [21] is electromagnetically coupled to a nearby conducting strip as shown in Figure 3.5, essentially the same phenomena as in the above structure of the aperture-coupled type that is observed and shown in Figure 3.6.

Note again that, in both coupling problems from the slit in the upper wall of the PPW in Chapter 2 and the open end slit of the flanged PPW, under consideration here, the height $h$ of the feeding waveguide is much smaller than the wavelength. In this case, even a narrow slit for both structures behaves like an open circuit when no nearby conductor is present. That is, the most incident power in the guide is reflected at the slit discontinuity and so the radiation through the slit becomes very small.

But when the nearby conducting scatterer is present, perfect matching can be achieved so that most incident power $P_i$ in the guide can be made to be radiated through the slit. The presence of the nearby conducting scatterer can be described as follows: it enhances the conductance of a small slit which would be much lower without scatterer's influence close to the characteristic admittance of the feeding guide, plus it provides the required inductive susceptance for resonance, i.e., cancellation of the reactive component as discussed above. Note that the $TM_{01}$ resonant mode underneath the nearby conducting strip also can be taken as a simple Fabry–Perot resonant field configuration.

### 3.1.4  Fano-type resonance in the microstrip antenna radiation

It is interesting to point out that the Fano-like transmission resonance curve is also observed in the aperture-coupled microstrip antenna structure as seen in Figure 3.4. In the figure, the unsymmetrical resonance curve which is composed of rapid change from the total transmission (impedance matching) to the transmission zero within a very narrow frequency range is clearly seen. This can be explained as follows: first, consider the case where the nearby conducting strip is absent. In this

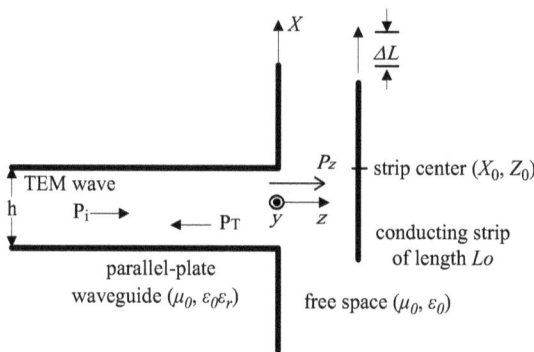

*Figure 3.5  Flanged parallel-plate waveguide structure coupled to nearby conducting strip*

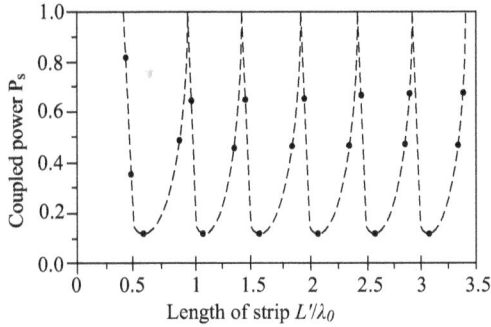

*Figure 3.6 Coupled power $P_s$ through the open end of the flanged parallel-plate waveguide for the case of $\varepsilon_r = 1.0, h = 0.012\lambda, L' = L + \Delta L, L = 0.438\lambda_0, x_0 = 0.01\lambda_0 + 0.5\Delta L,$ and $z_0 = 0.03\lambda_0 (\lambda_0$ : free space wavelength)*

case, only the narrow slit whose reactance can be canceled by use of the short stub behaves as a non-resonant type of radiator. On the other hand, the nearby strip corresponds to the resonant type of radiator. These two types of radiation interfere to result in the Fano-like transmission (radiation) characteristics.

## 3.2    Transmission resonances through a small rectangular slot in a thick conducting screen as 3-dimensional problem

Here we deal with the analysis method for the diffraction problem by a small rectangular slot in a thick conducting screen and look into the properties of the transmission resonance and compare these resonance properties with those for the infinitely long narrow slot problem in the previous Section 3.1 and describe the fundamental difference between 2-dimensional and 3-dimensional transmission resonance problems. Experimental verification for the existence of two kinds of transmission resonance phenomena is dealt with. The physical origin of the shortening factor for magnetic dipole (narrow slot) is also discussed.

### 3.2.1    Analysis method

Figure 3.7(a) shows the geometry under consideration which is composed of the narrow slot structure in a thick conducting screen. This issue is viewed as an expansion of the 2-dimensional problem discussed in the preceding Section 3.1, focusing primarily on the resonance phenomena during transmission in the context of the 3-dimensional rectangular slot problem.

For the purpose of analyzing the problem, the equivalence principle is employed to divide the original problem into three equivalent situations by placing appropriate equivalent magnetic current sheets in the shorted apertures, as shown in Figure 3.7(b).

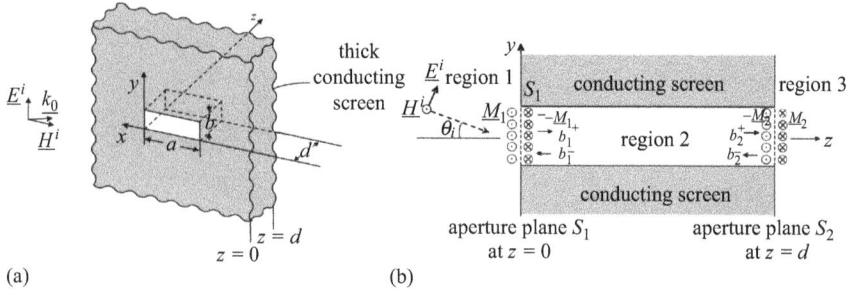

*Figure 3.7    Transmission resonance problem under consideration. (a) Geometry, (b) equivalent problem (reprinted from [22] with permission from IEEE)*

The modal expansion is used to express the fields in the waveguide region and then the tangential electric and magnetic fields just to the left of the right-side aperture $S_2(z = d)$ can be given as a superposition of the transverse electric (TE) mode set $\{e_{mn}^h, h_{mn}^h\}$ and transverse magnetic (TM) mode set $\{e_{mn}^e, h_{mn}^e\}$ as

$$E_{2t} = \sum_m \sum_n \left[ b_{2mn}^h e_{mn}^h(x,y) + b_{2mn}^e e_{mn}^e(x,y) \right] \tag{3.16}$$

$$H_{2t} = \sum_m \sum_n \left[ Y_{2mn}^h b_{2mn}' h h_{mn}^h(x,y) + Y_{2mn}^e b_{2mn}' e h_{mn}^e(x,y) \right] \tag{3.17}$$

Here the modal functions, the modal admittances, and the modal phase constant (appearing later) follow the usual meaning as in [23,24], the subscript 2 denotes the aperture planes $S_2$, $b_{2mn}^{h(e)}$ and $b_{2mn}'^{h(e)}$ mean $\lfloor b_{2mn}^{h(e)+} + b_{2mn}^{h(e)-} \rfloor$ and $\lfloor b_{2mn}^{h(e)+} - b_{2mn}^{h(e)-} \rfloor$. respectively, where $b_{2mn}^{h(e)+}$ $\left[ b_{2mn}^{h(e)-} \right]$ corresponds to the coefficient of the incident [reflected] electric modal field at the right aperture $S_2$. The scattered electromagnetic fields in the right half-space (and left half-space modal as well) are expressed in terms of the 3-dimensional Green's function as in [25]. In Figure 3.7 (b), the magnetic currents $M_i(i = 1, 2)$ and the tangential electric field $E$ on the apertures $S_i(i = 1, 2)$ are related by $M_i = E \times \hat{n}_i$. where $\hat{n}_i$ is a unit normal vector directing inwardly in each region on the aperture planes $S_i(i = 1, 2)$.

Next, imposing boundary conditions across the aperture $S_2$ and testing both sides of the resultant equation with the same set of modal functions as used in the guided region, we obtain the scattering matrix $[S]$ as a relationship between the coefficient matrix $[b_2^+]$ of the incident modal electric fields and $[b_2^-]$ of the reflected modal electric fields at the aperture $S_2$ through the use of $b_{2mn}^{h(e)} = b_{2mn}^{h(e)+} + b_{2mn}^{h(e)-}$ and $b'^{h(e)}_{2mn} = b_{2mn}^{h(e)+} - b_{2mn}^{h(e)-}$ as

$$\left[b_2^-\right] = \left(1 + [Y]^{-1}[M]^{-1}\right)^{-1}\left(1 - [Y]^{-1}[M]\right)\left[b_2^+\right] = [S]\left[b_2^+\right] \tag{3.18}$$

here

$$[Y] = \begin{bmatrix} Y^h & 0 \\ 0 & Y^e \end{bmatrix} \tag{3.19}$$

$$[M] = \begin{bmatrix} H & K \\ Q & E \end{bmatrix} \tag{3.20}$$

where $Y^{h(e)}$ is a submatrix comprising the modal admittance elements for TE(TM) mode as in [23,24] and the expressions for the elements in $[H]$, $[K]$, $[Q]$, and $[E]$ are given as follows:

$$
\begin{aligned}
H_{mnst} = &-\frac{j\omega\varepsilon_0}{2\pi} \int_{S_2}\int_{S_2} \left[\boldsymbol{h}_{mn}^h(x,y)\cdot \boldsymbol{e}_{st}^h(x',y')\right]\frac{e^{-jk_0\cdot R}}{R}\,ds\,ds' \\
&+\frac{j}{2\pi\omega\mu_0}\int_{S_2}\left[\nabla_t\cdot\boldsymbol{h}_{mn}^h(x,y)\right]\int_{S_2}\left[\nabla_t'\cdot\boldsymbol{e}_{st}^h(x',y')\times\hat{z}\right]\frac{e^{-jk_0\cdot R}}{R}\,ds\,ds'
\end{aligned} \tag{3.21}
$$

$$K_{mnst} = -\frac{j\omega\varepsilon_0}{2\pi}\int_{S_2}\int_{S_2}\left[\boldsymbol{h}_{mn}^h(x,y)\cdot\boldsymbol{e}_{st}^e(x',y')\right]\frac{e^{-jk_0\cdot R}}{R}\,ds\,ds' \tag{3.22}$$

$$Q_{mnst} = -\frac{j\omega\varepsilon_0}{2\pi}\int_{S_2}\int_{S_2}\left[\boldsymbol{h}_{mn}^e(x,y)\cdot\boldsymbol{e}_{st}^h(x',y')\right]\frac{e^{-jk_0\cdot R}}{R}\,ds\,ds' \tag{3.23}$$

$$E_{mnst} = -\frac{j\omega\varepsilon_0}{2\pi}\int_{S_2}\int_{S_2}\left[\boldsymbol{h}_{mn}^e(x,y)\cdot\boldsymbol{e}_{st}^e(x',y')\right]\frac{e^{-jk_0\cdot R}}{R}\,ds\,ds' \tag{3.24}$$

where $R = |\underline{R}| = |\underline{r} - \underline{r}'|$.

Generally, in the boundary value problem of the flanged rectangular aperture [25–29], the fields in the aperture plane possess singular behaviors due to the second derivative of the Green's function. Some difficulties may therefore arise due to these singularities in the computational procedure of matrix elements. These singular problems are efficiently dealt with by integrating parts after testing the integral equation with a set of modal functions inside the guided region, which is similar to the scheme in [25–29]. This method along with the transformation of rectangular to polar coordinates is employed in the integration procedure for calculating all the matrix elements in $[H]$, $[K]$, $[Q]$, and $[E]$.

Next, using phase relationship between $b_{2mn}^{h(e)\pm}$ and $b_{1mn}^{h(e)\pm}$, i.e., $b_{2mn}^{h(e)+} = b_{1mn}^{h(e)+}e^{-j\beta_{mn}d}$ and $b_{2mn}^{h(e)-} = b_{1mn}^{h(e)-}e^{j\beta_{mn}d}$ in (3.18) we express the scattering matrix $[S']$ on the left aperture plane $S_1$ in terms of the scattering matrix $[S]$ on the right aperture plane $S_2$ as $[S'] = \lfloor e^{-j\beta_{mn}d}\rfloor[S]\lfloor e^{+j\beta_{mn}d}\rfloor$.

Then imposing the continuity condition of the tangential magnetic field on the left aperture plane $S_1$ with the inclusion of the incident plane wave and going through the same procedure as that used in deriving, equation (3.18), we obtain the expression for the relation between $[b_1^\pm]$ and $[S^i]$ at the left aperture plane $S_1$ as

$$[b_1^+] = \left\{([Y] + [M]) - ([Y] - [M])[S']\right\}^{-1}[S^i] \tag{3.25}$$

$$[b_1^-] = \left\{([Y] + [M])[S'] - ([Y] - [M])\right\}^{-1}[S^i] \tag{3.26}$$

Here $[Y]$ and $[M]$ are the same matrices as those in (3.19) and (3.20), $\lfloor S^i \rfloor$ is a matrix comprising two submatrices $[S^h]$ and $[S^e]$ whose elements are given by

$$S_{mn}^{h(e)} = \int_{S_1} \left\{\boldsymbol{h}_{mn}^{h(e)}(x,y) \cdot \boldsymbol{H}^i(x,y)\right\} ds \tag{3.27}$$

where $\underline{H}^i(x,y)$ is the incident magnetic field on the aperture plane $S_1$ which is given by $\boldsymbol{H}^i(x,y,z) = -\hat{x}H_0 e^{jk_0(\sin\theta_i y - \cos\theta_i z)}$  ($H_0$ : constant amplitude, $\hat{x}$ : unit vector along the x-axis). By use of (3.25) and (3.26), transmitted power $P_t$ through a thick slot into region 3 is obtained as

$$P_t = \frac{1}{2}\text{Re}\sum_m\sum_n\left\{b_{mn}^h\left(Y_{mn}^h b_{mn}' h\right) + b_{mn}^e\left(Y_{mn}^e b_{mn}'^e\right)^*\right\} \tag{3.28}$$

where $\left\{b_{mn}^{h(e)}, b_{mn}'^{h(e)}\right\}$ mean either $\left\{b_{1mn}^{h(e)}, b_{1mn}'^{h(e)}\right\}$ or $\left\{b_{2mn}^{h(e)}, b_{2mn}'^{h(e)}\right\}$.

For the particular case of a narrow slot in the thick conducting screen, only the $TE_{10}$ mode can propagate through the slot region, and higher-order modes rapidly decay exponentially along the guide region when generated at the discontinuity of the left aperture $S_1$. In this case, the transmitted power through the slot into the region 3 is expressed approximately as follows:

$$P_t = \frac{1}{2}\text{Re}\left\{\int_{S_1(or\ S_2)} \boldsymbol{M} \times \boldsymbol{H}^* \cdot \underline{ds}\right\}$$

$$= \frac{1}{2}\text{Re}\left\{b_{10}^h\left(Y_{10}^h b_{10}' h\right)^*\right\}$$

$$= \frac{1}{2}\text{Re}\left[\frac{|S_{10}^h|^2 Y_{10}^{h*}\left\{\left(Y_{10}^h + M_{1010}\right)e^{j\beta_{10}d} + \left(Y_{10}^h - M_{1010}\right)e^{-j\beta_{10}d}\right\}}{\left|\left(Y_{10}^h + M_{1010}\right)^2 e^{j\beta_{10}d} - \left(Y_{10}^h - M_{1010}\right)^2 e^{-j\beta_{10}d}\right|^2}\right.$$

$$\left. \cdot\left\{\left(Y_{10}^h + M_{1010}\right)e^{j\beta_{10}d} - \left(Y_{10}^h - M_{1010}\right)e^{-j\beta_{10}d}\right\}^*\right] \tag{3.29}$$

Here, $\underline{M}$ is the magnetic current just to the right (left) of the aperture plane $S_1(S_2)$ in Figure 3.7(b). $M_{1010}\left[= G + jB = Y_L = \left(1 + b_{10}^+/b_{10}^-\right)/\left(1 + b_{10}^-/b_{10}^+\right)\right]$ means the radiation admittance of the flanged rectangular slot for the narrow slot case where the reflection coefficient becomes similar to that of a lossy magnetic

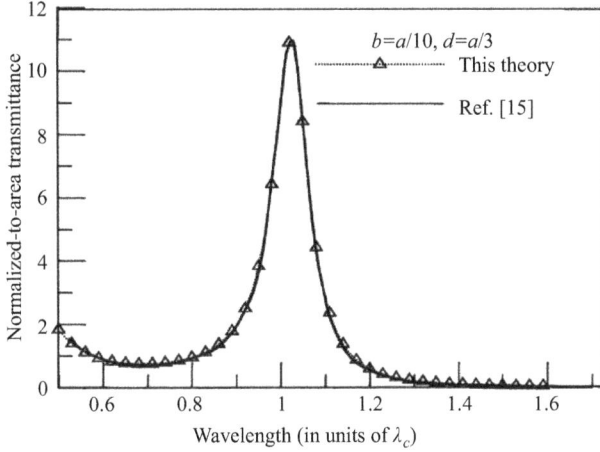

*Figure 3.8    Comparison of the normalized-to-area transmission between results obtained by the present method [22] and those in [23] (reprinted from [22] with permission from IEEE)*

wall as the guide height $b$ decreases as discussed in [14,30]. Using this relationship, $P_t$ is expressed in the simple form [31] as

$$P_t = \frac{1}{2}\mathrm{Re}\left\{ b_{10}^h \left( Y_{10}^h b_{10}'^h \right)^* \right\}$$

$$= \frac{1}{2}\mathrm{Re}\left\{ \frac{4\left|S_{10}^h\right|^2 \left(Y_{10}^h\right)^2 G}{\left| \left(Y_{10}^h + M_{1010}\right)^2 e^{j\beta_{10}d} - \left(Y_{10}^h - M_{1010}\right)^2 e^{-j\beta_{10}d} \right|^2} \right\} \qquad (3.30)$$

To validate the present method, some numerical results obtained by the use of this method are compared with those available that are obtained by use of the approximate spectral domain method [22] in Figure 3.8. The calculation was carried out for the normalized-to-area transmittance versus the wavelength (in units of the cutoff wavelength $\lambda_c = 2a$) for the case that $b = a/10$ and $d = a/3$. Here the normalized-to-area transmittance [22] has the same meaning as the transmission coefficient $T$ in Figure 3.9. A comparison shows good agreement, as displayed in Figure 3.8.

## 3.2.2    Diffraction property

We have investigated the transmission characteristics through a narrow rectangular slot in an arbitrarily thick conducting screen by use of (3.28). The representative numerical result of the transmitted power through a narrow slot as a function of the thickness of the screen (i.e., corresponding to the finite length of a rectangular waveguide of the slot in region 2) is given in Figure 3.9. The data have been

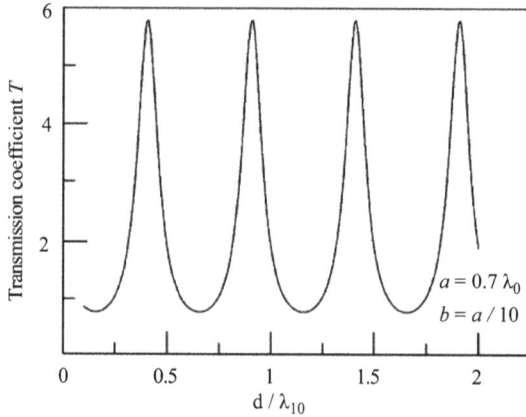

*Figure 3.9   Plots of transmission coefficient T versus $d/\lambda_{10}$ for the case that a = 0.7 $\lambda_0$ b = a/10 [Reprinted from [22] with permission from IEEE: Cho Y.K., Kim k.W., Ko J.H., Lee J.I. 'Transmission through a narrow slit in a thick conducting screen.' 2009; IEEE Trans. Antennas Propagat., AP-57, pp. 813–816]*

obtained for cases in which $a = 0.7\lambda_0$ and $b = a/10$ under the assumption that the power density $S$ of a normal incident wave is $S = 1 \text{ W/m}^2$. The transmission coefficient $T$ in Figure 3.9 is defined as the ratio of transmitted power $P_t$ through the thick slot to the incident power $S \cdot ab$ upon the actual area $ab$, that is, $T \equiv \frac{P_t}{S \cdot ab}$ as defined above. It is observed that the transmission coefficient $T$ reaches its maximum when the screen thickness $d$ is somewhat less than an integer multiple of half wavelengths of the $TE_{10}$ mode in region 2. Assuming that $\tan(\beta_{10}d)$ varies linearly in the vicinity of each zero, we have the resonant thickness $d = d_{\text{res}}$ where the transmission coefficient $T$ is maximized as

$$d_{\text{res}} \cong \left(n - \frac{2B}{\pi Y_{10}^h}\right) \frac{\lambda_{10}}{2} \qquad (3.31)$$

when the condition $\left(Y_{10}^h\right)^2 \gg G^2 + B^2$ has been used in the approximation.

As $b \to 0$, $T$ reaches the maxima at so-called resonant thicknesses that approach multiples of $\lambda_0/2$. as predicted by (3.31). This observation is the same as that in the prior work [1] on the maximum transmission (or resonance) phenomena through the (infinitely long) narrow slit in a thick conducting screen.

To better understand the underlying physics of the condition under which the maximum transmission through the slot in the conducting screen of an arbitrary thickness occurs, we have investigated transmission characteristics for the case of various combinations of the geometrical parameters ($a$, $b$, and $d$). Figure 3.10 shows transmission characteristics $T$ versus frequencies for the following three cases:

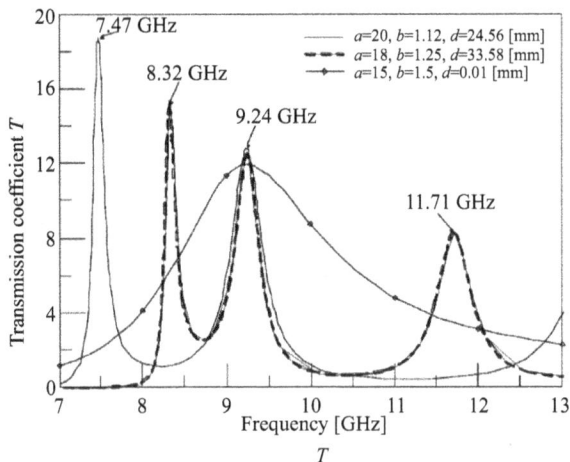

*Figure 3.10 Plots of transmission coefficient* T *versus frequency [Reprinted from [22] with permission from IEEE: Cho Y.K., Kim k.W., Ko J.H., Lee J. I. 'Transmission through a narrow slit in a thick conducting screen.' 2009; IEEE Trans. Antennas Propagat., AP-57, pp.813-816]*

Case (1) $a = 20$, $b = 1.12$, and $d = 24.56$

Case (2) $a = 18$, $b = 1.25$, and $d = 33.58$ and

Case (3) $a = 15$, $b = 1.5$, and $d = 0.01$ mm.

Here, small values of $b$ are chosen so that both slots $S_1$ and $S_2$ may behave like a lossy magnetic wall, which results in the formation of a lossy cavity in the thick slot region 2.

The solid line represents the transmission characteristics $T$ in Case (1), which shows two transmission peaks at frequencies of 7.47 and 9.24 GHz. The second peak at 9.24 GHz is observed as occurring when the screen thickness $d$ is somewhat less than an integer multiple of the half-guide wavelengths of the dominant $TE_{10}$ mode, as expressed in (3.11). This type of transmission peak is observed only when the lossy cavity is formed along the longitudinal direction (i.e., z-direction).

The dashed line represents the transmission characteristics for Case (2). Among three peaks at 8.32, 9.24, and 11.71 GHz for the Case (2), the peaks at 9.24 and 11.71 GHz are seen to belong to the type observed at 9.24 GHz in the above Case(1). This type of transmission peak will be designated as a transmission cavity resonance (TCR) type to distinguish between this and another type of transmission peak corresponding to the peaks at the lowest frequencies (7.47 and 8.32 GHz) in Cases (1) and (2).

Note that the frequencies 7.47 and 8.32 GHz, where the lowest transmission peaks occur are just below the cutoff frequencies (7.5 and 8.33 GHz) in Cases (1) and (2), respectively. Interestingly enough, this type of transmission peak is observed for an arbitrarily thick screen when the propagation constant $\beta_z$ along the

$z$-axis become almost zero as discussed in [22]. This type of transmission peak is therefore observed to occur irrespective of the screen thickness $d$. The irrelevancy of the transmission peak to the screen thickness suggests that, unlike the afore-mentioned TCR case, transverse resonance along the $x$ direction may contribute to the maximum transmission. It should also be noted that the frequency at which maximum transmission of this type occurs is observed to approach the exact cutoff frequency of the $TE_{10}$ mode as the screen thickness $d$ is increased. These types of transmission peaks due to the transverse resonance through a resonant slot in the screen of an arbitrary thickness can therefore be regarded as a generalized type of transmission resonance phenomena through a resonant slot in the infinitesimally thin conducting screen.

### 3.2.3   Equivalent circuit representation

As discussed in [22], when the guide height (corresponding to $b$) is much smaller than the wavelength $\lambda_{10}$. the flanged rectangular slot, seen from the guide, behaves like a lossy magnetic wall. Here, the loss is due to the small amount of radiation from the flanged aperture into the half-space and the magnetic wall means that the reflection coefficient at the aperture plane approaches positive unity because $Y_{10}^h$ becomes larger than $|G+jB|$ as $b \to 0$. The slot region therefore constitutes the lossy cavity, whose equivalent circuit is represented by a transmission line of the electrical length $(\beta_{10}d)$ with both the left and right ends terminated by the radiation admittance $Y_L(= G+jB)$ corresponding to the lossy magnetic wall of the flanged slots, as shown in Figure 3.11. This formation of the lossy cavity plays an important role in the transmission resonance phenomenon for cases in which the transmitted power $P_t$ through the thick slot reaches its maximum, as discussed later. Note also that for the values of $d$ in (3.31), the total admittance $Y_{in}$ in Figure 3.11 becomes purely real, that is, $Y_{in} = 2G$.

### 3.2.4   The condition for the maximum power transmission

The condition for the maximization of the transmitted power $P_t$ is found by investigating the minimum of the magnitude $f(\beta_{10}d)$ of the denominator in (3.30). That is, from the equation $f'(\beta_{10}d) = 0$ the minimum condition of $f(\beta_{10}d)$ is found

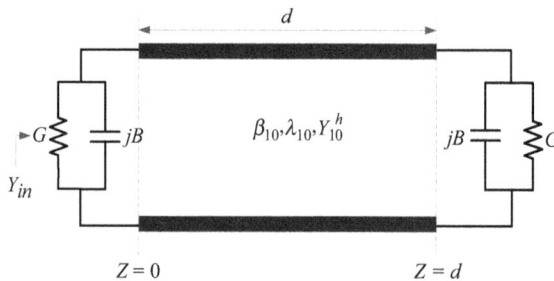

*Figure 3.11   Equivalent circuit for the present transmission resonance problem*

to be

$$\tan \beta_{10} d = \frac{2Y_{10}^h B}{G^2 + B^2 - \left(Y_{10}^h\right)^2} \tag{3.32}$$

which corresponds to the condition for maximum power transmission through the slot. It is interesting to note that this expression for the maximum transmission condition is exactly the same as that for the maximum radiation condition which is obtained by using the transmission line model in the rectangular microstrip patch antenna structures [32].

### 3.2.5 Fundamental difference of the diffraction properties between infinite long slit and rectangular slot problems

The difference in the diffraction properties between infinitely long and rectangular slot problems is summarized as follows: unlike the infinitely long slit problem in which case only the Fabry–Perot type resonance phenomenon is observed, there are two kinds of transmission resonance types—Fabry–Perot type and near-cutoff mode type in rectangular slot problem. Though the near-cutoff mode in the rectangular slot can be taken as the zeroth order of Fabry–Perot resonance, the transmission resonance phenomena are classified into the above two types due to the conceptual utility of the difference between the two types. It is worthwhile to mention that the near-cutoff mode occurs at a frequency slightly higher than the cutoff frequency and the group velocity becomes almost near zero as mentioned above. This near-cutoff mode can be used to simulate the epsilon near zero material [33]. What is more, if the rectangular ridged guide structure is considered under the condition of the near-cutoff mode, the length of the structure corresponding to the transmission cavity length can be reduced to zero without degradation of the transmission resonance. This helps to understand the relation between the nanoaperture such as ridged aperture in Chapter 4 and the transmission cavity resonance in Chapter 5 having the same cross section.

### 3.2.6 Physical origin of the shortening factor for narrow slot magnetic dipole

In analyzing a representative slot coupling problem, we have investigated the transmission characteristics of the narrow rectangular slot in a thin conducting screen for Case (3). The result of this case is represented by a solid line marked with circles in Figure 3.10. The transmission peak is observed at 9.24 GHz, for which the free space wavelength is 32.5 mm. As expected from a resonant slot whose reactive part of the admittance can be tuned to zero by making the slot length slightly less than $\lambda_0/2$. the transmission peak is observed when the slot length $a(=15$ mm$)$ is a little smaller than $\lambda_0/2(=16.25)$. This is analogous to the shortening factor introduced to eliminate the reactive component of the input impedance of the half-wavelength electric dipole antenna structure.

## 3.2.7   Experimental verification for the existence of two kinds of transmission resonance phenomena; Fabry–Perot type and near-cutoff mode (near-zero group velocity) type

A microwave waveguide setup is used for experimental verification of the presence of two kinds of transmission resonance phenomena.

Figure 3.12 shows the experimental setup, where a narrow slot structure is placed in a thick conducting screen inside a rectangular waveguide such that the narrow slot is centrally located in the larger rectangular waveguide with a cross-section of $a \times b$ mm$^2$. By employing this structure for the experimental verification of the two kinds of transmission resonances, the transmission resonance problem through a narrow slot in a thick conducting screen under the incidence of the plane wave is transformed into an impedance-matching problem through a thick rectangular window in a larger rectangular waveguide, which is much easier to deal with experimentally. As a reminder, the two types of transmission resonances, transverse resonance type and Fabry–Perot (FP) resonator type have been observed to occur in a narrow rectangular slot structure [31] in a thick, perfect, conducting screen under the incidence of plane wave.

Significant impedance mismatching can be expected to occur at the input (at $z = 0$) and the output (at $z = L$) junctions due to a sharp variation in the characteristic impedance (or admittance). In this case, a strong resonant cavity field may be established in the feeding guide region between the left coaxial probe of the standard WR-90 adapter and the input junction of the thick rectangular window, thereby significantly decreasing the power transmitted through the thick window. However, this was never found to be the case, as will be discussed next.

Figure 3.13 shows the experimental results for $|S_{11}|$ at port 1. Interestingly, at the four sharp dips, good impedance matching was observed to be achieved, implying that electromagnetic energy squeezing and smooth guiding occurred through the very narrow window (central rectangular waveguide) channel in the

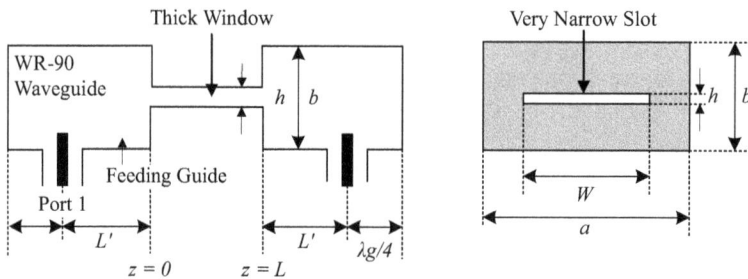

*Figure 3.12   Experimental setup and geometrical parameters: (a) lateral view and (b) cross-sectional view (a = 22.86 mm, b = 10.16 mm, W = 14 mm, h = 1 mm, L = 60 mm, L' = 100 mm)*

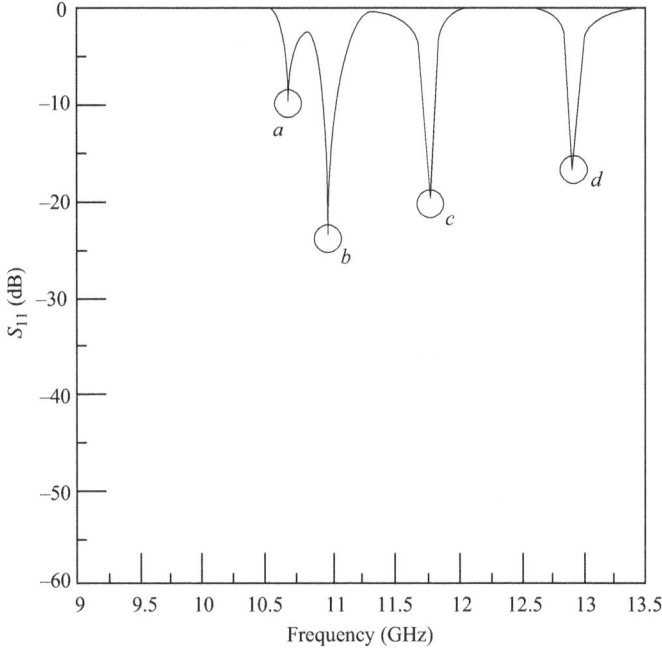

*Figure 3.13   Experimental and simulation results for $|S_{11}|$*

absence of any sharp variation in the waveguide's transverse cross section at the input junction at $z = 0$.

Figure 3.13 also compares the experimental results for $|S_{11}|$ with the simulation results over a frequency range between 9 GHz and 13.5 GHz, where the solid line represents the experimental results and the dotted line represents the simulation results. The two sets of results agree well with each other. The four experimental (simulation) sharp dips are marked with the letters a, b, c, and d, and the corresponding frequencies were 10.695 (10.738) GHz, 10.945 (10.991) GHz, 11.745 (11.782) GHz, and 12.91 (12.973) GHz, respectively. Here, the simulation was conducted using HFSS.

For the first sharp dip (point a), at the impedance matching point frequency of point a, the guide wavelength approached a very large value (almost infinite), implying that this matching frequency was very near to (and slightly higher than) the cutoff frequency of the lowest $TE_{10}$ mode in the central narrow waveguide. This means also that $\beta_z$ (phase constant along the guided direction) $\cong 0$ mimicking propagation through a medium with zero permittivity. This implies also that this type of transmission resonance (impedance matching) condition mainly depends upon the transverse slot length of the thick window irrespective of the longitudinal slot's length $L$. In this sense, the medium under the transmission resonance corresponding to this impedance-matching frequency is thought to be similar to the ENZ (permittivity or epsilon near zero) channel type [33].

At the frequencies corresponding to points b, c, and d, transmission resonance phenomena were observed to occur when the length $L$ of the central narrow waveguide (corresponding to the thick window) was somewhat less than integral multiples of half the guide wavelength of the lowest $TE_{10}$ mode in the central narrow waveguide. For reference, half the guide wavelengths corresponding to points *b*, *c*, and *d* were 67.1 mm, 31.2 mm, and 20.8 mm, respectively, and the length $L$ of the central narrow waveguide was 60 mm. The above values of 67.1 mm, 31.2 mm, and 20.8 mm were obtained by finding the dip frequencies corresponding to the impedance matching points b, c, and d from the experimental data in Figure 3.13. From a knowledge of the dip frequency half the guide wavelength, $\frac{1}{2}\lambda_g$. in the rectangular waveguide structure inside the thick slot region is obtained by use of the relation of $\lambda_g = \frac{\lambda_0}{\sqrt{1-\left(\frac{\lambda_0}{2w}\right)^2}}$, where $\lambda_0$ is the respective free-space wavelength at each of the three dip frequencies corresponding to the points b, c, and d in Figure 3.13.

The above experimental values correspond almost to the theoretical value of 1/2 $\lambda_g$ satisfying the typical expression for the longitudinal FP cavity length of $L = \left(n - \frac{2B}{\pi Y_{10}}\right)\frac{\lambda_g}{2}$ [31] for the three experimental dip frequencies. In this expression, $n(= 1,2,3\ldots)$ is an integer, $Y_{10}$ is the characteristic admittance for the lowest $TE_{10}$ mode of the rectangular waveguide structure inside the thick slot region, and $B$ is the junction susceptance at waveguide junction discontinuities where abrupt changes in the cross section areas occur at the input and the output junction at $z = 0$ and $z = L$ in Figure 3.12.

The values for $B$ are calculated for the three frequencies by using the commercial software HFSS. We note that this junction susceptance $B$ is different from the radiation susceptance $B$ in the transmission cross section (TCS) problem of a narrow slot in the thick conducting screen under the assumption of an incident plane wave, which is obtained by use of the method mentioned in [31].

First, let us consider the half-guide wavelength $\frac{\lambda_g}{2}$ (67.1 mm) corresponding to point b in Figure 3.8. If we calculate $B$ and $Y_{10}$ for the given frequency at point b and insert them into the expression for the relation between $L$ (=60 mm) and $\frac{\lambda_g}{2}$, the obtained value of $\frac{\lambda_g}{2}$ after solving the equation $L = \left[n - \frac{2B}{\pi Y_{10}}\right]\frac{\lambda_g}{2}$ for $\frac{\lambda_g}{2}$ as an unknown, is found to be very close to 67.1 mm for $n = 1$. Similarly, the value of $\frac{\lambda_g}{2}$ for points c and d are found to be almost equal to 31.2 mm and 20.8 mm, respectively, for $n = 2$ and $n = 3$. Thus, the transmission resonance phenomena corresponding to points b, c, and d appeared to be FP resonator-type phenomena as they occurred when the length $L$ was somewhat smaller than $n\frac{\lambda_g}{2}$. We should note again that the above calculation of the half-guide wavelength $\frac{\lambda_g}{2}$ for the transmission resonance (impedance matching) used the experimental values for the transmission resonance frequencies, as those frequencies agreed well with the simulation frequency results. The difference between the experimental and the theoretical value of $\frac{1}{2}\lambda_g$ for transmission resonance (impedance matching) is observed to be smaller as the guide height $h$ in the thick slot region in Figure 3.12 becomes smaller.

Returning to the original transmission resonance problem through a narrow rectangular slot in a thick conducting screen under an incident plane wave, the TCS characteristics versus the frequency were investigated in the case of a plane wave at normal incidence. Recall that the TCS was defined as the ratio of the transmitted power [Watts] through the slot into the opposite half space to the incident power density [Watts/m$^2$] for the resultant physical dimensions of the area, whose meaning, as a measure of the transmission efficiency, is clarified in Figure 3.3. Figure 3.14 shows the TCS characteristics versus the frequency. Four maximum peaks were observed at the points marked with the letters a, b, c, and d. The frequencies corresponding to points a, b, c, and d were 10.71 GHz, 10.93 GHz, 11.71 GHz, and 12.9 GHz, respectively. For example, the physical area of the narrow rectangular slot is $W \times h = 14$ mm$^2$. whereas its TCS is 205.1 mm$^2$ at the point of 10.71 GHz. This means that the incident power in the TCS (205.1 mm$^2$) much larger than the physical area (14 mm$^2$) is transmitted through the slot and is re-radiated into the half-space opposite to the incident side.

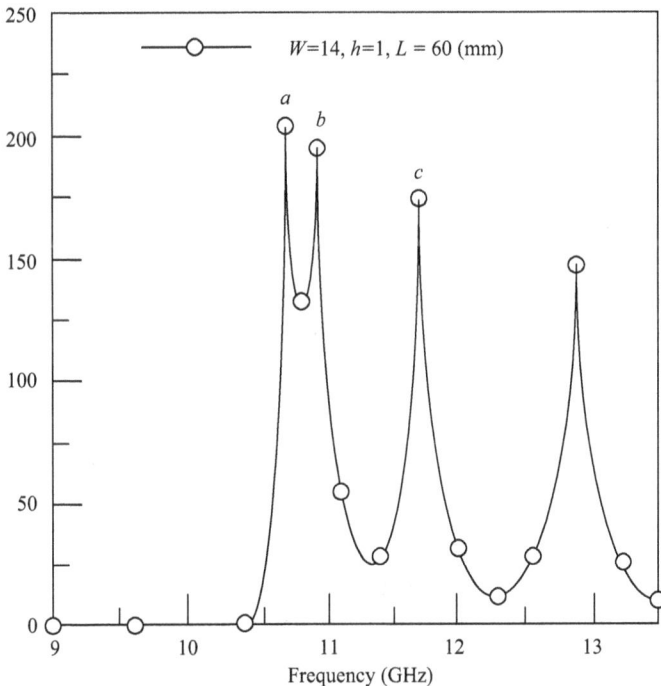

*Figure 3.14   TCS characteristics*

The guide wavelength for point a was quite long, which means that the frequency corresponded very closely to the cutoff frequency of the lowest $TE_{10}$ mode in the guide inside the thick slot region, and that $\beta_z \cong 0$.

For points b, c, and d, half the guide wavelengths were 69.4 mm, 31.8 mm, and 20.9 mm, respectively. In addition, the conducting screen thickness (corresponding to the cavity length) was 60 mm. In comparison with these values, the transmission resonances corresponding to points b, c, and d occur roughly when the cavity length $L$ is slightly less than integral multiples of half the guide wavelength of the lowest dominant $TE_{10}$ mode, similar to the abovementioned FP resonator type. Thus, the transmission resonances at point a are thought to be the transverse-resonance-type resonance while the transmission resonances at points b, c, and d appeared to be FP resonator-type resonance (the longitudinal resonance type). A comparison of the frequencies of the four sharp dips in Figure 3.13 and four maximum peaks in Figure 3.14 revealed a very close pair correspondence between the respective points a, b, c, and d. The reason for this is that a smaller guide height $b$ inside the slot region produces a stronger transmission resonance with a higher $Q$. In detail, when the lowest dominant $TE_{10}$ (TEM) mode in the rectangular (parallel plate) waveguide is incident upon the flanged open-ended aperture of a rectangular slot (infinitely-long slit), the reflection coefficient approaches 1 as the guide height becomes smaller [22]. This physical situation in which the reflection coefficient approaches 1 corresponds to the lossy magnetic wall which has been widely used as a cavity wall model along the microstrip patch antenna edge of a thin dielectric substrate. It is worthwhile to remember that the thinner the dielectric substrate of the microstrip patch antenna is, the higher the $Q$ of the microstrip patch antenna becomes. Thus, the impedance-matching frequencies in the thick rectangular window problem in the rectangular waveguide in Figure 3.12 appear to approach the maximum peak frequencies in the TCS problem in which the input/output feeding guides in the former problem are replaced by the respective half space. For the same reason, the graph of $|S11|$ in Figure 3.8 remains almost the same, regardless of the length $L'$ in Figure 3.12.

A very similar study [34] on multiple transmission resonance through a thick window in a waveguide was also conducted almost 40 years ago. However, the presence of two kinds of transmission resonance phenomena was not fully recognized at that time. It is also worthwhile to note that no connection (common feature and difference) has yet been established between previous studies of the 2-dimensional slit problem [1,4,9] and the 3-dimensional slot problem [31]. For the 2-dimensional infinitely long slit problem in a thick conducting screen, only FP resonator (or longitudinal resonance) type transmission resonance occurs as there is no cutoff for the lowest TEM mode in a 2-dimensional thick slit structure, where the inside corresponds to a parallel-plate waveguide. In contrast, for the 3-dimensional narrow slot problem, transverse resonance type, and longitudinal resonance (FP resonator) type transmission resonances are observed to occur as the transverse resonance type occurs at the cutoff frequency of the lowest $TE_{10}$ mode in a 3-dimensional narrow slot where the inside corresponds to a rectangular guide in a thick conducting screen.

So far, many research works have been done on the resonance phenomena of long slit and rectangular slot structures. It is worthwhile to note that not a few research results are thought to be divided into two kinds of resonance phenomena, transverse resonance phenomena and longitudinal FP resonance type. Some works [31,35] belong to the former type whereas other works [4,9,36] belong to the latter type.

## 3.2.8 *Physical origin of nanoaperture*

We have seen that two kinds of transmission resonance peak through the narrow rectangular slot in a thick conducting screen are observed. One is the longitudinal FP resonance type, and the other is the transverse resonance type occurring at near-cutoff frequency. Particularly the latter type of transmission peak is observed to occur irrespective of the screen thickness $d$. when the propagation constant $\beta_z$ along the longitudinal direction becomes almost zero. If the ridged waveguide is employed instead of the narrow rectangular waveguide with the small guide height retained, the cutoff frequency can be made to be significantly reduced. In this case too, the transmission peak which is almost irrespective of the conducting screen thickness can be found similar to the above rectangular waveguide case corresponding to the rectangular narrow slot in a thick conducting screen. Based upon this observation, the nanoapertures that belong to the transmission resonant aperture (TRA) such as H-shaped aperture and its variation can be designed. In other words, if the circular or rectangular ridged waveguide type for reducing the cross-section or lowering the cutoff frequency is employed instead of the above rectangular guide type, in the case of the transverse resonance type whose transmission resonance frequency is irrespective of the longitudinal length of the cavity corresponding to the conducting screen thickness, the structure determined in the limit of the zero thickness of the conducting screen can be taken as a nanoaperture. Note that this concept holds also for the TRA appearing in Chapter 4.

## References

[1]   Harrington R. F., Auckland O. 'Electromagnetic transmission through narrow slots in thick conducting screen'. *IEEE Trans. Antennas Propagat.* 1980, vol. 28(5), pp. 616–622

[2]   Park J. E., Cho Y. K. 'Comparison of transmission resonance phenomena through small coupling apertures between two kinds of transmission resonance structures'. Proc. URSI International Symposium on Electromagnetic Theory; 2010, pp. 898–902

[3]   Takakura Y. 'Optical resonance in a narrow slit in a thick metallic screen'. *Phys. Rev. Lett.* 2000, vol. 86(24), pp. 5601–5604.

[4]   Jiso X., Wang P., Tang L, Lu Y. Li Q., Zhang D. *et al.* 'Fabry-Perot-like phenomenon in the surface plasmons resonant transmission of metallic gratings with very narrow slits', *Appl. Phys. B* 2005, 80, pp. 301–305

[5]   Astilean S., Lalanne Ph., Palamaru M. 'Light transmission through metallic channels much smaller than the wavelength'. *Optics Comm.* 2000, vol. 175, pp. 265–273

[6]   Ruan Z., Qiu M. 'Enhanced transmission through periodic arrays of sub-wavelength holes: the note of localized waveguide resonances'. *Phys. Rev. Lett.* 2006, vol. 96(1–4), pp. 233901.

[7]   Porto J. A., Garcia-Vidal F. J., Pendry J. B 'Transmission resonances on metallic gratings with very narrow slits.' *Phys. Rev. Lett.* 1999, vol. 83(14), pp. 2845–2848

[8]   Yang F., Sambles J. R. 'Resonant transmission of microwaves through a narrow metallic slit.' *Phys. Rev. Lett.* 2002, vol. 89(6) pp. 063901-1–063901-3

[9]   Marguier F., Greffet J. J, Collin S., Pardo F., Pelouard J. R 'Resonant transmission through a metallic film due to coupled modes.' *Opt. Express* 2004, vol. 13(1), pp. 70–76

[10]  Collin S., Fardo F. Teissier R., Pelouard J.R., 'Horizontal and vertical surface resonances in transmission metallic grating.' *J. Opt. A: pure Appl. Opt.* 2002, vol. 4, pp. 5154–5160

[11]  Ebbesen, T.W. Lezec, H.J., Ghaemi, H.F., Thio, T., Wollf, P.A. 'Extraordinary optical transmission through subwavelength hole array.' *Nature (London).* 1998, vol. 391(6668), pp. 667–669

[12]  Harrington R.F. 'Small resonant scatters and their use for field measurements'. *IEEE Trans. Microwave Theory Tech.*, 1962, vol. 10, pp. 165–174

[13]  Harrington R.F. 'Resonant behavior of a small aperture backed by a conducting body'. *IEEE Trans. Antennas Propagat.* 1982 vol. 30(2), pp. 205–212

[14]  Cho, Y.K. 'On the equivalent circuit representation of the slitted parallel-plate waveguide filled with a dielectric'. *IEEE Trans. Antennas Propagat.* 1989, vol. 37(9), pp. 1193–1200

[15]  Lee J. I., Ko J. H., Cho Y. K. 'A note on diffraction and equivalent admittance properties of a transverse slit in a parallel plate waveguide filled with a homogeneous dielectric'. *IEICE Trans. Commun.* 2003, vol. E86-B(12), pp. 3600–3605

[16]  Garg R., Bhartia P., Bahl I., Ittipiboon A. *Microstrip Antenna Design Handbook.* Artech-House; 2001, pp. 90–102

[17]  Cho Y. K., Park J. E. 'Aperture-body resonance (ABR) and transmission cavity resonance (TCR) conditions in aperture coupling problems.' *Proc. European Microwave Conf.* 2007, vol. 37, pp. 424–427

[18]  Lee J.I., Cho Y.K., 'Electromagnetic transmission through a narrow slit backed by a nearby conducting strip.' *Proc. Intl. Symp. Antennas Propagat. (ISAP 2005)* 2005, pp. 1089–1092

[19]  Derneryd A. G. 'A theoretical investigation of the rectangular microstrip antenna element.' *IEEE Trans. Antennas Propagat.* 1978, vol. 26(4), pp. 532–535

[20]  Lee J.I., Cho Y.K. 'Maximum electromagnetic coupling to a nearby conducting strip through narrow and wide slits in a parallel plate waveguide'. *IEEE Trans. Antennas Propagat.* 2003, vol. 51(4), pp. 693–699

[21]  Lee J. I., Ko J. H., Cho Y. K. 'Coupling through a slit in a flanged parallel-plate waveguide with a conducting strip.' *Proc. Intl. Symp. Antennas Propagat. (ISAP'04)* 2004, pp. 589–592

[22]    Cho Y. K., Kim K. W., Ko J. H., Lee J. I. 'Transmission through a narrow slot in a thick conducting screen.' *IEEE Trans. Antennas Propagat.* 2009. 7, vol. 57(3), pp. 813–816

[23]    Marcuvitz N. *Waveguide Handbook.* McGraw-Hill; 1951. no. 21

[24]    Wade J. D., MacPhie R. H. 'Scattering at circular-to-rectangular waveguide junctions.' *IEEE Trans. Microwave Theory Tech.* 1986, vol. 34(11), pp. 1085–1091

[25]    Teodoridis V., Sphicopoulos T., Garndiol F. 'The reflection in open-ended rectangular waveguides.' *Mikrowellen Magaz in* 1985, vol. 11(3), pp. 234–238

[26]    MacPhie R. H., Zaghloul A. I. 'Radiation from a rectangular waveguide with infinite flange-exact solution by the correlation matrix method.' *IEEE Trans. Antennas Propagat.* 1980, vol. 28(4), pp. 497–503

[27]    Encinar J. A., Rebollar J. M. 'Convergence of numerical solutions of open-ended waveguide by modal analysis and hybrid modal-spectral techniques.' *IEEE Trans. Microwave Theory Tech.* 1986, vol. 34(7), pp. 809–814

[28]    Baudrand H., Tao J., Atechian J. 'Study of radiating properties of open-ended rectangular waveguides.' *IEEE Trans. Antennas Propagat.* 1988, vol. 36(8) pp. 1071–1077

[29]    Teodoridis V., Sphicopoulos T., Gardiol F.E. 'Radiation from an open-ended rectangular waveguide terminated by a layered dielectric medium.' *IEEE Trans. Microwave Theory Tech.* 1985, vol. 33(5), pp. 359–366

[30]    Kim K.W., Woo D. S., Cho Y. K. 'A conically coupled waveguide-to-coaxial line transmission in a reduced-height waveguide for compact transverse.' *Microw. Opt. Tech. Lett.* 2006, vol. 48(4), pp. 669–673

[31]    Garcia-Vidal F. J., Moreno E., Porto J. A., Martin-Moreno L. 'Transmission of light through a single rectangular hole.' *Phys. Rev. Lett.* 2005, vol. 95, pp. 103901-1–103901-4

[32]    Derneryd A.G. 'Linearly polarized microstrip antenna'. *IEEE Trans. Antennas Propagat.* 1976, vol. 24(6), pp. 846–851

[33]    Edwards B., Alu A., Young M.E., Silveirinha M., Engheta N. 'Experimental verification of epsilon-near-zero metamaterial coupling and energy squeezing using a microwave waveguide.' *Phys. Rev. Lett.* 2008, vol. 100, p. 033903

[34]    Luebbers R. J., Munk B.A. 'Analysis of thick rectangular waveguide windows with finite conductivity.' *IEEE Trans. Microwave Theory Tech.* 1959, vol. 21(7), pp. 461–468

[35]    H. R. Park, S. M. Koo, O. K. Suwal, Y. M. Park, and J. S. Kyung, M. A. Seo, S. S. Choi, N. K. Park, D. S. Kim, K. J. Ahn, 'Resonance behavior of single ultrathin slot antennas on finite dielectric substrates in terahertz regime.' *Appl. Phys. Lett.* 2010, vol. 96(21), p. 21109(3)

[36]    W. Lee, M. A. Seo, D. J. Park, S. C. Jeoung, Q. H. Park, Ch. Lienau, and D. S. Kim, 'Terahertz transparency at Fabry-Perot resonances of periodic slit arrays in a metal plate: experiment and theory.' *Optics Express*, 2006, vol. 14(26), p. 12637

*Chapter 4*

# Transmission resonance problem through small apertures and its dual resonant scatterer problem

## Synopsis

Employing a small circular aperture whose diameter $D$ is significantly smaller than the wavelength $\lambda$ in a conducting screen to obtain high transmission efficiency and spatial confinement is of importance in various application areas such as near-field scanning optical microscope (NSOM) probe design, optical data storage, etc. However, it is well known that, according to Bethe's theory, the transmission cross section (TCS) as a measure of the transmission efficiency of the small aperture is very small, i.e., TCS $\propto \left(\frac{D}{\lambda}\right)^4 \cdot D^2$. Here we deal with how to enhance the transmission efficiency (or transmission cross section) through the use of the transmission resonance phenomena by reforming the small apertures in shape. Various transmission resonant apertures are considered including C-shaped, H-shaped apertures, ridged apertures, and complementary split ring resonator (CSRR) type, etc. The physical condition of the transmission resonance is described along with applications to the NSOM probe design. In parallel, resonant scatterer structure as a dual problem is also discussed along with some applications.

## 4.1  Transmission resonant aperture (TRA) structure

Designing a small aperture in a conducting screen to obtain high transmission efficiency for high spatial resolution beyond the diffraction limit is very important. But when the lateral dimensions of an aperture are significantly smaller than half the wavelength, the transmission of the light through such a hole is typically very weak. It is usually believed to scale as the fourth power of an aperture diameter, for example, for small circular aperture [1]. Recently a kind of ridged aperture [2–5] that provides a significant enhancement of the transmission efficiency through the small aperture was proposed for applications of near-field optical areas such as optical data storage, nano-lithography, and nano-microscopy.

As typical structures that have been mainly investigated to this end, representatives are circular ridged aperture, H-shaped aperture, and C-shaped apertures. These structures have a common narrow gap structure to make up for the required capacitive

Shaded region−>conducting plane

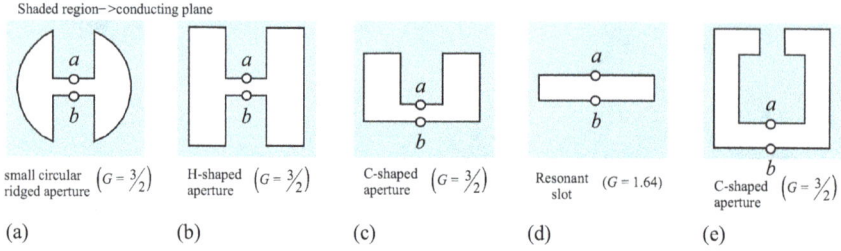

| small circular ridged aperture $\left(G = \frac{3}{2}\right)$ | H-shaped aperture $\left(G = \frac{3}{2}\right)$ | C-shaped aperture $\left(G = \frac{3}{2}\right)$ | Resonant slot $(G = 1.64)$ | C-shaped aperture $\left(G = \frac{3}{2}\right)$ |
|---|---|---|---|---|
| (a) | (b) | (c) | (d) | (e) |

*Figure 4.1   Various transmission resonant apertures*

susceptance for the transmission resonance. In the sense that the incident power on the significantly larger area than the physical aperture area can be transmitted through the small resonant aperture, these structures are named transmission resonant aperture (TRA). Because of sharing the above common narrow gap structure and the enhanced transmission efficiency, a resonant slot (called slot dipole) and a folded slot widely used as CSRR (complementary split ring resonator) [6] structure are also categorized into the TRA group as shown in Figure 4.1.

In Figure 4.1 the ports a−b mean input impedance (or admittance) port and $G$ means the aperture gain in the half-space of the incident side. Here the gain and directivity are interchangeably used. Note that under the plane wave incidence, the transmitted power at resonance through any TRA in Figure 4.1 is given by $P_i \times \frac{2G\lambda^2}{4\pi}$ irrespective of the physical area of aperture, where $P_i$ and $\lambda$ mean the incident power density and wavelength respectively.

There is another structure that provides a significant enhancement of transmission efficiency by employing the small aperture-to-cavity-to-small aperture coupling structure as discussed in [7]. As is well known, this structure has been widely used as a typical band pass filter structure in the microwave engineering area. The working principle of the transmission resonance in this structure is similar to that of the Fabry–Perot resonator structure. This structure is called transmission resonant cavity (TRC) structure in the next chapter to distinguish this from the transmission resonant aperture (TRA) type under discussion here.

The aim of this chapter is to study the transmission resonance phenomena through a small transmission resonant aperture with a main interest centering on the transmission cross section as a measure of transmission efficiency.

### 4.1.1   Transmission resonance condition and its equivalent circuit description

Here we are going to deal with the transmission resonance conditions for a transmission resonance aperture. First, we consider the small ridged aperture as an example of the transmission resonance aperture in Figure 4.2.

As mentioned already, when a plane wave is incident upon the small circular aperture whose diameter $D$ is much smaller than half the wavelength, the transmission efficiency through the aperture is very poor. The transmission efficiency

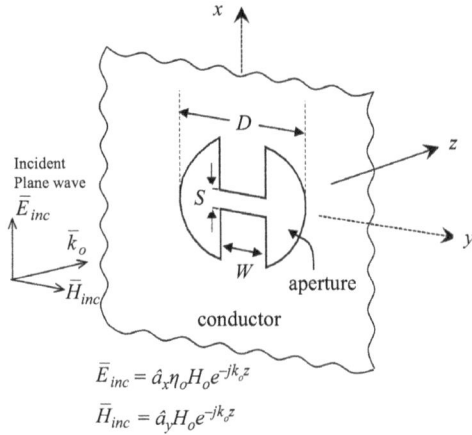

*Figure 4.2   A small ridged aperture in an infinite conducting screen*

can be, however, made to be significantly enhanced by introducing the ridge type modification in aperture shape as shown in Figure 4.2. To investigate the transmission resonance phenomena in which the maximum transmission occurs, we have calculated the transmitted power through the ridged aperture structure versus frequency, for the appropriately chosen geometrical parameters of separation between two ridges $S$ = 0.478 mm, ridge width $W$ = 3 mm and diameter of the circular aperture $D$ = 10 mm. Numerical analysis has been done by using the Rao–Wilton–Glisson (RWG) [8] method of moments. Figure 4.3 shows the variation of the transmitted power versus frequency. The transmission area $T$ is obtained by dividing the transmitted power [W] by incident power density [W/m²] such that its dimension may have that of area meaning the transmission cross section (TCS) of the aperture. For the numerical data, the incident power density from the left incident side is assumed to be 1 [W/m²].

The line marked with a small circle represents the transmitted power for the ridged circular aperture, which clearly shows the transmission resonance behavior. On the other hand, the transmitted power for the original circular aperture marked with small cross is seen to be much smaller than that near the resonance frequency of 7.442 GHz for the ridged circular aperture. When the transmission resonance for which the transmitted power through the aperture becomes maximum, the radiation admittance $G + jB (= 2G_{hs} + j2B_{hs})$ at input port a–b becomes real [3], as expected.

Here $G + jB$ is the total radiation admittance seen by the aperture port a–b into two (right and left) half-spaces in parallel, while $G_{hs} + jB_{hs}$ is the radiation admittance seen by the aperture port a–b looking into the single half-space region, i.e., $z < 0$ or $z > 0$ region. So at transmission resonance, the complex power $\left( 1/2 \int_S \vec{M} \cdot \vec{H}^* ds \right)$ transmitted through the ridged aperture into the right half-space (i.e., $z > 0$ region) becomes real.

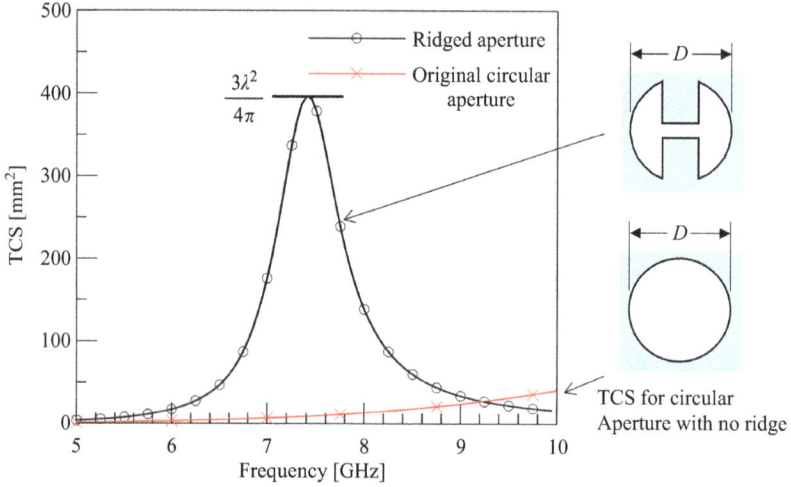

*Figure 4.3    Transmitted power versus frequency for the ridged circular aperture and the original circular aperture*

It has been observed [3] also that, at resonance, the phases of the dominant field components in the ridged aperture structure, $E_x$ and $H_y$ are almost the same and the phase difference between $E_x$ and $H_y$ is almost $0^0$. What is more, the intensity of $E_x$ component is seen to be the strongest in the gap between the two ridges. For this reason, the radiation source through the ridged aperture into the right half-space ($z > 0$ region) is approximated as uniform magnetic current $\left( \vec{M}_y = \vec{E}_x \times \vec{a}_z \right)$ over the rectangular area whose width is s (chosen to be the same as the gap $S$ between two ridges) and effective length is chosen to be $l_{eff}$ in the conducting plane of the opposite side at $z = z^+$ to the incident side.

Here the effective length $l_{eff}$ of the equivalent magnetic current source on the conducting plane of the $z^+$ side is determined such that, under the transmission resonance, the transmitted power through the ridged aperture into the right half-space may be equal to the radiated power from the above equivalent magnetic current source into the right half-space, i.e.,

$$\frac{1}{2}\int_{s(ra)} \vec{M} \cdot \vec{H}^* \, ds = \frac{1}{2}\mathrm{Re}\int_{s \times l_{eff}} \vec{M}_e \cdot \vec{H}_e^* \, ds \tag{4.1}$$

Here the integration range $s(ra)$ on the left side means the ridged aperture area and $\vec{M}_e$ and $\vec{H}_e$ mean, respectively, the above equivalent magnetic current source and the magnetic field due to $\vec{M}_e$. The $s \times l_{eff}$ on the right side means the area of the equivalent magnetic current source whose width is $s$ and the length $l_{eff}$. The operator Re on the left side has been dropped because of the real integration at resonance over the ridged aperture area. Note, on the other hand, that the operator Re on the right side should be retained because the above equivalent magnetic

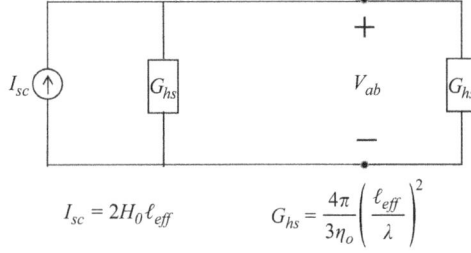

$$I_{sc} = 2H_0 \ell_{eff} \qquad G_{hs} = \frac{4\pi}{3\eta_0} \left( \frac{\ell_{eff}}{\lambda} \right)^2$$

*Figure 4.4    Equivalent circuit representation of the ridged aperture at resonance*

current source of the short length ($l_{eff}$) alone cannot radiate a real amount of energy but a complex amount of energy. When we calculate the radiated power from the above equivalent source into the right half-space, it is convenient to use the expression for the radiated power from the magnetic current element of $I_m l_{eff}$. Note that the equivalent magnetic current source is uniformly distributed over the rectangular area $s \times l_{eff}$ is expressed as a magnetic current element of $I_m l_{eff}$.

Here the magnetic current $I_m$ means $\int_b^a E_x dl = \int_b^a M_y \, dl$, which corresponds to the voltage $V_{ab}$ between port a–b in Figure 4.1(a) or the voltage across the narrow gap width $S$ of the equivalent magnetic current source in Figure 4.2. Because the radiated power from the magnetic current element $I_m l_{eff} = \left( V_{ab} l_{eff} \right)$ residing on the $z^+$ surface onto the right half-space is given as $\frac{2\pi}{3\eta_0} \left| \frac{V_{ab} l_{eff}}{\lambda} \right|^2$, setting this to be equal to the dissipated power $1/2 |V_{ab}|^2 G_{hs}$ in the conductance $G_{hs}$ [9] (the conductance seen by the aperture looking into the right half-space) leads us to the desired expression for $G_{hs}$ as follows:

$$G_{hs} = \frac{4\pi}{3\eta_0} \left( \frac{l_{eff}}{\lambda} \right)^2 \qquad (4.2)$$

Taking account of the total radiation conductance $2G_{hs}$ for left and right half-spaces and using the relationship $V_{ab} = \frac{I_{sc}}{2G_{hs}}$ between the voltage $V_{ab}$ and the short-circuit current $I_{sc} = 2H_0 l_{eff}$, we obtain the equivalent circuit description for the transmission resonance in the ridged aperture as shown in Figure 4.4.

From the above equivalent circuit, the dissipated power in the right $G_{hs}$ is obtained to be $\frac{1}{2} VI = \frac{1}{2} V^2 G_{hs} = \frac{1}{2} \eta_0 H_0^2 \frac{3\lambda^2}{4\pi}$ [W], which corresponds to the transmitted power into the right half-space through the ridged aperture under the transmission resonance. Because the incident power density is $\frac{1}{2} \eta_0 H_0^2$ [W/m²], the transmission area $T$ is found to be $\frac{3\lambda^2}{4\pi}$. This result is seen to be compatible with the numerical result at 7.442 GHz in Figure 4.3. Note that $V_{ab}$ and $I$ are real in the resonant circuit in Figure 4.4. Note that the gain $G$ of the small magnetic dipole which belongs to the TRA structure under consideration is 3/2. Here the gain $G$ and directivity $D$ can be used interchangeably under the assumption of the no-loss system. What is more the TCS of the aperture in the conducting screen is well known to be proportional to the directivity. For this reason we assume that the general expression for the TCS be chosen to be $\frac{2G\lambda^2}{4\pi}$ [m$^2$]. This general expression is validated for the single aperture whose TCS is larger than 3/2 in this chapter and even for the finite array of apertures of the non-resonant aperture in Chapter 8. So, the general expression for the TCS is thought to give an accurate result for both single TRA structure and the array of non-resonant aperture in parallel configuration. This has been a discussion on the transmission resonance phenomenon which can be used to design the transmission resonant aperture (TRA) for enhancing the transmission efficiency of the small aperture by modifying the aperture shape.

### 4.1.2  Various types of transmission resonant apertures

There are various types of transmission resonant aperture that have been mainly investigated to enhance the transmission cross section as a measure of the transmission efficiency for various applications such as optical data storage, nano-lithography, nano-microscopy, and meta-material and EMI-related areas.

As mentioned above, most of the TRA structures have been known to have one feature in common, the small gap structure, which is the key structure for providing high transmission efficiency and highly concentrated beam density. Here we deal with various TRAs of different shapes that belong to the ridged aperture [10] type, such as H-shaped aperture (double ridged), C-shaped aperture (single ridged), folded aperture of complementary split ring resonator (CSRR) type, and Jerusalem cross type of aperture as shown in Figure 4.5.

#### 4.1.2.1  H-shaped TRA

As discussed above, if geometrical parameters of the small aperture are appropriately chosen such that the input admittance between the port across the gap becomes real, then the incident power on the TCS much larger than the physical aperture area is transmitted through the aperture, amounting to the maximum of transmitted power, which is called transmission resonance. Under such a transmission resonance condition, incident power upon the TCS is funneled and transmitted through the aperture and reradiated into the back side half-space irrespective of the aperture shape and size.

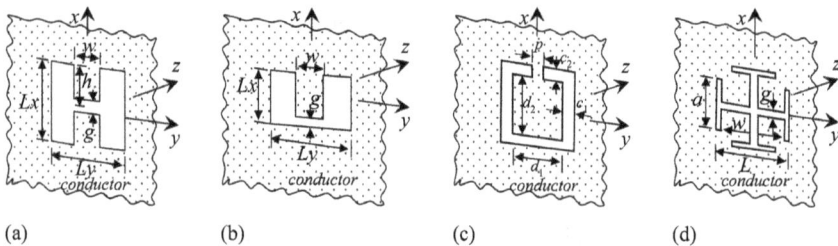

(a)                    (b)                    (c)                    (d)

*Figure 4.5   Transmission resonant apertures under consideration. (a) H-shaped, (b) C-shaped, (c) folded (CSRR) type, and (d) Jerusalem cross-type*

In order to investigate the transmission characteristics, a calculation was carried out for the case that $L_x = 10$ mm, $L_y = 10$ mm, $w = 3$ mm, and $g = 0.5$ mm. For numerical analysis, the Rao–Wilton–Glisson (RWG) method as a numerical method of moments (MoM) was used. Figure 4.6 shows the transmission resonance characteristics which are drawn as a solid line for H-shaped TRA of Figure 4.5(a).

In Figure 4.6, the TCS curve expressed by $\frac{2G\lambda^2}{4\pi}$ is also given by a dotted line and the TCS curve is seen to be tangent to the transmission characteristics curve at the peak frequency of 6.38 GHz. At this intersection of 6.38 GHz, the total input admittance across the gap becomes purely real and the transmission resonance occurs.

Under this condition, the TCS is observed to be 539 mm$^2$, which is much larger than the physical area of the aperture as expected from the funneling effect. For validation, the present numerical results are compared with those obtained by Microwave Studio® CST software. Note that the directivity $G$ of 1.53 used in Figure 4.6 is slightly larger than that of the ideal short dipole because of the slightly larger aperture size in comparison with the ideal short magnetic dipole.

### 4.1.2.2 C-shaped aperture

In the case of $L_x = 5$ mm, $L_y = 10$ mm, $w = 3$ mm, and $g = 0.25$ mm in the C-shaped aperture as shown in Figure 4.5(b) which has been obtained by bisecting the H-shaped aperture in Figure 4.5(a), we investigated the transmission characteristics, which is represented by a solid line along with the dotted line representing the TCS curve in Figure 4.7.

In comparison with the H-shaped aperture, a frequency shift to a higher frequency, for example, from 6.38 GHz to 7.16 GHz, is observed in the case of

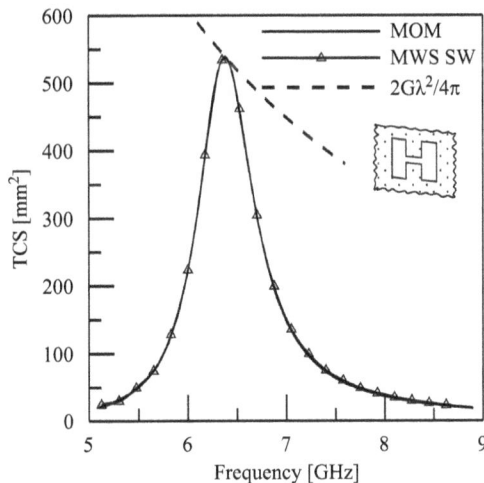

*Figure 4.6  Transmission cross section as transmission characteristics for the H-shaped aperture*

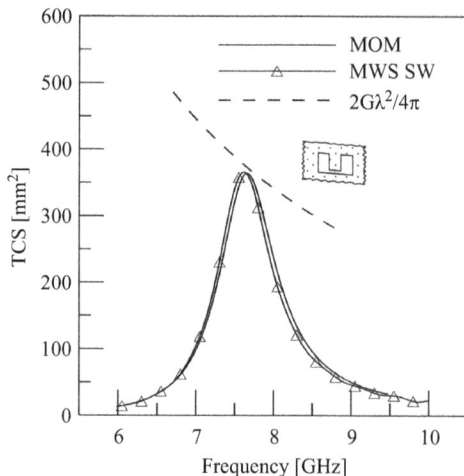

*Figure 4.7    Transmission cross section as transmission characteristics for C-shaped aperture*

C-shaped aperture. But in this C-shaped aperture case, by decreasing the gap $g$ somewhat, the transmission resonance frequency can be lowered to the previous resonant frequency of 6.4 GHz in the original H-shaped aperture structure, though not illustrated here.

So it is seen that the physical size of the TRA can be reduced roughly by half while keeping the TCS constant. This means that the TCS can be maintained constant irrespective of the physical aperture area of the TRA under the same transmission resonance condition of the real input impedance at the slot center.

### 4.1.2.3    Folded slot type aperture

As another example of TRA, the folded slot type of aperture which has been widely used as a complementary split ring resonator (CSRR) structure in the meta-material area is investigated with the main interest centering on the transmission resonance characteristics. In particular how the main horizontal slot length $d_1$ affects the transmission resonance is examined. Figure 4.8 shows a variation of the transmission characteristics with the main slot length $d_1$ as parameters for the case that $d_2 = 0.8$, $c_1 = 1.5, c_2 = 0.25$, and $p = 1.4$ [mm]. As expected, the transmission resonance frequency decreases as $d_1$ increases and every transmission resonance peak is seen to be slightly larger than the respective TCS values $\frac{3\lambda^2}{4\pi}$ at every transmission resonance peak frequencies. This means that the directivity of the folded slot dipole is slightly larger than that of the small slot dipole.

### 4.1.2.4    Jerusalem cross aperture

As a polarization-independent TRA, the transmission characteristics of Jerusalem cross type of aperture that consists of a pair of crossed vertical and horizontal

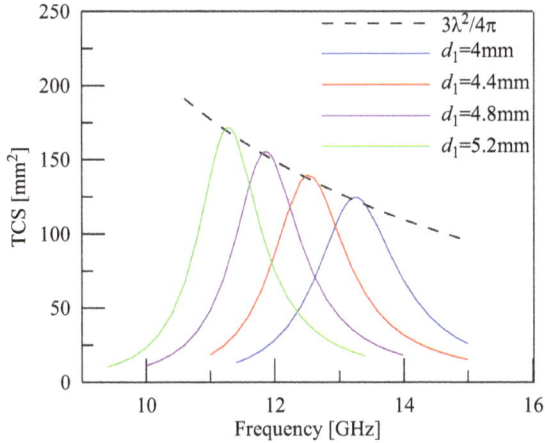

*Figure 4.8 Variation of the transmission characteristics of the folded slot dipole (CSRR type) with the main slot length d₁ as parameters*

H-shaped aperture as shown in Figure 4.5(d) are investigated. In more detail, the structure under consideration is made of crossing at right angles the two vertical and horizontal H-shaped apertures at the center intersection point. This cross-section point at each center of the above H-shaped apertures corresponds to the minimum loading points to each other. As a result, only the slight changes in the TCS value and its frequency are expected to result due to the connection in comparison with those of respective isolated H-shaped apertures.

For the case that $L = 10$ mm, $a = 7$ mm, $w = 8$ mm, and $g = 1$ mm in the Figure 4.5(d), the transmission characteristics of the Jerusalem cross type of aperture are presented along with the TCS curve in Figure 4.9.

Note that the directivity of the above four transmission resonant apertures ranges from 1.5 for small slot dipole to 1.53 for slightly larger aperture than the small slot dipole. For the case of the aperture having a larger directivity than 1.5, a general expression for the TCS, $\frac{2G\lambda^2}{4\pi}$ should be used for better accuracy.

## 4.1.3 Comparison of TCS between circular ridged aperture and circular aperture

It is interesting to compare the TCS of the circular ridged aperture as a TRA and that of the circular aperture with no ridge. As expected from the foregoing discussion, the TCS of the circular ridged aperture is much larger than that of the circular aperture. Figure 4.10 shows the comparison of TCS between resonant (ridged) and non-resonant (circular) aperture cases.

The transmission resonance characteristics of the circular ridged apertures have been investigated for the gap width of $g = 0.5$ mm. The numerical result is illustrated in comparison with that for the circular aperture with no ridge in Figure 4.10. From the figure it is seen that, in the case of the ridged aperture, the

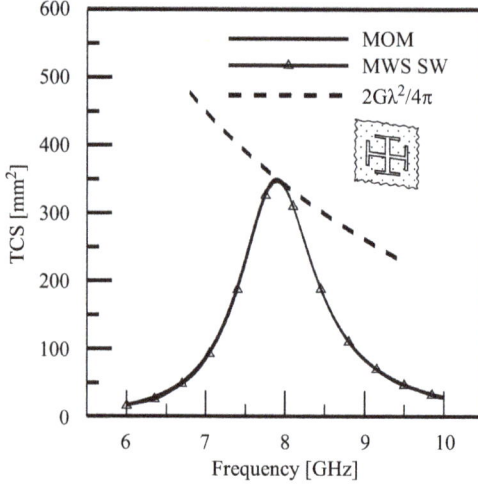

*Figure 4.9    Transmission characteristics of the Jerusalem cross type of aperture*

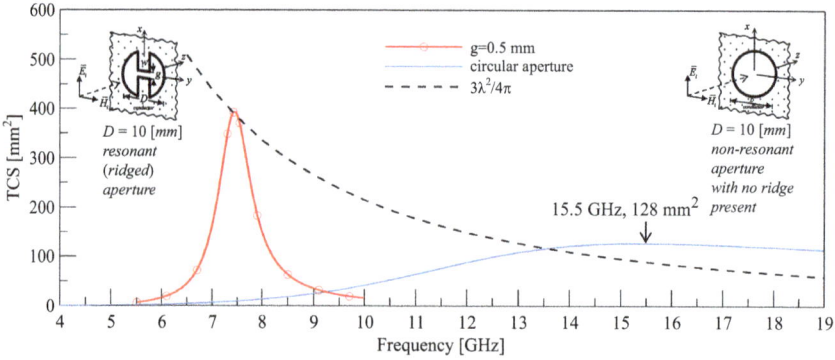

*Figure 4.10    Comparison of TCS between circular ridged aperture and circular aperture*

transmission resonance frequency is lowered while the resonance peaks will be in contact with the TCS curve. Here $G = 1.5$ for the directivity is used for the TCS curve. As expected, the TCS of the circular ridged aperture is observed to be much larger than that of the circular aperture with no ridge. The maximum TCS for the circular aperture is observed at 15.5 GHz at which frequency the TCS is seen to be 128 mm$^2$. This TCS is somewhat larger than the physical aperture area of 78.5 mm$^2$.

It can be said that transmission resonance of very low $Q$ occurs at 15.5 GHz, which roughly corresponds to the frequency (15 GHz) of the maximum effective aperture of the half-wavelength dipole antenna whose overall length is 10 cm. So, by employing the TRA structure we can significantly lower the transmission resonance frequency and increase the quality factor $Q$.

It is interesting to point out that the TCS of the circular aperture with no ridged is approximately two times the effective aperture of the half-wavelength dipole antenna. The directivity of the circular aperture is found to be 2.147 for reference.

### 4.1.4 Near-field probe application

As is well-known, near-field microscopy employs a probe of a subwavelength size and an object that is mounted in the near-field of the probe, so that the spatial resolution is determined by the probe size [11] rather than by the wavelength.

As a near-field probe, a non-resonant type of aperture such as a small circular aperture has been widely used from the early stage [11] of the near-field scanning microscope technology. Recently a narrow resonant slit type [12] of aperture as shown in Figure 4.11 was proposed as a near-field probe which was motivated by advantages such as high transmission efficiency and low Q (wide bandwidth) over conventional non-resonant small aperture.

The sensing probe in Figure 4.11 is composed of a narrow slit cut in the end plate of the rectangular waveguide. The length of the slit is approximately equal to the half wavelength in the free space, while the width may be exceedingly small.

In comparison with the resonant narrow slit type of probe, the TRA type of probe under consideration is thought to be more advantageous from the viewpoint of the better transmission beam confinement for better resolution, because the transmission area of the EM energy can be reduced significantly by confining EM beam passage within the small gap of the H-shaped aperture for example.

Consider both the $\lambda/2$ narrow slot dipole and the H-shaped aperture in the thin infinite conducting plane. As discussed above, the cross section area of passing EM beam can be made significantly small in the H-shaped aperture, because most EM energy is confined within the small gap. So the H-shaped (ridged) aperture is better from the viewpoint of the spatial confinement of the passing power whether it is used as a coupling aperture in the end plate of a guiding structure or on the conducting plane. This is the main reason why the H-shaped (or ridged) aperture has been chosen as the most promising nanoaperture in the near-field optical applications such as optical data storage, nanolithography, and nano microsopy.

As expected from the viewpoint of the resolving power of the probe, good transmission efficiency and good impedance matching characteristics should be

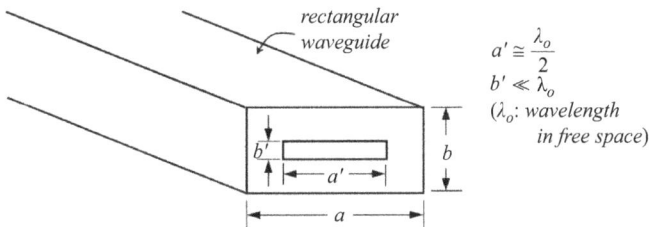

*Figure 4.11   Resonant narrow slit type of aperture*

maintained without any significant deterioration while the coupling hole size is kept as small as possible for better resolution.

The probe which comprises a small resonant circular ridged aperture cut in the end plate of the rectangular waveguide is thought to meet such requirements and to be a good choice.

### 4.1.4.1   Impedance characteristics of the TRA type of probe

We first obtain geometrical parameters such as ridge width $S$, gap $G$, and circular aperture radius $RA$ in Figure 4.12(a) for the given resonance frequency of the transmission resonance problem through the aperture in the infinite conducting plane by using the numerical method such as Rao–Wilton–Glisson (RWG) method.

Next, we investigate the impedance matching characteristics, i.e., scattering parameter $|S_{11}|$ of the circular ridged aperture in the end plate of the rectangular waveguide as in Figure 4.12(b).

Figure 4.13 shows the simulation and experiment of $|S_{11}|$ for the circular ridged aperture type.

When the TRA like the circular ridged aperture is installed in the end plate of the rectangular waveguide as above, the transmission resonance frequency is expected to shift somewhat from the aperture coupling problem in the infinite conducting plane, note that the greater the aperture's Q value, the less significant the frequency shift becomes.

In comparison with the conventional non-resonant aperture type of probe, the TRA type under consideration has several advantages for miniaturized apertures: (i) high transmission efficiency for good impedance matching, (ii) small spatial confinement of electromagnetic energy for better resolution, (iii) low Q factor in comparison with the transmission resonant cavity type following in the next chapter, enabling to transmit short pluses and to achieve good temporal resolution.

### 4.1.4.2   Field configuration in transmission resonant aperture

Field enhancement is intentionally tried to improve the resolution of the NSOM probe [13] and for use in heat-assisted surface modification [14] as well. For these purposes, the optical antenna of the bow-tie type has been used to realize a

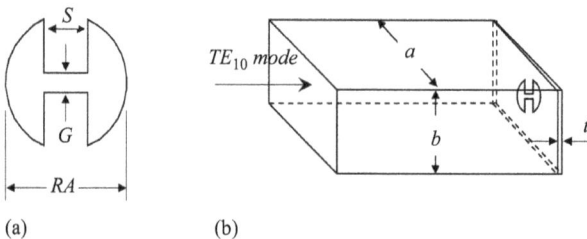

(a)                    (b)

*Figure 4.12   The probe of circular ridged aperture type. (a) Circular ridged aperture (S = 1.9, G = 0.4, and RA = 6.9), unit [mm], (b) feeding waveguide (a = 22.86, b = 10.16, and t = 0.2), unit [mm]*

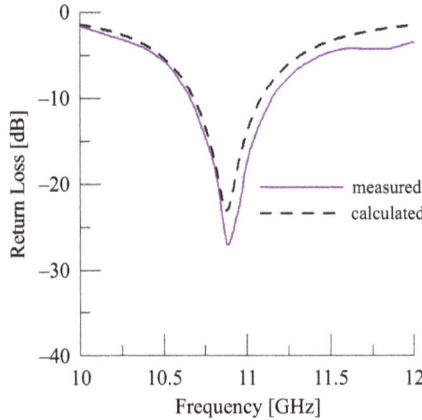

Figure 4.13　*Simulation and experiment of $|S_{11}|$ for the circular ridged aperture type*

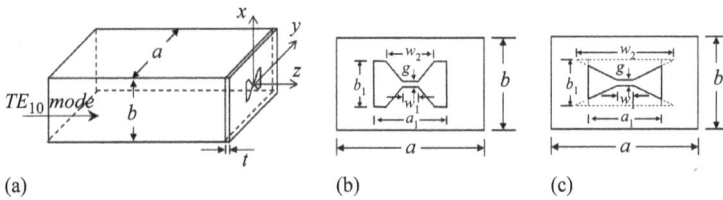

(a)　　　　　　　　　(b)　　　　　　　　　(c)

Figure 4.14　*Transmission resonant aperture of the double gradual ridged aperture (bow-tie type). (a) Feeding WR 90 rectangular waveguide ($a = 22.86, b = 10.16$, and $t = 0.2$) unit [mm]. (b) Double ridged aperture of bow-tie type ($w_1 = 2, w_2 = 10, g = 0.4, a = 8$, and $b = 4$) unit [mm]*

near-field optical probe that combines spatial resolution well below the diffraction limit with transmission efficiency approaching unity. The bow-tie type of antenna structure corresponds to the double gradual ridged aperture. So, we investigate the field configuration for the double gradual ridged aperture type. The structure of the gradually double ridged aperture and feeding guide are given in Figure 4.14. The geometrical parameters of the aperture are $w_1 = 2$ mm, $w_2 = 10$ mm, $g = 0.4$ mm, conductor thickness $t = 0.2$ mm, $a_1 = 8, b_1 = 4$, and the rectangular waveguide specification is WR90, i.e., $a = 22.86$ mm, $b = 10.16$ mm.

To investigate the reflection characteristics versus frequency and field distribution just over the aperture center for the geometrical parameters given above, we have calculated them by use of HFSS. Figure 4.15 shows the scattering coefficient $|S_{11}|$ versus frequency and the electric field distribution just over the aperture.

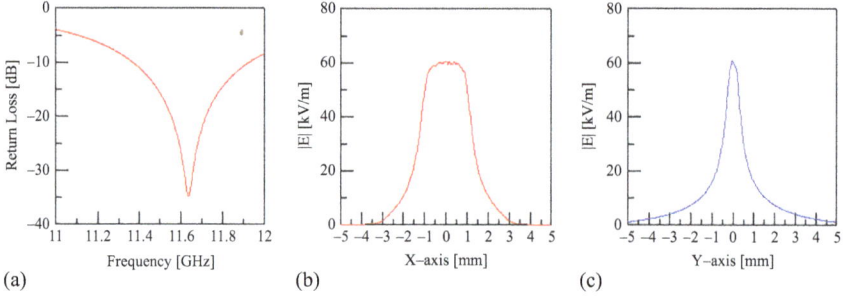

*Figure 4.15   Scattering coefficient $|S_{11}|$ (reflection coefficient) versus frequency
and the y-component electric field distribution in the TRA type of
probe. (a) Scattering coefficient $|S_{11}|$ (reflection coefficient) versus
frequency. (b) Electric field (y-component) along the x-axis at
$z = 0.2$ mm. (c) Electric field (y-component) along the y-axis at
$z = 0.2$ mm.*

From Figure 4.15, the probe whose aperture is of bow-tie type is seen to offer the advantages in both high transmission efficiency for good impedance matching and small spatial confinement of electromagnetic energy through the small aperture for better resolution. Another advantage of this TRA type is its low Q factor in comparison with the transmission resonant (TRC) type appearing later, enabling it to transmit short pulses and achieve relatively better temporal resolution. This bow-tie type of aperture is essentially the same as the H-shaped aperture. The H-shaped can be taken as a ridged waveguide whose longitudinal length is reduced to zero as a limiting case working near the cutoff state as discussed in Chapter 3.

As is well known in microwave engineering, the ridged waveguide is a broadband device and has a property that its cutoff wavelength in the $TE_{10}$ mode can be much larger than twice that of the rectangular waveguide size. So the bow-tie type of aperture can provide both a high transmission efficiency and a highly concentrated near-field spot size. It is worthwhile to note that the above bow-tie type of aperture is superior to the halfwave length slot dipole in terms of the smaller near-field spot size whereas the two apertures have almost the same transmission efficiency.

As mentioned above the TRA type of near-field probe can achieve lateral spatial resolution on the order of the aperture size of the probe because the electromagnetic field energy is squeezed through a tiny aperture and is collimated over a small confined region. So it is interesting to investigate the near-field distribution over the aperture. Figure 4.15(b) and (c) shows the y-component electric field respectively along the x and y axes at $z = 0.2$ [mm] at the impedance matching frequency of 11.64 GHz in Figure 4.15(a). In Figure 4.15(b) and (c), the near-field spot size of FWHM (full width at half maximum) is seen roughly to correspond to the gap size along the respective x and y axes.

## 4.2 Resonant scatterer structure as a dual structure to the transmission resonant aperture (TRA)

So far, we have dealt with the transmission resonance phenomena through the small resonant aperture which is called TRA. Under such a transmission resonance, the transmitted power through the aperture becomes maximum. Under this condition, the TCS is given by $\frac{3\lambda^2}{4\pi}$ if the resonant small aperture is approximated by a small magnetic dipole. Here we investigate the scattering resonance problem as a complementary structure to the resonant circular ridged aperture. As a word of caution, do not confuse the scattering resonance (or resonant scattering) here with the scattering resonance in Chapter 8. The scattering resonance here means the case in which the scattering cross section becomes maximum, whereas the scattering resonance in Chapter 8 means the case where the interaction becomes maximum between the outer incident wave and the leaky wave mode inside the guiding structure.

### 4.2.1 Small resonant scatterer as a complementary structure to the resonant circular ridged aperture

In the case of the small resonant scatterer like a short resonant dipole as a complementary structure to the above small circular ridged aperture problem as shown in Figure 4.16, the echo area or radar cross section has been well known to be $\frac{9\lambda^2}{4\pi}$ [m$^2$] under the usual assumption that the scatterer reradiates omnidirectionally.

If we assume that the small resonant scatterer radiates like a Hertzian dipole whose $\theta$-dependence is given by $\sin^2\theta$, the scattering cross section is found to be $\frac{3\lambda^2}{2\pi}$ [m$^2$].

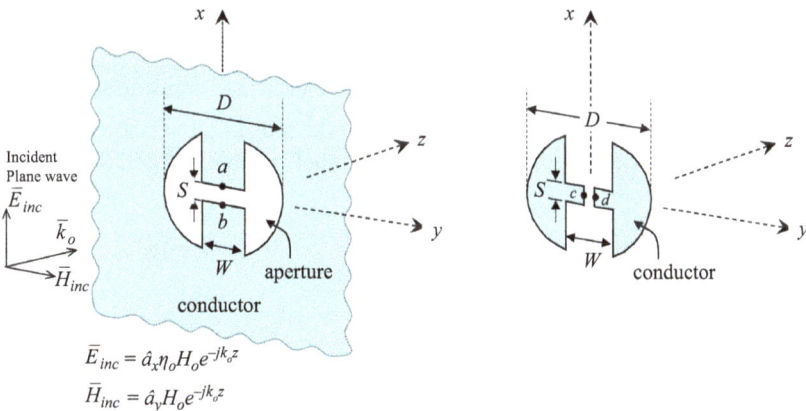

$$\bar{E}_{inc} = \hat{a}_x \eta_o H_o e^{-jk_o z}$$
$$\bar{H}_{inc} = \hat{a}_y H_o e^{-jk_o z}$$

*Figure 4.16   Resonant circular ridged aperture and its complementary structure. (a) Resonant circular ridged aperture, (b) resonant scatterer as a complementary structure to the resonant circular ridged aperture*

Interestingly this corresponds to the maximum scattering cross section of a shorted resonant antenna (under the condition $R_t = 0$, and $X_T + X_A = 0$ in Figure 4.17), which is four times as great as its maximum effective aperture as an absorber of energy (when $R_T = R_A = R_r$ and $X_T + X_A = 0$ in Figure 4.17).

Note that the maximum effective aperture is defined as a ratio of the maximum power transfer $\left(\frac{V^2}{4R_r}\right)$ to the antenna load $R_T$ to the incident power density $P_i$. Note that the maximum effective area of a short Hertzian dipole is given by $\frac{3\lambda^2}{8\pi}$ $[\text{m}^2]$ when the Hertzian dipole as a receiving antenna is assumed to be polarization-matched to the incident wave field and to be impedance-matched.

This is easily proved by the use of the equivalent circuit approach [15]. The scattering cross section of $\frac{3\lambda^2}{2\pi}$ $[\text{m}^2]$ also can be obtained from the above equivalent circuit approach by setting the load impedance, i.e., input impedance looking into the load port of the receiving antenna to be zero. It is interesting to point out also that this scattering cross section of $\frac{3\lambda^2}{2\pi}$ $[\text{m}^2]$ corresponds to the limit value of the maximum cross section of a single atom for a typical electric dipole transition [16,17]. For reference, when a small resonant scatterer is a single atom in a three-dimensional (3D) vacuum, its maximum scattering cross section is $(2l+1)\frac{3\lambda^2}{2\pi}$ at the atomic resonant frequency, where $l$ is the total angular momentum of the atomic transition involved [16,17].

The small resonant scatterer [18] under consideration is a kind of good scatterer [17] (though small)—scatterers with such an enhanced optical cross section that they are larger than the physical size of the scatterer.

If one-half of the scattered power $\left(\frac{3\lambda^2}{2\pi} \times P_i\right)$ by the small resonant scatterer is assumed to be reradiated into the half-space of the incident side, the total scattered power backwardly into the half-space of the incident side becomes identical to the maximum transmitted power through the small (subwavelength) resonant aperture, if and only if the two incident waves are the fields from conjugate sources.

The transmission cross section $\left(\frac{3\lambda^2}{4\pi} [\text{m}^2]\right)$ of the small transmission resonant aperture is much larger than the physical area of the transmission resonant aperture. As a result, a sort of electromagnetic energy "funnel" is obtained and the collection

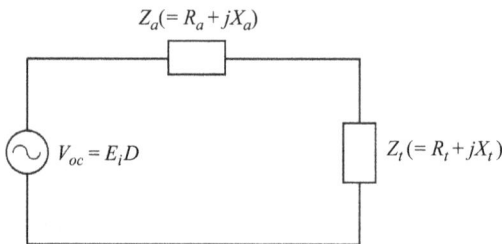

$$Z_a (= R_a + jX_a)$$

$$V_{oc} = E_i D \qquad Z_t (= R_t + jX_t)$$

*Figure 4.17   Equivalent circuit for the receiving mode of the small resonant scatterer*

of the electromagnetic energy into the small resonant aperture is favored as discussed in the report on light transmission through metallic channels much smaller than the wavelength [16,17]. That is, under the resonance condition, a significant amount of the incident electromagnetic energy outside the resonant aperture is funneled through the aperture.

Also in the case of the small resonant scatterer as discussed above, the scattering cross section $\frac{3\lambda^2}{2\pi}$ [m$^2$] is much larger than the geometrical cross sections in the scattering by a single atom [19–21]. A natural question [19–21] arises: "How can the small resonant scatterer interact with more than the electromagnetic energy incident on it?"

As expected from the funneling mechanism in the small resonant aperture problem, the small resonant scatterer illuminated by a plane wave disturbs the flow of electromagnetic energy in its neighborhood such that the field lines of the Poynting vector converge strongly near the small resonant scatterer. In this way the incident electromagnetic energy which interacts with the small resonant scatterer and is scattered is significantly increased, whereby the superscattering feature—scattering with such an enhanced optical scattering cross section (OSCS) that the enhanced OSCS is larger than the physical size of the scatterer—results. This wave picture is appropriate for accounts for the well-known large effective scattering cross sections, i.e., the superscattering feature, compared to the geometrical sections, of the atomic system, while the particle picture certainly fails to do so as discussed in [21].

The phenomenon of the transmission resonance through a small aperture is useful for applications to bandpass filter, small aperture antenna and array, FSS, and probe for NSOM (near-field scanning optical microscope). On the other hand, the phenomenon of the above resonant scattering by the small scatterer [22] can be applied to the field measurement by use of the resonant scatterer, band rejection filter and FSS, and power limiter design [23].

### 4.2.2 Small resonant scatterer as a dual structure to the circular ridged aperture

The structure of the small resonant scatterer is determined automatically by choosing the complementary structure to the circular ridged resonant aperture structure. Note that when we compare the transmitted field through the aperture and the back-scattered field from the resonant scatterer the two incident fields for the former and latter cases are assumed to be from conjugate sources, as shown in Figure 4.16.

When the small resonant conducting scatterer is assumed to be in receiving mode, the equivalent circuit is given as shown in Figure 4.17. In the figure, $Z_a$ means the radiation impedance and $Z_t$ means the input impedance looking into the receiver circuit between port c – d.

If we assume that $Z_t = Z_a^*$ , i.e., $R_a = R_t$ and $X_a = -X_t$ , then the dissipated power in $R_t$ is calculated to be $80\pi^2 \left(\frac{D}{\lambda}\right)^2$, which corresponds to the received power by the antenna. Dividing this result by incident power density gives the expression for the effective area of the Hertzian dipole, $\frac{3\lambda^2}{8\pi}$ [m$^2$].

When the terminal impedance becomes zero, the scattering power of the resonant scatterer becomes maximum. That is, when $R_t + jX_t = 0$ and complex conjugate matching condition $X_a = 0$ is also met, the maximum scattered (reradiated) power is calculated to be $\frac{3\lambda^2}{2\pi}$ [m²]. So the maximum scattered (reradiated) power as a resonant scatterer with the feeding port c−d shorted is four times as great as its maximum effective aperture as an absorber of energy.

In the present case where the small resonant scatterer is assumed to be located in the planar form in the $x$−$y$ plane, the radiation pattern is symmetrical about the $x$−$y$ plane at $z = 0$, so the half of total scattered power $P_i \times \frac{3\lambda^2}{2\pi}$ is scattered into $z < 0$ half-space and the $z > 0$ half-space, respectively.

So the backscattered power $P_i \times \frac{3\lambda^2}{4\pi}$ from the small resonant scatterer becomes the same as the transmitted power through the resonant aperture $P_i \times \frac{3\lambda^2}{4\pi}$, if and only if the two incident waves are the fields from conjugate sources as illustrated in Figure 4.16.

## 4.2.3   Resonant aperture and scatterer structures as a dual problem

The resonant aperture structure and the complementary structure constitute the typical dual problems whose radiation conductance $G(= 2G_{hs})$ and radiation resistance $R_r$ are multiplied to be $\eta_0^2/4$. Here $\eta_0$ means the intrinsic impedance of the free space.

It is straightforward to prove such a relationship. In the case of a resonant aperture such as a circular ridged aperture as shown in Figure 4.16(a), the radiation conductance under the resonance condition of cancellation of the imaginary part of input impedance at feeding port a−b is

$$G = 2G_{hs} = \frac{8\pi}{3\eta}\left(\frac{D}{\lambda}\right)^2 \tag{4.3}$$

On the other hand, the radiation resistance of the small scatterer under resonance at port c−d as shown in Figure 4.16(b) is

$$R_r = 80\pi^2 \frac{D^2}{\lambda^2} \tag{4.4}$$

Using the expressions for $G$ and $R_r$ in the ratio $R_r/G$ leads to the following desired expression:

$$\frac{R_r}{G} = \frac{R_r}{2G_{hs}} = \frac{\eta_0^2}{4}, \tag{4.5}$$

which is a general expression for impedance (admittance) relation between two dual problems. This constitutes also an indirect confirmation of the validity of formulations of the two dual problems.

### 4.2.4 Small resonant scatterer for use in field measurements

Resonant scatterers are found to have several advantages over nonresonant scatterers when used for field measurements [18,22]. To this end, widely used structures are loaded dipole and loaded loop. The echo areas of these two small resonant scatterers, under the resonance condition that the imaginary part of the total input impedance be canceled to be zero, are obtained to be $\sigma \cong \frac{9}{4\pi}\lambda^2$. This has been obtained under the assumption that the small resonant scatterer reradiates omnidirectionally.

If we assume that the small resonant scatterer radiates like a Hertzian dipole whose $\theta$ dependence is given by $\sin^2\theta$, the scattering cross section is calculated to be $\frac{3\lambda^2}{2\pi}$ [m$^2$]. This corresponds also to the maximum cross section of the small loaded dipole when $R_t = 0$ and $X_t + X_a = 0$ as a scatterer of energy. This is also four times as great as its maximum effective aperture as an absorber of energy as already mentioned.

One half $\left(\frac{3\lambda^2}{4\pi}P_i\right)$ of the total scattered power $\left(\frac{3\lambda^2}{2\pi}P_i\right)$ is backward (or forward) scattered. So this backward scattered power $\left(\frac{3\lambda^2}{4\pi}P_i\right)$ is equal to the transmitted power $\left(\frac{3\lambda^2}{4\pi}P_i\right)$ through the resonant aperture as a dual problem.

From the discussions that the small resonant scatterer for use for field measurement satisfies the same scattering resonance condition as the dual structure of the transmission resonant aperture, it is also thought of as a kind of dual structure of the foregoing TRA structure.

The transmission resonant aperture, along with its dual structure known as resonant scattering, finds applications in various areas, including the design of compact-sized filters and FSS. In particular, the TRA structure can be used in the design area of NSOM probe and the resonant scatterer structure can be used in the design of the power limiter [23] for low-power radar.

### References

[1]   Bethe H. 'Theory of diffraction by small holes.' *Phys. Rev.* 1944, vol. 66(7 and 8), pp. 163–182

[2]   Shi X., Hesselink L., Thornton R. 'Ultrahigh light transmission through a C-shaped nanoaperture.' *Opt. Lett.* 2003, vol. 28(15), pp. 1320–1322

[3]   Park J. E., Cho Y. K. 'Comparison of transmission resonance phenomena through small coupling apertures between two kinds of transmission resonance structures.' *URSI Int. Symp. Electromagn. Theory Proc.* 2010, pp. 744–747

[4]   Park J. E., Yeo J., Lee J. I., KO J. W., Cho Y. K. 'Resonant transmission of an electrically small aperture with a ridge.' *J. Electromagn. Waves Appl.* 2009, vol. 23, pp. 1981–1990

[5]   Chen Y.c., Feng J.Y., Tien C.H., Shieh H.P.D 'High-transmission hybrid-effect-assisted nanoaperture.' *Opt. Lett.* 2006, vol. 31(5), pp. 655–657

[6]   Bahrami H., Hakkak M., Pirhadi A. 'Analysis and design of highly compact bandpass filter utilizing complementary split ring resonators (CSRR).' *Prog. Electromagn. Res.* 2008, vol. 80, pp.107–122

[7]   Leviatan Y., Harington R., Mautz J. 'Electromagnetic transmission through apertures in a cavity in a thick conductor.' *IEEE Trans. Antennas Prapagat.* 1982, vol. 30(6), pp. 1153–1165

[8]   Rao S. M., Wilton D. R., Glisson A. W. 'Electromagnetic scattering by surfaces of arbitrary shape.' *IEEE Trans. Antennas Prapagat.*, 1982, vol. 30 (3), pp. 409–418

[9]   Harington R. F. 'Resonant behavior of a small aperture backed by conducting body.' *IEEE Trans. Antennas Propagat.* 1982, vol. 30(2), pp. 205–212

[10]  Eric X. J., Xianfan Xu. 'Finite-difference time-domain studies on optical transmission through planar nano-aperture in metal film.' *Jpn. J. Appl. Phys.*, 2004, vol. 43(1), pp. 407–417

[11]  Ash E. A., Nicholls G. 'Super-resolution aperture scanning microscope.' *Nature*, 1972, vol. 237, pp. 510–512

[12]  Golosovsky M., Davidov D. 'Novel millimeter-wave near-field resistivity microscope.' *Appl. Phys. Lett.* 1996, vol. 68(11), pp. 1579–1581

[13]  Grober R. D., Schoelkopf R. J., Prober D. E. 'Optical antenna: towards a unity efficiency near-field optical probe.' *Appl. Phys. Lett.*, 1997, vol. 70 (11), pp. 1354–1356

[14]  Challener W.A. 'Heat-assisted magnetic recording by a near field transducer with efficient optical energy transfer.' *Nat. Photonics*, 2009, vol. 3, pp. 220–224

[15]  Kraus J. D. *Antennas.*1950, chapter 3, pp. 41–56

[16]  Hamam R. E., Karalis A., Joannipoulos J. D., Soljacic M. 'Coupled-mode theory for general free space resonant scattering of waves.' *Phys. Rev. A*, 2007, vol. 75, pp. 053801-1–053801-5

[17]  Ruan Z, Fan S. 'Superscattering of light from subwavelength nanos-tructures.' *Phy. Rev. Lett.*, 2010, vol. 105, pp. 013901

[18]  Harrington R. F. 'Small resonant scatterers and their use for field measure-ments.' *IEEE Trans. Microwave Theory Tech.*, 1962, vol. 10, pp. 165–174

[19]  Poul H., Fischer R. 'Light absorption by a dipole.' *Sov. Phys. Usp.* 1983, vol. 26(10), pp. 923–926

[20]  Tribelsky M. I. 'Resonant scattering of light by small particles.' *Sov. Phys. JETP.* 1984, vol. 59(3), pp. 534–536

[21]  Tribelsky M. I., Luk'yanchuk B. S. 'Anomalous light scattering by small particles.' *Phys. Rev. Lett.* 2006, vol. 97(26), pp. 263902

[22]  Ryerson J. L. 'Scatterer echo area enhancement.' *Proc. IRE*, 1979, vol. 50, pp. 1979–1980

[23]  Cho Y. K., Park J. S., Yeo J. H., Lee J. I., Kim K. C. 'Compact microwave waveguide limiter.' *IEICE Electron. Express*, vol. 13(23), pp. 1–9

# Chapter 5

# Transmission resonance problem through transmission cavity structures

## Synopsis

We deal with the transmission resonant cavity (TRC) structure which shows almost the same transmission characteristics as that of the transmission resonant aperture (TRA) in Chapter 4. Various TRC structures are considered with an emphasis on the working principle of the structures. The common feature between transmission resonance phenomena by TRA and TRC structure is discussed with the main interest centering on the TCS characteristics. The scattering phenomenon is unique to the TRC structure, as a difference between scattering by TRA and TRC structures is also described. An application example is considered for near-field scanning optical microscopy (NSOM) probe design for microwave and millimeter range.

## 5.1  Transmission resonant cavity structure

So far, we have dealt with the transmission resonant aperture (TRA) to enhance the transmission efficiency of the small aperture having innately poor transmission efficiency. There is another method for enhancing the transmission efficiency of the small coupling aperture.

That is, by use of the transmission resonant cavity (TRC), such a mission can be achieved. Figure 5.1 shows how a transmission cavity may be composed by taking an example of the TRC structure which employs the small aperture-to-cavity-to-small aperture system [1,2]. The irises form the closed ends of the transmission cavity and the small holes in the centers of the irises allow coupling into and out of the TRC. The transmission frequency for this type of TRC structure is mainly determined by the length of the transmission cavity.

Though the above TRC structure is to be considered in this chapter, for a more complete discussion, there is one more transmission resonance structure that deserves mentioning. It is the transmission structure through Bethe's small hole in a metal waveguide screen [3]. In this case, different from the TRC structure the cavity structure is not required. So instead of employing the transmission resonant cavity structure, the transmission resonant structure composed of a Bethe's small coupling hole at the center of a metal waveguide screen [3] is used. For this reason,

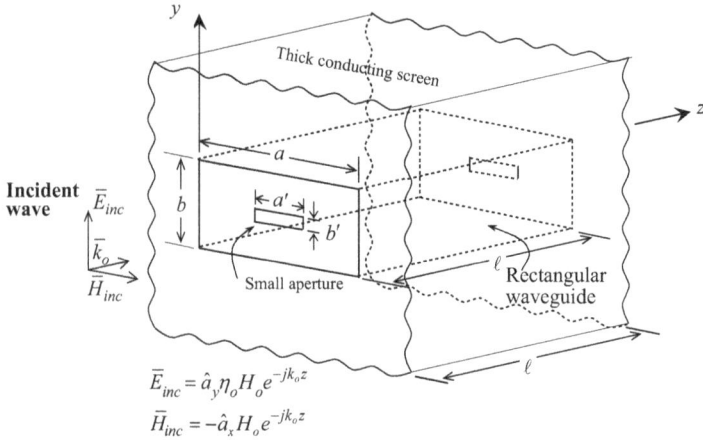

*Figure 5.1    Transmission resonant cavity (TRC) system*

the transmission resonance frequency is mainly determined by the waveguide height along which direction the incident electric field component exists, i.e., a periodic image array of magnetic current source over the hole is aligned in the parallel configuration [4] as discussed in Chapter 8. In more detail, the transmission resonance is observed just before the frequency where the interval between two adjacent images becomes the free space wavelength corresponding to the onset frequency of the lowest grating lobe. This array of images is thought to constitute a basic unit of a single linear chain of holes as a minimal system showing extra-ordinary optical transmission (EOT) phenomenon [5].

In a broad sense, extraordinary transmission refers to transmitting much more power than the power impinging on the actual aperture area of an incident side. Then the extraordinary transmission phenomena are thought to be present in the case of the single isolated aperture like the TRC here as well as the TRA in Chapter 4. It is worthwhile to mention that the cases of small holes over the metal waveguide screen and a finite array of the magnetic current sources of the holes in parallel configuration as discussed later in Chapter 8 can be categorized into this extraordinary transmission phenomenon.

Among the above transmission types, the single TRA and TRC structures corresponds to the waveguide resonance type and the infinite periodic image array of Bethe's small hole coupling problem inside the guide corresponds to the reso-nance type related to the periodicity in terms of the terminology used in the prior work on the extraordinary optical transmission (EOT) phenomena [6]. There have been discussions regarding the distinction between transmission resonance in a transmission cavity, which is currently under consideration, and that occurring through Bethe's small hole in the metal waveguide screen. However, these dis-cussions have slightly deviated from the main topic of the TCR.

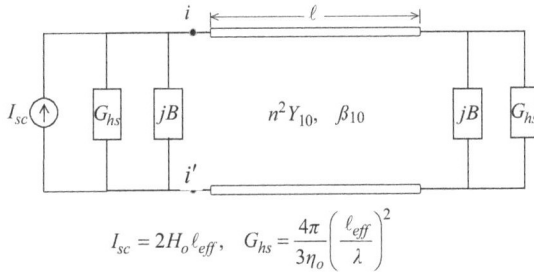

*Figure 5.2 Equivalent circuit for the transmission resonant cavity system*

Let us go back to the original topic of the TRC. The equivalent circuit for the above aperture-to-cavity-to-aperture system under consideration is given in the following Figure 5.2.

In Figure 5.2, $G_{hs}$ means the radiation conductance seen by the aperture looking into single (left or right) half-space, $B$ means the susceptance component seen by the aperture taking account of the total reactive (stored) energy outside and inside of the cavity region near each aperture, $n$ is a turns ratio of the ideal transformer which is introduced in the input and output ports of the equivalent circuit for the transmission resonant cavity [1], $Y_{10}$ and $\beta_{10}$ mean the characteristic admittance and propagation constant of the dominant $TE_{10}$ mode in the rectangular waveguide region. In the above equivalent circuit, if the cavity length $l$ is chosen such that the transformed value of the right load admittance $(G_{hs} + jB)$ along the transmission line of length $l$ may give $(G_{hs} + jB)^*$, then the total admittance at input port $i - i'$ becomes real as $2G_{hs}$, which is identical to that in the case of the circular ridged aperture which was dealt with as a kind of transmission resonant aperture (TRA).

Because $B$ is highly inductive for the small aperture, the length $l$ of the transmission line is found to be somewhat less than half the rectangular waveguide length of $TE_{10}$ mode. For the above reason, the transmission resonant cavity under consideration also has a maximum peak of the transmission area $\frac{3\lambda^2}{4\pi}$ [1,2] under the transmission resonance condition just like the TRA case such as the foregoing circular ridged aperture.

Here the definition of the transmission area is the same as that of the transmission cross section in Chapter 3. So these two can be used interchangeably.

As a numerical example, we have investigated the transmission characteristics of the structure of Figure 5.1 versus frequency for the case of $a = 22$ mm, $b = 6$ mm, $a' = 8$ mm, $b' = 1$ mm, and $l = 24$ mm.

The numerical result for the transmission area is given in Figure 5.3. Here the numerical analysis has been done by use of the mode matching technique.

The peak value of the curve is observed to be in good agreement with the result for the transmission cross section of $\frac{3\lambda^2}{4\pi}$.

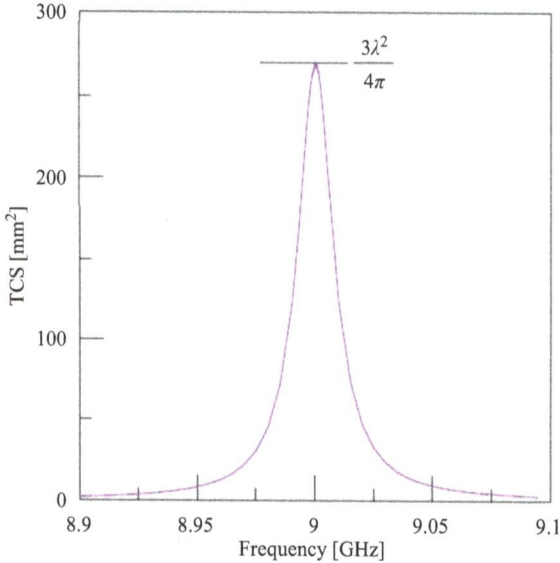

*Figure 5.3    Transmission area* T *versus frequency*

To validate further the result of $\frac{3\lambda^2}{4\pi}$ as the transmission cross section for the TRC type of structure, we have compared the results of $\frac{3\lambda^2}{4\pi}$ as a function of frequency with those numerically obtained in Figure 5.4.

In Figure 5.4, the results of $\frac{3\lambda^2}{4\pi}$ are compared with those obtained by use of the mode matching method for the structure in Figure 5.1. The results of the curve represented by the solid line have been obtained by calculating the maximum transmitted power for the waveguide cavity length $l$ determined such that maximum power transmission occurs at every frequency over 7 to 11 GHz. A comparison shows a uniform difference between the two results, though small. Note that the analytical expression of $\frac{3\lambda^2}{4\pi}$ for the TCS of the small aperture assumes that the coupling aperture is so small that the equivalent aperture source may be replaced by the magnetic Hertzian dipole whose directivity corresponds to 1.5. The difference tells us that the coupling aperture size used in numerical analysis is somewhat larger than the above magnetic Hertzian dipole and its directivity is larger than 1.5. So, if the smaller aperture is used in the numerical analysis or more accurate directivity for the aperture is used, the two results are expected to show better correspondence.

## 5.2    Various types of transmission resonant cavity (TRC) structure

We have discussed a typical transmission resonant type which is widely used as a bandpass filter in microwave and millimeter band. It is mainly composed of the

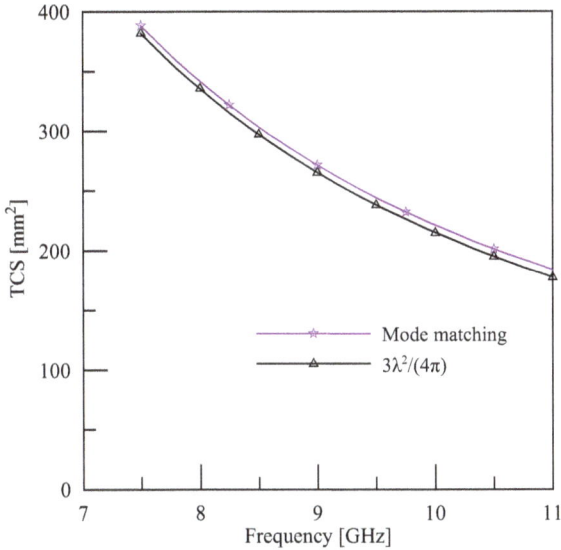

*Figure 5.4* *Transmission area* T *versus resonant frequency for the cavity length l determined such that maximum transmission occurs in the TRC structure with other geometrical parameters fixed*

iris-coupled transmission cavity structure, i.e., a small inductive aperture-coupled transmission cavity structure for both input and output coupling ports.

The possible TRC structures having almost the same TCS can be considered in terms of four combinations of input/output irises as shown below.

As shown in Figure 5.5, a transmission resonant cavity can be implemented either in the symmetrical form or in the unsymmetrical form. Figures 5.5(a) and (b) show the symmetrical form where both the input and output aperture are the same shape as the aperture, small hole (inductive) type, or capacitive iris type.

Unsymmetrical forms of transmission resonant cavity are illustrated in Figure 5.5(c) and (d), in which case if the input (incident) side aperture is of inductive small hole type, the output side aperture is given to be a capacitive type of aperture and vice versa. The unsymmetrical type of transmission resonant cavity has an advantage over the symmetrical type when used as an NSOM probe structure as discussed later.

In particular, for the case where intense local field enhancement [7,8] is required for resolution beyond the diffraction limit, the unsymmetrical TRC structure, featuring an output iris made up of a gradual ridge-type capacitive aperture, is preferred. It deserves mentioning that this type of output aperture belongs to the typical optical antenna structure [7].

For convenience, according to the combination of the input (incident) side and output side aperture types, we call L-L type, C-C type, C-L type, and L-C type.

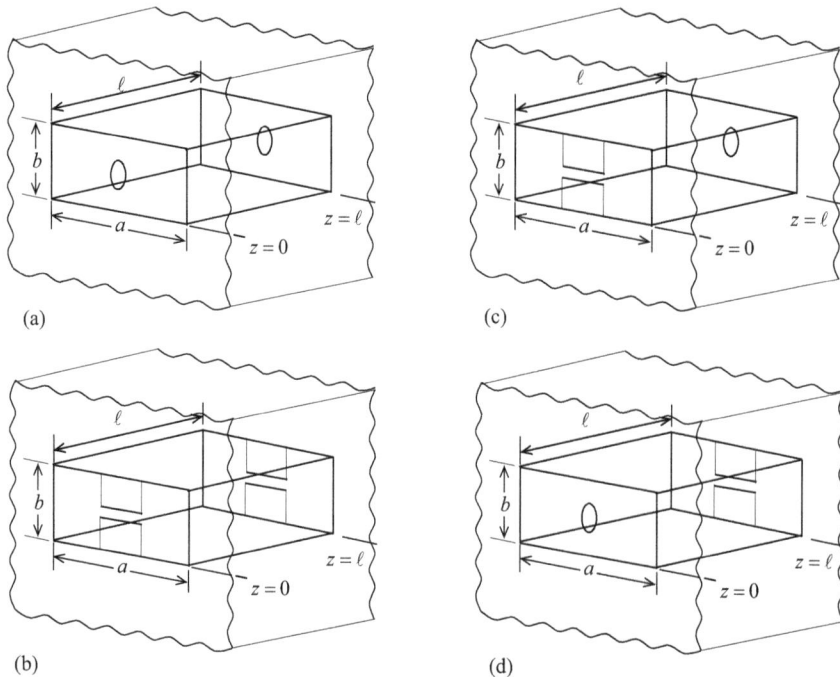

Figure 5.5   Four kinds of TRC structures

From the Smith chart, it is easy to see how the transmission resonance condition where the transmitted power becomes maximum is all satisfied for the above four cases.

Consider first the L-L case. In general, the radiation conductance $G_{hs}$ of the radiation admittance $Y_R(= G_{hs} + jB)$ at the output port is much smaller than the radiation susceptance $B(= -\frac{1}{\omega L})$. So, the admittance point $Y_R$ is located on the fourth quadrant somewhere near the short point as shown in Figure 5.6. There the radiation admittance $Y_R(= G_{hs} + jB)$ is normalized such that the real part may be unity, i.e., $Y_R = 1 + jb$ for convenience.

Transforming a line length somewhat smaller than the half-wavelength to the complex conjugate point and adding the normalized susceptance $\left(b = \frac{B}{G_{hs}}\right)$ to the normalized transformed admittance $Y_R{'}(= 1 - jb)$ makes the total admittance at the i-i' port in Figure 5.2 pure real. Under this condition, maximum transmission through the transmission cavity occurs. Note that $Y_L = Y_R$ for the symmetrical case [9,10].

From Figure 5.6, it is seen that for L-L type, the line of the TRC is less than a half-wavelength in length. Similarly, it is seen that for C-C type, the line is somewhat longer than a half-wavelength and that for C-L type the line is almost the same as the half-wavelength.

$L - L$ *type* $(b < 0)$

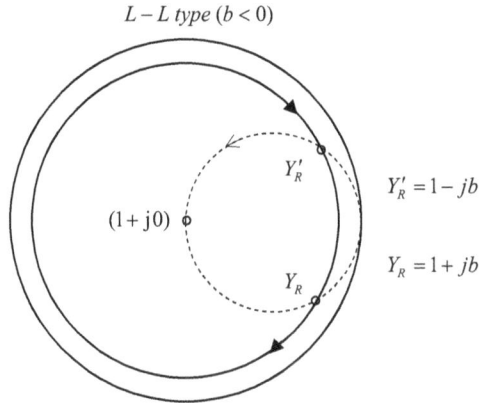

$Y'_R = 1 - jb$

$Y_R = 1 + jb$

$Y'_R$

$(1 + j0)$

$Y_R$

*Figure 5.6    Admittance locus for changing line position in Figure 5.2*

The present transmission resonant cavity problem is simply different from the waveguide band pass filter in that input and output ports are connected to respective half-space in the present problem whereas input and output ports of the waveguide band pass filter are connected to the waveguide whose transverse dimension is same to that of the bandpass filter structure [10]. If the TRC structure in Figure 5.5 (a) is incorporated in a rectangular waveguide run, it constitutes a bandpass filter, which is a well-known transmission cavity filter structure for microwave and millimeter band use. The remaining three structures can be also used for the design of the bandpass filter. When we deal with the waveguide filter problem, we have only to replace the radiation conductances $G_{hs}$ for left and right half-spaces in Figure 5.2 with the characteristics admittance $Y_c$ of the rectangular waveguide of the same cross section as the transmission resonant cavity along with some modification in $B$. Note that the incident plane wave for the TRC problem under consideration is modified to the dominant TE$_{10}$ mode for the waveguide filter problem.

Aside from the waveguide bandpass filter, the TRC structure under consideration can be applied to the design of the NSOM probe by installing the TRC structure in the end section of the rectangular waveguide. From the viewpoint of equivalent circuit description, the NSOM probe structure corresponds to the equivalent circuit in Figure 5.2 with the radiation conductance $G_{hs}$ for left (incident side) half-space replaced with the characteristic admittance of the feeding rectangular waveguide. The transmission resonance frequencies do not change appreciably for the three cases of TRC structure even when the cavity is connected to the outer structure for applications to the bandpass filter or NSOM probe if the unloaded $Q$ of the TRC is moderately high. The loading effect of the outer connecting structure to the transmission resonant cavity decreases as the quality factor $Q$ of the isolated TRC increases.

## 5.3    Common features between TRA and TRC

As mentioned already, the TRC is divided into two kinds, symmetrical TRC and unsymmetrical TRC. In the case of the symmetrical TRC whose input and output port iris are of the same inductive or capacitive type, the TCS is determined only by the aperture directivity $G$ of the incident side. Under the transmission resonance condition where the transmitted power through the transmission cavity becomes maximum, the TCS is given by $\frac{2G\lambda^2}{4\pi}$ [m$^2$]. This holds also for the TRA structure. So any symmetrical TRA structure can be implemented in terms of symmetrical TRC structures.

This can be taken as a common feature of the transmission characteristics between TRA and the symmetrical TRC structures. Such an equivalence for the TCS characteristics between TRA and TRC is clearly seen if we compare the results [1,8,11] for the two structures that were made by use of the small aperture coupling.

It is worthwhile to mention that the longitudinal length of the TRC cavity for all the types, irrespective of whether it is symmetrical type or unsymmetrical type, is determined such that the reactive component of the input and output port are summed to be canceled after transforming the output port admittance along the line length of the cavity to the location of input port admittance.

So, it can be said that the role of the TRC from the viewpoint of the equivalent circuit is to cancel the total sum of input port susceptance and the output port susceptance transformed along the cavity length.

This can be thought of as a common feature that the longitudinal cavity of all the TRC satisfies like the Fabry–Perot resonator under the transmission resonance.

As a result, the total admittance at the input port, where the equivalent current source $(= 2H_0)$ is applied, becomes real. Under this condition, the total admittance is found to be approximately two times the radiation conductance seen by the input aperture port looking into the left single half-space, or the radiation conductance seen by the output aperture port looking into the right single half-space. Because the radiation conductance is the same for both the left half-space and right half-space, the total radiation conductance becomes $2G_{hs}$, which is the same as the TRA case as discussed above.

## 5.4    Transmission property unique to transmission resonant cavity structures

To investigate the effect of the input coupling aperture shape on the TCS for the symmetrical and unsymmetrical TRC structures, we have looked into the symmetrical case where the input and output apertures are the same as each other in shape as narrow vertical slots as shown in Figure 5.7(a) and an unsymmetrical case where the input and output apertures are different from each other as shown in Figure 5.7(b).

Figure 5.8(a) and (b) shows the results for TCS for the two cases of Figure 5.7 (a) and (b), respectively.

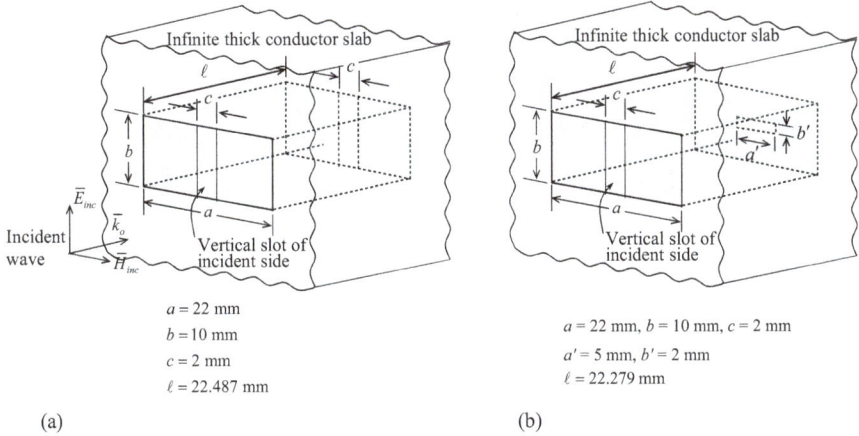

a = 22 mm
b = 10 mm
c = 2 mm
ℓ = 22.487 mm

(a)

a = 22 mm, b = 10 mm, c = 2 mm
a′ = 5 mm, b′ = 2 mm
ℓ = 22.279 mm

(b)

*Figure 5.7   (a) Symmetrical case where input and output aperture shapes are the same to each other as vertical slots. (b) Unsymmetrical case where the input and output aperture shapes are different from each other.*

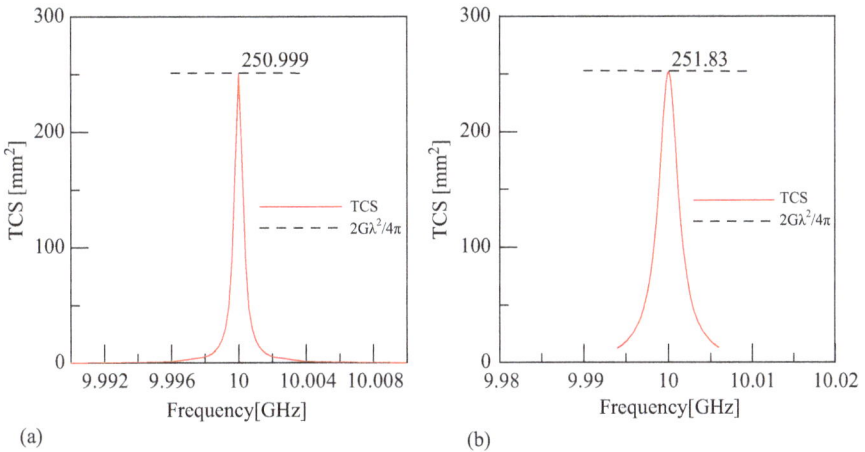

(a)

(b)

*Figure 5.8   TCS curves for and unsymmetrical TRCs. (a) TCS as a function of frequency for the case of Figure 5.7(a). (b) TCS as a function of frequency for the case of Figure 5.7(b).*

From a comparison between Figure 5.8(a) and (b), the TCS of the transmission resonant cavity is seen to remain almost the same if the same input (incident) side aperture shape is kept unchanged under the condition that the slot width is much smaller than the wavelength, even though the output aperture shape is changed if the transmission cavity still remains to be set up.

This means that the transmission efficiency such as TCS is mainly determined by the incident (input) side aperture shape and its gain, i.e., the output side aperture shape has almost no effect on the transmission characteristics. However, the beaming (radiation pattern) can be controlled by modifying the output side aperture. This observation is analogous to the result [12] that optical transmission through a single subwavelength slit is mainly governed by the structure of the incident side aperture surface and giving some perturbation on the output side for example by adding a grating to it does not pull more light energy through the slit although it shapes the emerging radiation pattern. Here it is worthwhile to add the comment that the above discussion so far holds only under the condition that the transmission cavity of high $Q$ is set up.

To check whether only the incident side aperture shape contributes to enhancing the transmission efficiency, we have investigated the TCS characteristics also for the case that the input aperture is changed to the finite horizontal slot type, i.e., that input and output apertures in Figure 5.7(b) are interchanged as shown in Figure 5.9.

The result of TCS for the transmission cavity structure in Figure 5.9 is given in Figure 5.10.

From Figure 5.10, it is seen that the change in the input coupling aperture shape results in a significant change in the transmission efficiency characteristics such as TCS. Note that the TCS (=219) is significantly decreased from that of the previous case (251) in the case of Figure 5.7(b). This observation backs up the foregoing argument again that the transmission characteristics of the transmission resonant cavity are mainly determined by the incident side aperture shape.

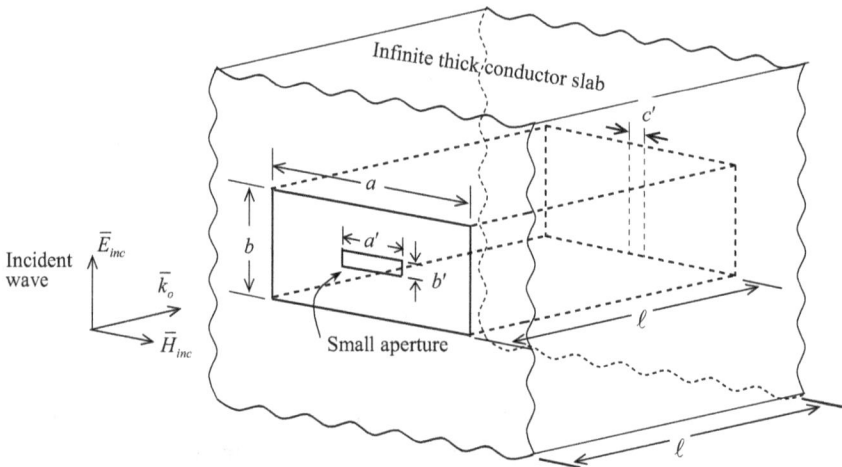

*Figure 5.9    Case where input and output apertures in Figure 5.7(b) are interchanged*

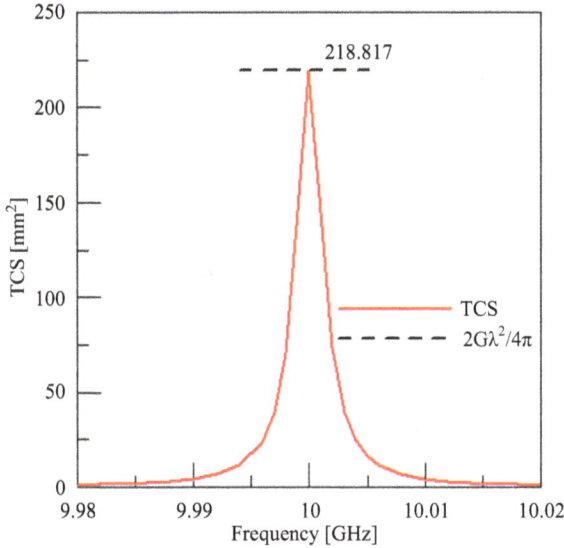

*Figure 5.10   TCS versus frequency for the case of Figure 5.9*

## 5.5   Review of the two types of transmission resonant cavity

In this chapter, we have worked with a transmission resonant structure in which two transverse walls, each with a small hole serving as input and output ports, are positioned some distance apart to create the desired cavity. In this structure, the cross section of the cavity region is chosen to be almost the same or comparable to that of the usual rectangular waveguide. In this structure, as a coupling hole becomes smaller, the aperture impedance seen by the aperture into the outer half-space $Z_L$ is much smaller than the characteristic impedance $Z_C$ of the guide inside the cavity, i.e., $Z_L \ll Z_C$ and so the reflection coefficient at the transverse wall with a small hole, $\frac{Z_L - Z_C}{Z_L + Z_C}$ approaches to (-1), which corresponds to the electric wall. In more detail, $Z_L$ means the input impedance looking at the coupling hole into the outside half-space of the cavity and $Z_C$ means the characteristic impedance corresponding to the lowest guided mode along the guide inside the cavity.

As discussed in Chapter 3, there is another transmission resonant cavity structure. Recall the perforated type of aperture in a thick conducting screen whose cross section of perforated aperture is that of a rectangular waveguide. When the guide height is much smaller than the wavelength, the radiation admittance $Y_L$ looking at the aperture center into the outside half-space is much smaller than the characteristic admittance $Y_C$ of the perforated aperture guide region. Note that the reflection coefficient $\Gamma$ at the aperture is given by $\Gamma = \frac{Y_C - Y_L}{Y_C + Y_L}$ and that as the guide height decreases, the characteristic admittance $Y_C$ of the aperture guide region becomes much larger than the aperture admittance $Y_L$, i.e., $Y_C \gg Y_L$, and so $\Gamma$ approaches +1, which corresponds to the lossy magnetic wall [13].

In this case, the structure in itself constitutes the transmission cavity if the cavity length is chosen to be roughly an integer multiple of half the guide wavelength. Here the transmission cross section is given by $\frac{2G\lambda^2}{4\pi}$, where gain $G$ is roughly 1.64 corresponding to that of half wave dipole as discussed already in Chapter 3.

So the transmission resonance phenomena occur for both types of transmission cavities; one is the type with both ends terminated by a lossy electric wall (transverse conducting wall having a small coupling hole) and the other with both ends terminated by a lossy magnetic wall (rectangular aperture itself). Here the loss means the radiation due to electromagnetic aperture coupling to the left (incident) side and the right side half-spaces.

The only difference between these two kinds of transmission resonant cavities is that, according to the shape of the coupling aperture, the transmission cross-section $\frac{2G\lambda^2}{4\pi}$ [m$^2$] becomes different, i.e., for the transmission resonant cavity having small coupling holes which is very small in comparison with the wavelength, so the gain ($G = 1.5$) for the small Hertzian dipole is used in the expression $\frac{2G\lambda^2}{4\pi}$ for the transmission cross section. In this case, the transmission cross section becomes $\frac{3\lambda^2}{4\pi}$ [m$^2$], which is also the case with the transmission resonant aperture of the circular ridged aperture type whose coupling aperture becomes a highly confined small area between two ridged gaps. On the other hand, for the transmission resonant cavity composed of lossy magnetic walls, the coupling aperture is of $\frac{\lambda}{2}$ slot dipole shape and the gain ($G = 1.64$) for $\frac{\lambda}{2}$ dipole is used in the calculation of the transmission cross section. For this reason, the transmission cross section for the latter case is somewhat larger than that for the former case.

Since the work of Ebbesen *et al.* [14] was reported as showing an extraordinary optical transmission (EOT) problem through a 2-D array of subwavelength holes perforated in thick metallic film, there has been a renewed interest in analyzing resonant transmission properties of a single subwavelength slot [15] as well as an array [6] of subwavelength slots in thick conducting screen. The foregoing transmission resonance problem for the transmission cavity composed of lossy magnetic walls is thought to be essentially the same as the single subwavelength slit problem, which has been widely investigated in connection with the EOT problem [16,17]. If we limit the slot to the narrow rectangular aperture and the slit to the infinitely long narrow slit, then the only difference in the transmission resonance between the two structures is due to the presence of the cutoff frequency in the former as discussed in Chapter 4. It is worthwhile to add a comment that the narrow slot type belongs to the small aperture category. Here the small aperture category means the small hole type inside of which the lowest guide mode is supportable even in the limit of the zero diameter of the hole corresponding to the zero physical aperture area. The cylindrical hole structure inside the plasmonic medium whose diameter approaches zero also belongs to this category. The half-wavelength rectangular slot whose height approaches zero as a counterpart in the perfect electric conducting (PEC) medium also belongs to this category.

## 5.6 Linkage of the near field optical probe using bow-tie optical antenna to the TRC type

Recently to overcome the low transmission efficiency of the conventional type of tapered fiber probe, a new type of probe employing an optical antenna structure was proposed [7]. The detailed configuration is illustrated in Figure 5.11.

As shown in Figure 5.11, as an optical antenna, the bow-tie type of antenna with its feeding point open-circuited is used to concentrate the beam for high spatial resolution.

For even larger transmission efficiency, the separation $S$ between the waveguide open-end aperture and bow-tie antenna is chosen to be approximately half wavelength, in which case the in-between region constitutes a transmission cavity with relatively low $Q$.

As mentioned above, by employing the optical bow-tie antenna type, the desired EM energy confinement near the gap can be achieved, but transmission efficiency through the transmission cavity system between the waveguide open end and the bow-tie antenna structure still remains to be improved further.

If we implement the above near-field optical probe system in terms of the waveguide structure, we can improve the transmission efficiency by increasing the $Q$ of the transmission cavity.

A possible microwave version of such a desired high $Q$ structure can be implemented, as shown in Figure 5.12 by incorporating the TRC structure into a guiding structure run, for example, a microwave waveguide or millimeter waveguide. A detailed discussion of this structure is given in Section 5.7.2.

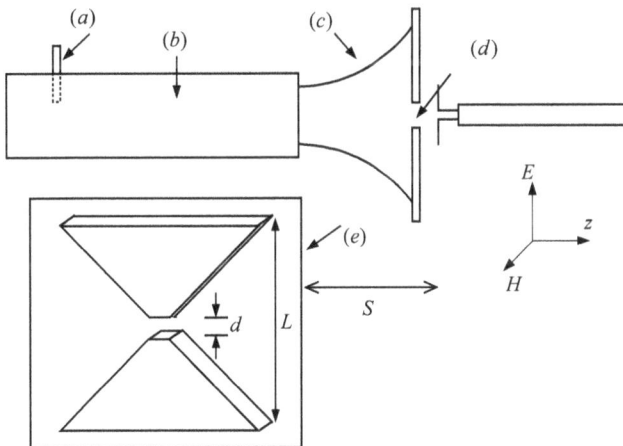

*Figure 5.11*  *Near field optical probe using bow-tie optical antenna for X-band use: (a) microwave source, (b) waveguide for X-band, (c) illumination beam, (d) bow-tie optical antenna gap, and (e) bow-tie optical antenna*

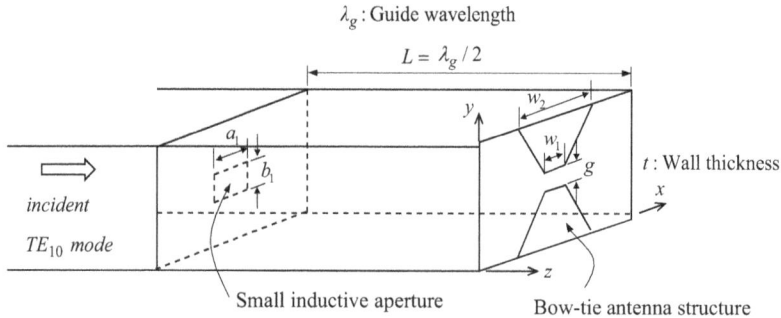

*Figure 5.12    Integrated structure of the near-field optical probe using bow-tie optical antenna type in the transmission resonant cavity (TRC)*

## 5.7    Enhanced electric field distribution for transmission resonant cavity case

Dividing the NSOM probe structures according to the technical trend, they can be categorized into two kinds, one is an aperture-based near-field structure, and the other is the type that employs the optical antenna structure [7].

As mentioned already, aperture-based near-field techniques employ a probe of subwavelength coupling aperture size and an object that is mounted in the near field of the probe so that the spatial resolution is determined by the coupling aperture size of the probe rather than by the wavelength. Several designs of the near-field probe have been used such as a circular aperture [18], rectangular aperture [19,20], and open waveguide [21] in microwave bands for example.

A smaller circular coupling aperture provides better resolution, although this is not very practical because the transmitted power is very small as is well known.

So, to overcome such a low transmission efficiency problem, the above two kinds of resonant types of near-field probe structures have been proposed, resonant aperture-based probe structure and optical antenna-inspired structure.

Interestingly these two structures can be explained in terms of the TRA and TRC concepts which have been described here.

For the sake of assessing the probe from the viewpoint of field enhancement, we are going to deal with the aperture field distribution for both TRA and TRC cases and compare the field distributions for both cases.

### 5.7.1    Transmission resonant aperture type

The most important single element in any form of scanning probe microscopy is the probe itself [22]. The spatial resolution of (microwave) imaging depends on how the electromagnetic field energy is confined and focused [22]. We first design geometrical parameters of the bow-tie type of aperture in an infinite conducting plane so that the transmission resonant frequency may be located roughly in the X-band range (8.2–12.4 GHz) by using the numerical analysis of Rao–Wilton–Glisson

(RWG) method [23]. Next, the guiding structure such as a rectangular waveguide as a feeding structure is terminated by the TRA designed above.

In Chapter 4, as a representative TRA type of probe, the probe structure which consists of the rectangular waveguide terminated by an end plate with a bow-tie type of aperture as shown in Figure 4.14 has been investigated with the main concern on the impedance matching and the near field distribution over the aperture.

From the results on the return loss and field distribution in Figure 4.15, the bow-tie type of TRA is seen to offer advantages in both high transmission efficiency for good impedance matching and small spatial confinement of electromagnetic energy through the small aperture for better resolution.

The present aperture-based near-field probe structures such as the TRC type under consideration as well as the foregoing TRA type in Chapter 4 can achieve lateral spatial resolution on the order of the aperture size of the probe because the electromagnetic field energy is squeezed as a result of the funneling mechanism through a tiny aperture and is collimated over a small confined region. So as seen in Figure 4.15, the near-field spot size, i.e., full width at half maximum (FWHM) of the electric field along the $x$ and $y$ directions are seen roughly to correspond to the gap size along respective directions. In the design procedure of the TRA type of probe, the input impedance into the gap is chosen to be comparable to the characteristic impedance level of the feeding guide. On the other hand in the TRC case, because the cavity should be set up, the input impedance of the output iris should be chosen to be quite different from the characteristic impedance of the guide inside the cavity. This is a fundamental difference between TRA-type and TRC-type probes.

## 5.7.2    Transmission resonant cavity type

To look into to what extent the field enhancement can be achieved in the TRC structure in Figure 5.12, we have investigated the return loss characteristics and field distribution over the aperture. Figure 5.13 shows such numerical data for the geometrical parameters that $a = 22.86, b = 10.16, a_1 = b_1 = 6, g = 0.4, w_1 = 2, w_2 = 5$, and $t = 0.2$ [mm].

Figure 5.13(a) shows the return loss (impedance matching characteristics). As seen from the comparison between Figure 4.15(a) in Chapter 4 and Figure 5.13(a), the cavity $Q$ for the latter TRC is much larger than the aperture $Q$ for the former TRA structure. The TRA structure of the low $Q$ has the advantage of relatively better temporal resolution for the transmission of shorter pulses.

As expected, as the longitudinal length of the TRC decreases, the transmission resonance frequency increases. To look into the field enhancement and energy confinement in comparison with the TRA case, we have investigated those characteristics for the TRC case. Figure 5.14 shows the highly confined electric field distribution which is required for better resolution.

It is worthwhile to note that the large E-field confinement to the small area near the gap corresponding to the subwavelength aperture is a combined result of the transmission cavity resonance effect and a lightning rod of tip effect.

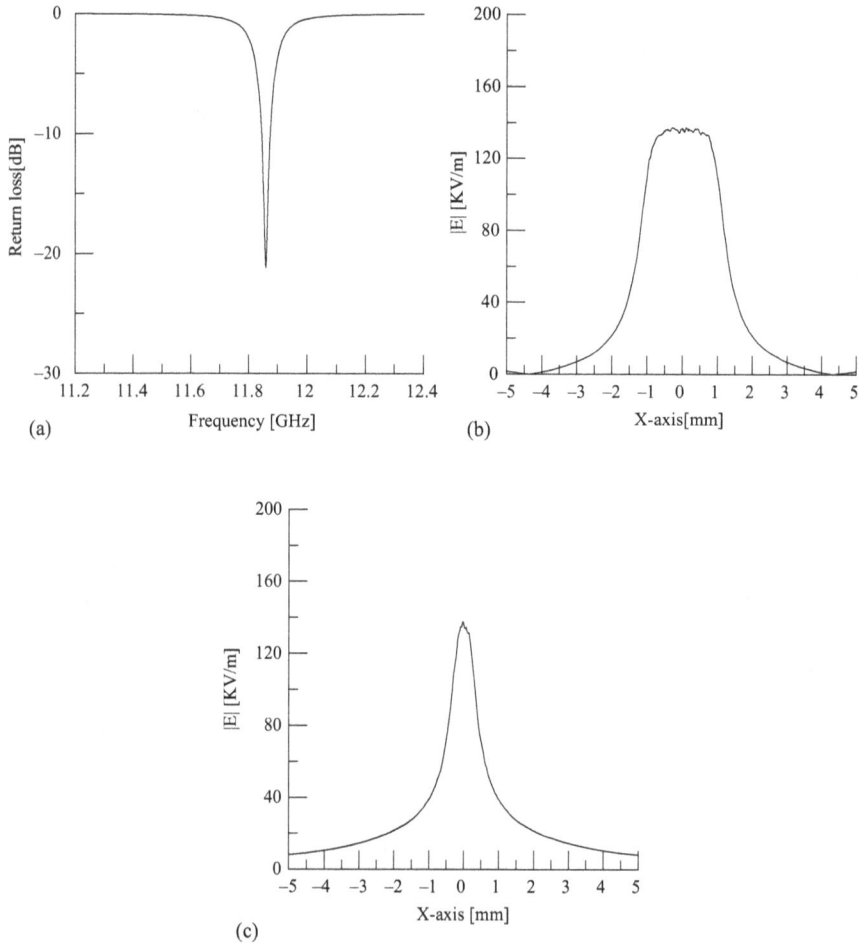

*Figure 5.13    Scattering coefficient $|S_{11}|$ (reflection coefficient) versus frequency and the y-component electric field distribution in the TRC type of probe. (a) Scattering coefficient $|S_{11}|$ (reflection coefficient) versus frequency. (b) Electric field (y-component) along the x-axis at $z = 0.2$ [mm]. (c) Electric field (y-component) along the y-axis at $z = 0.2$ [mm].*

Because electric field enhancement is useful for better active interaction between fields and matter as an object, it is interesting to compare the electric field enhancement for TRA and TRC cases. Figure 5.14 shows the electric field distribution of the $y$-component along the $z$-axis (longitudinal direction) at the center of the cross section for TRA (in Figure 4.14 in Chapter 4) and TRC cases. Note that the field distribution in the case of the TRC undergoes an abrupt change from the minimum point of the electric field to the maximum point as a combined result of a

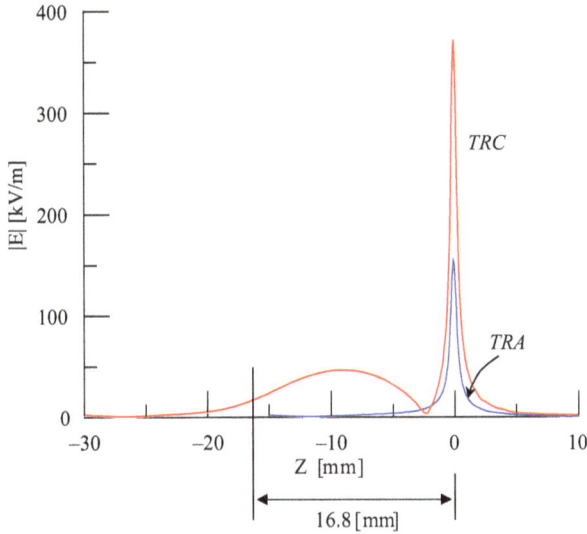

*Figure 5.14    Electric field distribution of y-component along the z-axis at the center of the waveguide for TRA and TRC cases*

resonant electric mode field inside the cavity and the divergent field across the gap at the output port. Field enhancement is seen to be better for the TRC type, which can be attributed to the transmission cavity effect and lightning-rod effect. So the TRC type is thought to apply to highly confined small areas [24] as well.

## References

[1] Park J.E., Cho Y.K., 'Comparison of transmission resonance phenomena through small coupling apertures between two kinds of transmission resonance structures' *URSI International Symposium on Electromagnetic Theory, Proc.* Berlin, Germany, 2019, pp. 744–747

[2] Leviatan Y., Harrington R.F., Mautz J.R., 'Electromagnetic transmission through small coupling apertures in a cavity in a thick conductor.' *IEEE Trans. Antennas Propagat.*, 1982, 30(6), pp. 1153–1165

[3] Pang Y., Hone A.N., So P.P.M., Gordon R., 'Optical transmission through a small hole in a metal waveguide screen.' *Opt. Express.* 2009, 17(6): 4433–4441

[4] Carter P.S., 'Circuit relations in radiating systems and applications to antenna problem.' *Proc. of IRE.*, 1932, vol. 20(6), pp. 1004–1041

[5] Bravo-Abad J., Garcia-Vidal F.J., Martin-Moreno L., 'Resonant transmission of light through finite chains of subwavelength holes in a metallic film.' *Phys. Rev. Lett.*, 2004, vol. 93(22), pp. 227401–1(4)

[6]   Ruan Z., Qiu M., 'Enhanced transmission through periodic arrays of sub-wavelength holes: The role of localized waveguide resonance.' *Phys. Rev. Lett.*, 2006, vol. 96(23), pp. 233901(1)–(4)

[7]   Grober R.D., Schoelkopf R.J., Prober D.E., 'Optical antenna: Towards a unity efficiency near-field optical probe.' *Appl. Phys. Lett.* 1997. vol. 70(11), pp. 1354–1356

[8]   Crozier K.B., Sundaramurthy A., Kino G.S., Quate C.F., 'Optical antennas: Resonators for local field enhancement. *Appl. Phys.* 2003, vol. 94(7), pp. 4632–4642

[9]   Thomassen K.I., *Introduction to Microwave Field and Circuits*. Prentice-Hall, 1971, pp. 153–158

[10]  Collin R.E. *Foundation for Microwave Engineering*. 2nd edn, New York: McGraw-Hill, 1992

[11]  Harrington R.F., 'Resonant behavior of a small aperture backed by a conducting body.' *IEEE Trans. Antennas Propagat.*, 1982, 30(2), pp. 205–212

[12]  Garcia-Vidal F.J., Lezec H.J., Ebbesen J.W., Martin-Moreno L., 'Multiple paths to enhance optical transmission through a single subwavelength slit.' *Phys. Rev. Lett.*, 2003, 30 pp. 213901–1(4)

[13]  Cho Y.K., Kim K.W., Ko J.H., Lee J.I., 'Transmission through a narrow slot in a thick conducting screen.' *IEEE Trans. Antennas Propagat.*, 2009, 57(3), pp. 813–816

[14]  Ebbesen T.W., Lezec H.J., Ghaemi H.F., Thio T., Wolff P.A., 'Extraordinary optical transmission through subwavelength hole arrays.' *Nature (London)*, 2005, 391, pp. 667–669

[15]  Garcia-Vidal F.J., Moreno E., Porto J.A., Martin-Moreno L., 'Transmission of light through a single rectangular hole.' *Phys. Rev. Lett.*, 2005, 95, pp. 103901(1)–(4)

[16]  Takakura Y., 'Optical resonance in narrow slit in a thick metallic screen.' *Phys. Rev. Lett.*, 2001, 86(24), pp. 5601–5603

[17]  Yang F., Sambles J.R., 'Resonant transmission of microwaves through a narrow metallic slit.' *Phys. Rev. Lett.*, 2002, 89(6), pp. 063901(1)–(2)

[18]  Ash A., Nicholls G., 'Super-resolution aperture scanning microscope.' *Nature*, 1972, vol. 237, pp. 510–512

[19]  Golosovsky M., Davidov D., 'Novel millimeter-wave near-field resistivity microscope.' *Appl. Phys. Lett.*, 1996, 68(11), pp. 1579–1581

[20]  Golosovsky M., Lann A.F., Davidov D., Frenkel A., 'Microwave near-field imaging of conducting objects of a simple geometric shape.' *Rev. Sci. Instrum.*, 2000, 71(10), pp. 3927–3932

[21]  Wang M.S., Borrogo J.M., 'Determination of shallow acceptor concentration in Si–GaAs from steady state and transient microwave photo-conductivity measurement.' *J. Electrochem. Soc.*, 1990, vol. 137(11), pp. 3648–3652

[22]  Betzig E., Trautman J.K. 'Near-field optics: microscopy. Spectroscopy, and surface modification beyond the diffraction limit.' *Science*, 1992, vol. 257, pp. 195–198

[23]   Rao S.M., Wilton D.R., Glisson A.W., 'Electromagnetic scattering by surfaces of arbitrary shape.' *IEEE Trans. Antennas Propagat.*, 1982, 30, pp. 418–419

[24]   Challener W.A. *et al.*, 'Heat-assisted magnetic recording by a near field transducer with efficient optical energy transducer.' *Nat. Photonics*, 2009. 3, pp. 220–224

*Chapter 6*

# Absorption resonance problem

## Synopsis

The electromagnetic coupling problem through a small aperture backed by a lossy cavity is considered. In particular, under the maximum electromagnetic coupling condition of an incident plane wave through an input coupling aperture, the absorption cross section as a measure of the absorption efficiency is almost the same as the transmission cross section of the transmission resonant cavity in Chapter 5.

Here the absorption cross section (ACS) is defined as the ratio of the penetrated through the coupling aperture and absorbed power into the cavity enclosed by a lossy wall over the incident power density (Poynting vector magnitude) such that the physical dimension may be that of the area. This concept is essentially the same as that of the transmission cross section (TCS) which has been introduced to assess the transmission efficiency of the transmission resonant cavity. Based upon the similarity of the expressions for the maximum values and frequency responses between the ACS and TCS, the absorption resonance means the resonance characteristics associated with the frequency characteristics of the absorption near the ACS (maximum absorption efficiency). A kind of absorption frequency selective surface (AFSS) is considered by composing a two-dimensional array of the above aperture coupled lossy cavity structure and then eliminating all the side walls except lossy end walls. The resultant AFSS structure made of the small aperture array over the lossy ground is investigated with the main interest centering on the absorption resonance phenomena as a possible application to the RAM (radar absorbing material). A comparison between this absorption resonance and the former transmission resonance is discussed.

## 6.1 Absorption resonant cavity structure and its resonance condition

As discussed in Chapters 4 and 5, since extraordinary optical transmission (EOT) phenomena through subwavelength apertures in optically thick metal film by Ebbesen group [1] was reported, huge amounts of research have been done on the fundamental physics of the transmission resonance relevant to the EOT phenomena.

In parallel with the transmission resonance phenomena associated with the EOT, the absorption resonance phenomena have been the subject of intensive research in recent years due to potential applications in photovoltaics [2] such as solar cell devices and photodetectors [3]. Here we are going to consider under what condition the maximum absorption occurs when the coupling aperture is backed by a closed lossy rectangular cavity [4]. This absorption resonance problem is different from the previous transmission resonance problem in that the main concern was the maximum transmission phenomena through the transmission structures such as the transmission resonant apertures and transmission resonance cavity in the latter case, whereas the main concern here is the maximum absorption phenomena. Figure 6.1 shows an absorption resonant cavity structure under consideration in the present problem.

In the figure, the inside walls of the absorption cavity backing the small input coupling aperture are assumed to be lossy due to the finite conductivity of the inside wall material.

We consider the problem of a plane wave incident normally on an infinite conducting plane containing a small coupling aperture of rectangular type which is backed by a rectangular cavity as depicted in Figure 6.1. Here restricting the incident angle of the incident wave to the normal incidence case does not result in loss of generality of this approach because the small aperture coupling of the incident side tends to make the absorption efficiency significantly less sensitive to the incident angle. By use of the method of moments (MoM) which is borrowed from the prior work [4–7], we analyze the electromagnetic coupling problem with the main interest centering on the condition for the maximum absorption into the lossy cavity.

Here a plane wave is assumed to be normally incident so incident $y$-component electric field and $x$-component magnetic field are, respectively, given by

$$\mathbf{E}^i = \mathbf{a}_y E_0 e^{-jk_0 z} \tag{6.1a}$$

$$\mathbf{H}^i = -\mathbf{a}_x H_0 e^{-jk_0 z} \tag{6.1b}$$

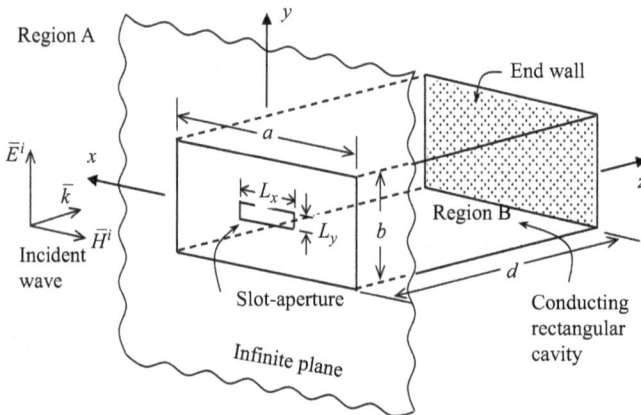

*Figure 6.1   Absorption resonant cavity structure under consideration*

where $a_x(a_y)$ is the unit vector in the *x*-(*y*-) direction and $E_0(H_0)$ is the magnitude of incident electric(magnetic) field amplitude. The infinite plane separates the half-space $(z < 0)$ and the interior of the cavity $(z > 0)$. We use the equivalence principle to divide the problem into two regions, the half-space (region a, $z < 0$) and the interior of the lossy cavity (region b, $z > 0$). The field in region A remains unchanged if the aperture is closed by a conductor and the equivalent surface magnetic current

$$\boldsymbol{M} = \boldsymbol{E} \times \boldsymbol{n} = V_0 \boldsymbol{M}_0 \tag{6.2}$$

is placed over the aperture region, where $\underline{E}$ is the electric field in the aperture of the original problem and $\boldsymbol{n}(= -\boldsymbol{a}_z)$ is the unit normal vector pointing outward from the cavity and $V_0$ is an unknown coefficient to be determined. $\boldsymbol{M}_0$ may be chosen to be a single term of the entire basis function or a set of expansion functions. Here the single entire basis function is first dealt with because a single-term expansion gives quite good results compared with those obtained by using a set of expansion functions and the next multi-term expansion case is briefly dealt with.

The field in region "A" is that produced by the incident wave and due to $\boldsymbol{M}$ over the aperture with the aperture covered by a perfect electric conductor. To determine the coefficient $V_0$, use is made of the requirement that the tangential component of the magnetic field is continuous across the aperture. This leads to an integral equation that can be converted into a single scalar equation by employing a single-term entire basis function expansion and Galerkin's testing. Following this approach, i.e., using $\boldsymbol{M}_0$ as the testing function, we obtain

$$\left(Y^a + Y^b\right)V_0 = I \tag{6.3}$$

where $Y^a$ and $Y^b$ are admittances at the aperture respectively, of the half-space and of the lossy cavity and $I$ is the current source

$$I = 2\iint_{ap} \left(\boldsymbol{M}_0 \cdot \boldsymbol{H}_o^i\right)\mathrm{d}s, \tag{6.4}$$

where the subscript "ap" means the surface integral area over the aperture and $\boldsymbol{H}_o^i$ is the incident magnetic field over the aperture in free space. The admittance looking into the half-space is evaluated as follows:

$$Y^a = -2\iint_{ap} \boldsymbol{M}_0 \cdot \boldsymbol{H}(\boldsymbol{M}_0)\mathrm{d}s \tag{6.5}$$

Here $\boldsymbol{H}(\boldsymbol{M}_0)$ is the magnetic field produced by $\boldsymbol{M}_0$ radiating into free space, and the admittance $Y^b$ looking into the cavity side is expressed by Liang and Cheng [4],

$$Y^b = -j\sum_i A_{oi}^2 Y_i \frac{\cot k_i d - (1-j)r_{si}}{1 + (1-j)r_{si}\cot k_i d}, \tag{6.6}$$

which has been obtained by following Galerkin's approach in the expression for

$$Y^b = \iint_{ap} M_0 \cdot H(-M_0) ds$$

Here $\underline{H}(-M_0)$ is the generalized magnetic field intensity produced by $-M_0$ radiating into the cavity.

In (6.6), $Y_i$ and $k_i$ denote, respectively, the characteristic admittance and wave number of the $i$th eigenmode in the cavity and $r_{si} = R_S Y_i$ is the normalized surface resistance for the $i$th mode and $R_S$ means the surface impedance of the lossy cavity wall appearing later in (6.12). In (6.6),

$$A_{oi} = \iint_{ap} M_o \cdot a_z \times e_i ds \tag{6.7}$$

where $e_i$ is the normalized $i$th mode transverse electric field vector [8]. For the detailed procedure for the expression for (6.6), refer to [4]. Because only the fundamental (dominant) mode is assumed to be propagating in the waveguide region b, all $Y_i$ and $k_i$ are imaginary except $Y_1$ and $k_1$. We have

$$Y_i = jB_i, (i \neq 1) \tag{6.8}$$

$$k_i = -j\alpha_i, (i \neq 1) \tag{6.9}$$

For $i = 1$, $Y_i$ is the real characteristic admittance of the dominant mode, and

$$k_1 = \beta_1 - j\alpha_1 \tag{6.10}$$

where $\beta_1$ is the phase constant which is related to the guide wavelength $\lambda_{g1}$ of the dominant mode $\left(\beta_1 = \frac{2\pi}{\lambda_{g1}}\right)$ and $\alpha_1$ is the dominant mode attenuation constant due to power loss in cavity walls, which is found by the use of the perturbation method [9] under the assumption of low loss. Based on the above, we can construct an equivalent circuit for the cavity-backed coupling aperture, i.e., the absorption resonant cavity structure in Figure 6.1. The equivalent circuit consists of a current source I in parallel with $Y^a$, input admittance looking into the left half-space, together with a susceptance

$$jB = j\sum_{i \neq 1} A_{oi}^2 B_i \coth(\alpha_i d) \tag{6.11}$$

representing the equivalent susceptance due to an evanescent higher-order mode, and a terminated transmission line coupled through an ideal transformer with a turns-ratio $A_{o1} = n_2/n_1$ as in Figure 6.2.

For reference, the admittance $Y^b$ in (6.6) is expressed as the sum of the admittance of the dominant mode term $\left(Y^b|_{i=1} = G^b + jB^b\right)$ and the higher-order mode $\left(Y^b|_{i \neq 1} = jB\right)$ as $Y^b = G^b + jB^b + jB$.

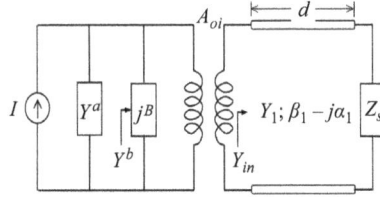

*Figure 6.2 Equivalent circuit representation for the absorption resonant cavity structure*

The termination $Z_s$ represents the surface impedance of the end wall

$$Z_s = (1 + j)R_S = (1 + j)\sqrt{\frac{\pi f \mu}{\sigma}} \tag{6.12}$$

where $\sigma$ is the conductivity of the cavity material.

Referring to Figure 6.2, the total input admittance $Y_0$ looking rightwardly from the current source is

$$\begin{aligned} Y_0 &= Y^a + j^B + A_{01}^2 Y_{in} \\ &= G^a + j(B^a + B) + A_{01}^2 Y_{in} \end{aligned} \tag{6.13}$$

which can be rewritten as

$$Y_0 = G^a + jB' + \text{Re}\{A_{01}^2 Y_{in}\} + j\text{Im}\{A_{01}^2 Y_{in}\} \tag{6.14}$$

Here $B'(= B^a + B)$ is the sum of the input susceptance looking into the left half-space of region "a" and the transformed susceptance due to the evanescent higher-order modes in region "b". $B'$ can be either a positive or a negative quantity depending on the coupling aperture size and shape. The real and imaginary part of $A_{01}^2 Y_{in}$ can be obtained by expanding the input admittance $Y_{in}$ of the terminated lossy transmission line

$$Y_{in} = Y_1 \left[ \frac{1 + jz_1 \tan(\beta_1 d - j\alpha_1 d)}{z_s + j\tan(\beta_1 d - j\alpha_1 d)} \right] \tag{6.15}$$

where $z_s = Z_S Y_1 = (1 + j)R_S Y_1 = (1 + j)r_s$ is the normalized terminating impedance and $r_s = R_S Y_1$.

In the calculation procedure of (6.15), because the cavity length becomes roughly half the waveguide wavelength and attenuation constant $\alpha_1$ and the normalized load impedance of the end wall $z_s$, $\tan\beta_1 d$, and $\alpha_1 d$, are much smaller than unity, equation (6.15) is approximated to be, after some algebraic manipulation,

$$Y_{in} \cong Y_1 \frac{r_d - jr_e}{r_d^2 + r_e^2} \tag{6.16}$$

where

$$r_d = r_s + a_1 d \tag{6.17}$$

$$r_e = r_s + \tan\beta_1 d \tag{6.18}$$

both of which are very small compared to unity.

So the real and imaginary parts of the input admittance of the dominant (fundamental) mode inside the cavity region b are expressed as

$$G^b = \text{Re}\left(A_{o1}^2 Y_{in}\right) = A_{o1}^2 Y_1 \frac{r_d}{r_d^2 + r_e^2} \tag{6.19}$$

$$B^b = \text{Im}\left(A_{o1}^2 Y_{in}\right) = -A_{o1}^2 Y_1 \frac{r_e}{r_d^2 + r_e^2} \tag{6.20}$$

When $r_e = r_d$, $\left|B^b\right|$ has a maximum value as

$$\max\left|B^b\right| = A_{o1}^2 Y_1 \frac{1}{2r_d} \tag{6.21}$$

We define aperture-cavity resonances as those corresponding to $\text{Im}(Y_0) = 0$, i.e.,

$$\text{Im}\left(A_{o1}^2 Y_{in}\right) = -B' \tag{6.22}$$

If we introduce a new parameter $\xi$ such that

$$\xi = \frac{A_{o1}^2 Y_1}{2r_d |B'|} = \frac{r_d^2 + r_e^2}{2r_d r_e} \tag{6.23}$$

From (6.21), (6.22), and (6.23), we can see that aperture-cavity resonances can be obtained only if

$$\xi \geq 1 \tag{6.24}$$

From (6.23), we can get the expression for $(r_e/r_d)$

$$\frac{r_e}{r_d} = F(\xi) = \xi \pm \sqrt{\xi^2 - 1} \tag{6.25}$$

The lossy cavity length d at which aperture-cavity resonances, i.e., maximum absorption occurs is obtained by use of the (6.25). In particular, for the negative $B'$ to which case the small coupling aperture under consideration belongs, we have

$$\tan(\beta_1 d) = -r_s - r_d \left(\xi \pm \sqrt{\xi^2 - 1}\right), \tag{6.26}$$

which reveals clearly that a pair of aperture-cavity resonances exist through the use of the graphical method in [4]. The plus and minus signs in (6.25) are chosen to correspond to the first and second resonances, respectively. For low-loss cavities, $\xi$

is usually much greater than one, so $F(\xi)$ is expressed by

$$F(\xi) \cong 2\xi \text{ or } \frac{1}{2\xi} \tag{6.27}$$

Given this along with (6.18), the two solutions for the absorption cavity length $d$ which satisfies the absorption resonance condition, (6.22) were found to be, at the first resonance,

$$d_1 = \frac{\lambda_{g1}}{2}\left[ n - \frac{1}{\pi}\tan^{-1}(r_s + 2\xi r_d) \right] \cong \frac{\lambda_{g1}}{2}\left[ n - \frac{1}{\pi}\tan(2\xi r_d) \right] \tag{6.28}$$

and at the second resonance

$$d_2 = \frac{\lambda_{g1}}{2}\left[ n - \frac{1}{\pi}\tan^{-1}\left( r_s + \frac{1}{2\xi}r_d \right) \right] \cong \frac{\lambda_{g1}}{2}\left[ n - \frac{1}{\pi}\tan^{-1}(r_s) \right] \tag{6.29}$$

Approximations have been made because $r_s \ll 2\xi r_d$ and $r_s \gg \frac{1}{2\xi}r_d$ in (6.28) and (6.29), respectively. However, since $r_s$ is itself very small, the cavity length $d_2$ for the second aperture-cavity resonance is seen to be only very slightly less than $n\lambda_{g1}/2$.

## 6.2   Characteristics of double aperture-cavity resonances

We now examine the characteristics of the two kinds of aperture-cavity resonance. At the first resonance, the use of (6.23) and (6.26) with the plus sign in (6.19) yields

$$G_1^b = \frac{2|B'|\xi}{4\xi^2 + 1}, \tag{6.30}$$

where the subscript 1 means the first resonance. The magnitude of the voltage in the aperture at the first resonance is

$$|V_{R1}| = \frac{I}{G^a + G_1^b} = \frac{I}{G^a}\left[ \frac{1}{1 + \frac{2\xi Q_a}{4\xi^2 + 1}} \right] \tag{6.31}$$

where

$$Q_a = \frac{|B'|}{G^a} \tag{6.32}$$

Because $d_2$ at the second resonance is very close to the cavity length at natural resonance and $r_e \ll r_d$. The combined use of (6.19) and (6.23) leads to

$$G_2^b = 2|B'|\xi \tag{6.33}$$

where the subscript 2 means the second resonance. So the magnitude of the voltage in the aperture at the second resonance is

$$|V_{R2}| = \frac{I}{G^a}\left[\frac{1}{1+2\xi Q_a}\right]$$
(6.34)

Since $\xi$ is much greater than one in (6.31) and (6.33), it is seen that

$$|V_{R1}| \gg |V_{R2}|$$
(6.35)

and their ratio is

$$\frac{|V_{R1}|}{|V_{R2}|} \cong \frac{4\xi^2 Q_a}{2\xi + Q_a}$$
(6.36)

In general, the peak electric field inside the lossy cavity at the first resonance is much larger than that at the second resonance, that is, a much stronger mode field is set up at the first resonance, which plays an important role in the maximum power absorption due to the surface current on the lossy wall of the cavity as discussed later. In addition, the peak electric field strengths in the aperture show the same trend but interestingly the difference in the field strengths in the aperture is far more than that in the peak electric fields in the cavity at the two resonances as seen in (6.36), which is reminiscent of the funneling phenomena which was observed in the transmission resonance through the small holes.

## 6.3   Maximum power absorption

Between the two solutions for the length d for the absorption resonance in which the absorbed power through the coupling aperture becomes maximum, only the first solution $d_1$ is related to the maximum absorption resonance as discussed in [4]. So we deal with the power transmitted through the input coupling aperture and absorbed inside the cavity for this case. To this end, we use the expressions for the transverse electromagnetic fields $E_{t1}$ and $H_{t1}$ for the dominant mode inside the cavity which are given as

$$E_{t1} = V_o A_{o1} e_1 \frac{\sin k_1(d-z) + (1-j)r_s\cos k_1(d-z)}{\sin k_1 d + (1-j)r_s\cos k_1 d}$$
(6.37)

$$H_{t1} = -jV_o A_{o1} Y_1 a_z \times e_1 \frac{\cos k_1(d-z) - (1-j)r_s\sin k_1(d-z)}{\sin k_1 d + (1-j)r_s\cos k_1 d}$$
(6.38)

For the case of the first solution $d_1$, the power transmitted through the aperture and absorbed inside the cavity is

$$P_1 = \frac{1}{2}\mathrm{Re}\iint_{ap}(E \times H^*)\cdot \underline{ds} = \frac{1}{2}\mathrm{Re}\iint_{a\times b}(E_{t1} \times H_{t1}^*)|_{z=0^+}\cdot a_z \mathrm{ds}$$
(6.39)

where 'ap' and a × b mean respectively 'aperture' and 'cross section' of the lossy cavity side at $z = 0^+$. A detailed explanation of the second equation in (6.39) in connection with the funneling phenomenon is given later in Section 6.9. Substituting (6.37) and (6.38) in (6.39) leads to the following expression:

$$P_1 = \frac{I^2 G_1^b}{2\left(G^a + G_1^b\right)^2} \tag{6.40}$$

where the approximation of

$$\frac{\cos k_1 d - (1 - j)r_s \sin k_1 d}{\sin k_1 d + (1 - j)r_s \cos k_1 d} \cong \frac{1}{\tan k_1 d + (1 - j)r_s}$$

has been made, and the identities $\tan k_1 d + (1 - j)r_s = r_e - jr_d$, $e_1 \times (a_z \times e_1) \cdot a_z = e_1 \cdot e_1$, the normalization condition $\iint_{a \times b} e \cdot e = 1$, and the relation between the current source I and $V_o$ at resonance in (6.3) have been used. It is seen clearly that $P_1$ reaches a maximum when $G^a = G_1^b$ and the maximum is

$$\max \quad P_1 = \frac{I^2}{8G^a} \tag{6.41}$$

where the subscript 1 in $G_1^b$ means "at the first resonance." Interestingly this maximum condition is observed to be met when the imaginary part of the total admittance seen by the current source I is cancelled and the input conductance $G_1^b$ inside the cavity is matched to the radiation conductance $G^a$ seen by a coupling aperture into the left half-space. This is essentially the same as the maximum transmission condition in the previous transmission resonance problem case.

## 6.4 Admittance parameters for half-space region

The magnetic field in the half-space region due to the equivalent magnetic current M over the aperture is expressed as

$$H_F(r) = -j\omega\varepsilon F(r) - \frac{j}{\omega\mu}\nabla(\nabla \cdot F(r)) \tag{6.42}$$

$$= -\frac{j}{\omega\mu}\left(k^2 + \nabla(\nabla \cdot)\right)\iint_{s'} M(r_t')\varphi(R)ds', \tag{6.43}$$

where F means the electric vector potential due to the equivalent magnetic current source after imposing the image effect of the infinite conducting plane and so

$$F(\underline{r}) = \frac{1}{2\pi}\iint_{s'} M(r_t)\frac{e^{-jkR}}{R}ds' \tag{6.44}$$

Here $\varepsilon$ and $k$ follows the usual meaning in the free space and the subscript $t$ in $\underline{r_t}$ means the tangential component $(x', y')$ over the aperture, $ds'$ means the source (aperture) region, $R = \sqrt{(x - x')^2 + (y - y')^2 + z'^2}$ and $\nabla \cdot \boldsymbol{F}$ is chosen to be $-j\omega\mu\phi_m$, where $\phi_m$ means the magnetic scalar potential [10]. For a more general case of multi-term representation of the equivalent magnetic current source over the aperture than the previous single-term case as in (6.2), the unknown magnetic current source $\boldsymbol{M}$ is approximated by a finite number of the entire basis function $\boldsymbol{M}_n \left( = \boldsymbol{a}_x \cos \frac{(2n-1)\pi x}{L_x} \right)$ as

$$\boldsymbol{M} = \sum_{n=1}^{N} V_n \boldsymbol{M}_n \tag{6.45}$$

For this multi-term representation of the equivalent magnetic current source over the aperture, the admittance parameters are given in matrix form by

$$[Y^a]_{N \times N} = -2 \iint_{ap} \boldsymbol{M}_m \cdot \boldsymbol{H}(\boldsymbol{M}_n) \mathrm{d}s, \ m, n = 1, 2, \ldots, N \tag{6.46}$$

$$[Y^b]_{N \times N} = \iint_{ap} \boldsymbol{M}_m \cdot \boldsymbol{H}(-\boldsymbol{M}_n) \mathrm{d}s, \ m, n = 1, 2, \ldots, N, \tag{6.47}$$

where $\boldsymbol{H}(\boldsymbol{M}_n)$ means the magnetic field due to the equivalent source $\boldsymbol{M}_n$ over the aperture in the half-space region, and $\boldsymbol{H}(-\boldsymbol{M}_n)$ means the magnetic field due to the equivalent source $-\boldsymbol{M}_n$ over the aperture inside the cavity region. Recall that, for the case of the single-term representation of the unknown magnetic current distribution over the aperture, the integral equation is converted into a single scalar equation different from the present matrix equation.

The detailed expressions for the admittance matrix $[Y^a]_{N \times N}$ in the half-space region are obtained by employing Galerkin's testing scheme, i.e., inserting (6.43) into (6.46) as follows:

$$[Y^a]_{NN} = \frac{j}{\omega\mu} k^2 \iint_{ap} \boldsymbol{M}_m(\underline{r}) \cdot \left\{ \iint_{ap} \boldsymbol{M}_n\left(\underline{r_t'}\right) \varphi(R) \mathrm{d}s' \right\} \mathrm{d}s$$
$$+ \iint_{ap} \iint_{ap'} \boldsymbol{M}_m \cdot \nabla\left(\nabla \cdot \left(\boldsymbol{M}_n\left(\underline{r_t'}\right) \varphi(R)\right)\right) \mathrm{d}s \mathrm{d}s', \ m, n = 1, 2, \ldots, N \tag{6.48}$$

where "ap" means the aperture area in terms of field coordinate and "ap" means the aperture area in terms of the source coordinate. Note that the factor 2 in (6.5) has been taken into account in (6.48).

In the second term of (6.48), the operation of $\nabla\nabla \cdot \left(\boldsymbol{M}_n\left(\underline{r_t'}\right)\varphi(R)\right)$ includes the third order of singularity, i.e., $R^{-3}$, which may cause the difficulty or inaccurate result in numerical integration procedure. Interestingly this difficulty can be overcome by use of the following identity [11]:

$$\iint_s \iint_{s'} \boldsymbol{M}_m(\underline{r}) \cdot \nabla\left(\nabla \cdot \left(\boldsymbol{M}_n\left(\underline{r_t'}\right)\varphi(R)\right)\right) \mathrm{d}s' \mathrm{d}s$$
$$= -\iint_s \iint_{s'} (\nabla \cdot \boldsymbol{M}_m(\underline{r})) \left(\nabla' \cdot \boldsymbol{M}_n\left(\underline{r_t'}\right)\right) \varphi(R) \mathrm{d}s' \mathrm{d}s \tag{6.49}$$

For reference, the detailed procedure for proof of identity is given in Appendix A. From (6.49), it is seen that the singularity problem in the integral procedure has been significantly alleviated. Using the expression for the cosinusoidal expansion for the magnetic current distribution $M_n$ in (6.45) and the equality of (6.49) in (6.48), we obtain the detailed expression for the admittance matrix element $[Y^a]_{N \times N}$ as follows:

$$[Y^a]_{N \times N} = \frac{j}{\omega\mu} \left[ \begin{array}{c} k^2 \iint_{\Delta s} \iint_{\Delta s'} M_m\left(r_{\underline{\iota}}\right) \cdot M_n\left(r_{\underline{\iota}}'\right) \varphi(R) ds' ds \\ - \iint_{\Delta s} \iint_{\Delta s'} \left(\nabla \cdot M_m\left(r_{\underline{\iota}}\right)\right) \left(\nabla' \cdot M_n\left(r_{\underline{\iota}}'\right)\right) \varphi(R) ds' ds \end{array} \right]$$

$$= \frac{1}{2\pi\omega\mu} \left[ \begin{array}{c} k^2 \int_{-\frac{L_x}{2}}^{\frac{L_x}{2}} \int_{-\frac{L_y}{2}}^{\frac{L_y}{2}} \int_{-\frac{L_x}{2}}^{\frac{L_x}{2}} \int_{-\frac{L_y}{2}}^{\frac{L_y}{2}} \cos\left[\frac{(2m-1)\pi x}{L_x}\right] \cos\left[\frac{(2n-1)\pi x'}{L_x}\right] \frac{e^{-jk\sqrt{(x-x')^2+(y-y')^2}}}{\sqrt{(x-x')^2+(y-y')^2}} dy dx dy' dx' \\ - \frac{(2m-1)(2n-1)\pi^2}{L_x^2} \int_{-\frac{L_x}{2}}^{\frac{L_x}{2}} \int_{-\frac{L_y}{2}}^{\frac{L_y}{2}} \int_{-\frac{L_x}{2}}^{\frac{L_x}{2}} \int_{-\frac{L_y}{2}}^{\frac{L_y}{2}} \sin\left[\frac{(2m-1)\pi x}{L_x}\right] \sin\left[\frac{(2n-1)\pi x'}{L_x}\right] \frac{e^{-jk\sqrt{(x-x')^2+(y-y')^2}}}{\sqrt{(x-x')^2+(y-y')^2}} dy dx dy' dx' \end{array} \right]$$

$$,m,n=1,2,\cdots,N$$

$$(6.50)$$

For numerical integration of the expression in (6.50), the integration variables $x$ and $y$ of the observation coordinate are transformed into the polar coordinates as

$$x - x' = \rho\cos\theta, y - y' = \rho\sin\theta \qquad (6.51)$$

In this case, inside the integration region over the aperture of the observation coordinate, the components of the source coordinates $(x', y')$ are taken as constants so the Jacobian $J$ for the present transformation is

$$J = \frac{\partial(x,y)}{\partial(\rho,\theta)} = \rho \qquad (6.52)$$

From the use of the integral in the transformed polar coordinate, the admittance matrix element $[Y^a]_{N \times N}$ is cast into the more convenient expression for numerical integration as follows:

$$[Y^a]_{N \times N} = \frac{j}{2\pi\omega\mu} \left[ \begin{array}{c} k^2 \int_{-\frac{L_x}{2}}^{\frac{L_x}{2}} \int_{-\frac{L_y}{2}}^{\frac{L_y}{2}} \int_{0}^{2\pi} \int_{0}^{\rho_c(\theta)} \left\{ \cos\left[\frac{(2m-1)\pi(x'+\rho\cos\theta)}{L_x}\right] \cos\left[\frac{(2n-1)\pi x'}{L_x}\right] e^{-jk\rho} \right\} d\rho d\theta dy' dx' \\ - \frac{(2m-1)(2n-1)\pi^2}{L_x^2} \int_{-\frac{L_x}{2}}^{\frac{L_x}{2}} \int_{-\frac{L_y}{2}}^{\frac{L_y}{2}} \int_{0}^{2\pi} \int_{0}^{\rho_c(\theta)} \left\{ \sin\left[\frac{(2m-1)\pi(x'+\rho\cos\theta)}{L_x}\right] \sin\left[\frac{(2n-1)\pi x'}{L_x}\right] e^{-jk\rho} \right\} d\rho d\theta dy' dx' \end{array} \right]$$

$$m, n = 1,2,\cdots,N,$$

$$(6.53)$$

where the distance $\rho_c(\theta)$ from $(x', y')$ to the periphery of the aperture varies according to the angle $\theta$ as shown in Figure 6.3.

For reference, the distance $\rho_c(\theta)$ from $(x', y')$ to the periphery of the aperture of the aperture is expressed by

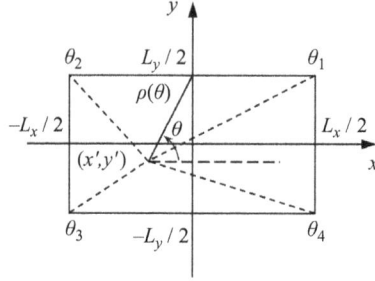

*Figure 6.3    Polar coordinate representation in the rectangular aperture*

$$
\begin{aligned}
\rho_c(\theta) &= \frac{L_x/2 - x'}{\cos\,\theta}\ (\theta_4 < \theta \leq \theta_1),\ \theta_1 = \arctan\left[\frac{L_y/2 - y'}{L_x/2 - x'}\right] \\[2mm]
\rho_c(\theta) &= \frac{L_y/2 - y'}{\sin\,\theta}\ (\theta_1 < \theta \leq \theta_2),\ \theta_2 = \arctan\left[-\frac{L_y/2 - y'}{L_x/2 + x'}\right] \\[2mm]
\rho_c(\theta) &= -\frac{L_x/2 + x'}{\cos\,\theta}\ (\theta_2 < \theta \leq \theta_3),\ \theta_3 = \arctan\left[\frac{L_y/2 + y'}{L_x/2 + x'}\right] \\[2mm]
\rho_c(\theta) &= \frac{L_x/2 + y'}{\sin\,\theta}\ (\theta_3 < \theta \leq \theta_4),\ \theta_4 = \arctan\left[-\frac{L_y/2 + y'}{L_x/2 - x'}\right]
\end{aligned}
\tag{6.54}
$$

## 6.5  Admittance parameters in the lossy cavity region

Here we aim at specifying the admittance parameters in the lossy cavity region. For that purpose, we first obtain the dominant-mode field expressions in the cavity by use of the simple transmission line theory [4]. Consider the lossy transmission line shown in Figure 6.4 that has a characteristic impedance (admittance) $Z_1(Y_1)$ and wavenumber $k_1$ and is terminated in an arbitrary lossy load $Z_s$ corresponding to the end wall of the lossy cavity under consideration.

Let the incident electric field $E_i^i$ of the $i$th mode corresponding to the generated field due to the aperture discontinuity at $z = 0$ be

$$
E_i^i = Ke_i(x,y)e^{jk_i(d-z)}
\tag{6.55}
$$

where $K$ is a constant that is determined according to the equivalent magnetic current $-M$ inside the lossy cavity, $e_i$ means the transverse vector, whose transmitted power through the cross section is normalized to unity, and $k_i$ means the wave number of the $i$th mode as given in (6.9) and (6.10). If the reflection coefficient is $\Gamma$ at $z = d$, the reflected electric field $E_i^r$ of the $i$th mode is

$$
E_i^r = K\Gamma e_i(x,y)e^{-jk_i(d-z)}
\tag{6.56}
$$

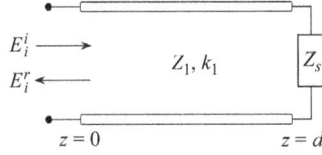

*Figure 6.4   Lossy transmission line with an arbitrary lossy load*

The total electromagnetic fields of the *i*th mode are given at an arbitrary *z*-point as

$$\boldsymbol{E}_{ti} = \boldsymbol{E}_i^i + \boldsymbol{E}_i^r = K\boldsymbol{e}_i(x,y)\left[e^{jk_i(d-z)} + \Gamma e^{-jk_i(d-z)}\right]$$

$$= K_i\boldsymbol{e}_i(x,y)\frac{[(1-\Gamma)\sin k_i(d-z) - j(1+\Gamma)\cos k_i(d-z)]}{(1-\Gamma)\sin k_i d - j(1+\Gamma)\cos k_i d} \tag{6.57}$$

$$\boldsymbol{H}_{ti} = -jK_iY_i\boldsymbol{a}_z \times \boldsymbol{e}_i\frac{[(1-\Gamma)\sin k_i(d-z) - j(1+\Gamma)\cos k_i(d-z)]}{(1-\Gamma)\sin k_i d - j(1+\Gamma)\cos k_i d} \tag{6.58}$$

where

$$K_i = K[(1+\Gamma)\cos k_i d + j(1-\Gamma)\sin k_i d] \tag{6.59}$$

In the above $\boldsymbol{e}_i$ is a transverse electric field, so the subscript $t$ in $\boldsymbol{E}_{ti}$ and $\boldsymbol{H}_{ti}$ mean the transverse. As a next step, we express the reflection coefficient $\Gamma$ at the ending wall in terms of the surface impedance $Z_S$ normalized for the characteristic admittance (impedance) $Y_i(Z_i)$ of the *i*th mode. For that purpose, we use the following relation:

$$\Gamma = \frac{Z_s - Z_i}{Z_s + Z_i} = -1 + \frac{2Z_s}{Z_s + Z_i} \cong -1 + 2Z_{si}, \tag{6.60}$$

where the normalized surface impedance $Z_{si}$ is given by

$$Z_{si} = Z_s Y_i = (1+j)\frac{R_s}{Z_i} = (1+j)r_{si} \tag{6.61}$$

here $R_S = \sqrt{\frac{\pi f \mu}{\sigma}}$. Recall that in the perturbation approach under consideration where the low loss is assumed, $Z_s \ll Z_i$, i.e., $z_{si} \ll 1$.

The characteristic admittance $Y_i$ of the *i*th mode used to normalize the surface impedance is defined according to the transverse magnetic (TM) or transverse electric (TE) mode as

$$Y_i = \frac{r_i}{j\omega\mu} \quad \text{for TE mode} \tag{6.62}$$

$$Y_i = \frac{j\omega\varepsilon}{r_i} \quad \text{for TM mode,} \tag{6.63}$$

where, in the expression of $r_i = jk_i$, $k_i$ means the wave number of the *i*th mode and is given, in terms of the phase constant $\beta_i$ and attenuation constant $\alpha_i$, by

$$k_i = \beta_i - j\alpha_i \tag{6.64}$$

Because all the higher-order modes except the lowest fundamental mode are cutoff, $k_i = -j\alpha_i$ for higher-order modes of both TE and TM modes, whereas $k_1 = \beta_1 - j\alpha_1$ for the lowest fundamental mode of the TE mode which is assumed to be only the propagating $TE_{10}$ mode. Here the attenuation constant $\alpha_1$ is the dominant mode $(TE_{10})$ attenuation constant due to power loss in cavity walls, which is found to be, based on the perturbation method [9],

$$\alpha_1 = \frac{R_S}{a^3 b\beta k\eta_0} \left(2b\pi^2 + a^3 k^2\right) \quad N_p/m \tag{6.65}$$

where $\beta$ is the propagation constant of the rectangular waveguide whose cross-section dimension is $a \times b$, $\eta_0$ and $k$ mean the intrinsic impedance $\sqrt{\frac{\mu_0}{\varepsilon_0}}$ of the air filling the lossy cavity and the wave number of the free space, respectively. Using the expression for $\Gamma$ in (6.60) in the electromagnetic field expressions of the $i$th mode in (6.57) and (6.58), we obtain the following expressions for electromagnetic fields which are related to the ending wall resistance in the lossy cavity region:

$$E_{ti} = K_i e_i(x,y) \frac{[\sin k_i(d-z) + (1-j)r_{si}\cos k_i(d-z)]}{\sin k_i d + (1-j)r_{si}\cos k_i d} \tag{6.66}$$

$$H_{ti} = -jK_i Y_i a_z \times e_i(x,y) \frac{[\cos k_i(d-z) - (1-j)r_{si}\sin k_i(d-z)]}{\sin k_i d + (1-j)r_{si}\cos k_i d} \tag{6.67}$$

Extending the above expression for the $i$th mode to all the modes inside the lossy cavity region that include only one fundamental (propagating) mode and all the higher-order (evanescent) modes due to the magnetic current $-M$ of the cavity side, we obtain the desired expression for electromagnetic fields inside the lossy cavity to be

$$E_t(-M) = \sum_i K_i e_i(x,y) \frac{[\sin k_i(d-z) + (1-j)r_{si}\cos k_i(d-z)]}{\sin k_i d + (1-j)r_{si}\cos k_i d} \tag{6.68}$$

$$H_t(-M) = -j\sum_i K_i Y_i a_z \times e_i(x,y) \frac{[\cos k_i(d-z) - (1-j)r_{si}\sin k_i(d-z)]}{\sin k_i d + (1-j)r_{si}\cos k_i d}$$
$$\tag{6.69}$$

where the unknown constant $K_i$ remains to be determined. These constants can be found by use of the orthonormal relation between the transverse vector mode function $e_i(x,y)$ and each entire vector basis function as follows:

$$K_i = \iint_{ap} M \cdot [a_z \times e_i(x,y)] ds$$
$$= \sum_{n=1}^{N} V_n \iint_{ap} M_n \cdot [a_z \times e_i(x,y)] ds \tag{6.70}$$
$$= V_1 A_{1i} + V_2 A_{2i} + \cdots + V_N A_{Ni}$$

where the multi-term expansion of the magnetic current in (6.45) has been used and

$$A_{ni} = \iint_{ap} \mathbf{M}_n \cdot [\mathbf{a}_z \times \mathbf{e}_i(x,y)]\,ds \tag{6.71}$$

The detailed procedure for the derivation of $K_i$ is listed in Appendix B. As seen in (6.70), $A_{ni}$ corresponds to the amplitude of the normalized electric field mode $\mathbf{e}_i(x,y)$ which the magnetic current density $-\mathbf{M}$ of the cavity side produces.

From the (6.70) and (6.71), the total transverse magnetic field $\mathbf{H}_t(-\mathbf{M}_n)$ due to $-\mathbf{M}_n$ is seen to be expressed by

$$\mathbf{H}_t(-\mathbf{M}_n) = -j\sum_i A_{ni} Y_i \mathbf{a}_z \times \mathbf{e}_i(x,y) \frac{[\cos k_i(d-z) - (1-j)r_{si}\sin k_i(d-z)]}{\sin k_i d + (1-j)r_{si}\cos k_i d} \tag{6.72}$$

Following the Galerkin testing in (6.72) as in (6.47), we get the desired admittance matrix equation for the lossy cavity region as

$$[Y^b]_{N\times N} = -j\sum_i A_{mi}A_{ni}Y_i\left[\frac{\cot k_i d - (1-j)r_{si}}{1 + (1-j)r_{si}\cot k_i d}\right], m,n = 1,2,3,\cdots,N \tag{6.73}$$

For most metals whose surface resistance $R_S$ is very small and so $r_{si} \ll 1$, the expression in (6.73) is approximated to be

$$[Y^b]_{N\times N} = -j\sum_i A_{mi}A_{ni}Y_i\cot k_i d, m,n = 1,2,3,\cdots,N \tag{6.74}$$

Note that (6.73) should be used for the case that the conductivity is not so high but for higher-order modes in which case the attenuation constant $\alpha_i$ becomes significantly large, the above approximation expression can be used even when the conductivity is not so high. $Y_i$ in (6.74) means the characteristic admittance of the $i$th mode and differently given according to TE or TM modes as mentioned above in relation to (6.62) and (6.63). Recall that the characteristic admittance $Y_i$ was used there for normalizing the surface impedance $Z_s$ of the ending wall. But here for the calculation of the admittance matrix element, we should divide into fundamental mode and higher-order (evanescent) mode and deal with the characteristic admittance for TE mode case as follows:

$$Y_i = \frac{r_i}{j\omega\mu} = \begin{cases} \dfrac{\beta_1}{\omega\mu}, & \text{for fundamental mode } (i = 1) \\[2mm] -j\dfrac{\alpha_i}{\omega\mu}, & \text{for higher order modes } (i \neq 1) \end{cases} \tag{6.75}$$

where $r_i = \alpha_i + j\beta_i$, $r_1 = j\beta_1 = j\sqrt{k^2 - k_{c1}^2}$ for $i = 1$ and $\alpha_i = \sqrt{k_{ci}^2 - k^2}$ and $\beta_i = 0$ for $i \neq 1$.

On the other hand, for the TM mode case where all the modes are assumed to be cutoff, the characteristic admittance $Y_i$ is given by

$$Y_i = \frac{j\omega\varepsilon}{\alpha_i} \tag{6.76}$$

The normalized transverse electric field $\mathbf{e}_i$ which is used in the present perturbation method for calculation of the admittance matrix element in (6.74) are expressed by, for TE mode,

$$
\begin{aligned}
e_i(x,y) = a_x \frac{j\omega\mu\pi B}{k_{c,AB}^2 b} C_{AB}\cos\left[\frac{A\pi\left(x+\frac{a}{2}\right)}{a}\right]\sin\left[\frac{B\pi\left(y+\frac{b}{2}\right)}{b}\right] \\
- a_y \frac{j\omega\mu\pi A}{k_{c,AB}^2 a} C_{AB}\sin\left[\frac{A\pi\left(x+\frac{a}{2}\right)}{a}\right]\cos\left[\frac{B\pi\left(y+\frac{b}{2}\right)}{b}\right]
\end{aligned}
\tag{6.77}
$$

and for TM mode,

$$
\begin{aligned}
e_i(x,y) = a_x \frac{-\alpha_{AB}\pi A}{k_{c,AB}^2 a} D_{AB}\cos\left[\frac{A\pi\left(x+\frac{a}{2}\right)}{a}\right]\sin\left[\frac{B\pi\left(y+\frac{b}{2}\right)}{b}\right] \\
+ a_y \frac{-\alpha_{AB}\pi B}{k_{c,AB}^2 b} D_{AB}\sin\left[\frac{A\pi\left(x+\frac{a}{2}\right)}{a}\right]\cos\left[\frac{B\pi\left(y+\frac{b}{2}\right)}{b}\right],
\end{aligned}
\tag{6.78}
$$

where $TE_{01}$ mode and $TM_{11}$ mode are the lowest modes respectively in the TE and TM modes and $C_{AB}$ and $D_{AB}$ are found to be, by use of the normalization condition of the mode power, i.e., $\iint_{a\times b}\left(e_x^2 + e_y^2\right)ds = 1$

$$
C_{A0} = \pm j\frac{k_{c,A0}}{\omega\mu\pi A}\sqrt{\frac{2a}{b}} \text{ for } B = 0
\tag{6.79}
$$

$$
C_{0B} = \pm j\frac{k_{c,AB}^2}{\omega\mu\pi B}\sqrt{\frac{2b}{a}} \text{ for } A = 0
\tag{6.80}
$$

$$
C_{AB} = \pm j2\frac{k_{c,AB}}{\omega\mu}\sqrt{\frac{1}{ab}} \text{ for } A, B \neq 0
\tag{6.81}
$$

and

$$
D_{AB} = \pm 2\frac{k_{c,AB}}{\alpha_{TM,AB}}\sqrt{\frac{1}{ab}} \text{ for } A, B \neq 0
\tag{6.82}
$$

here $k_{c,AB} = \sqrt{\left(\frac{A\pi}{a}\right)^2 + \left(\frac{B\pi}{b}\right)^2}$ and $\alpha_{TM,AB} = \sqrt{\left(\frac{A\pi}{a}\right)^2 + \left(\frac{B\pi}{b}\right)^2 - k^2}$

To calculate the admittance matrix elements inside the lossy cavity in (6.73), the expression for $A_{ni}$ in (6.70) still remains to be derived. Desired expression for $A_{ni}$ is obtained by using (6.71) and is summarized according to TE and TM mode as follows:

For the TE mode, independent of whether it is propagating mode $(k \geq k_{c,AB})$ or evanescent mode $(k \leq k_{c,AB})$, the expression for $A_{ni}$ is divided into three cases as

i) for the first case that $B=0$, the expressions for $A_{ni}$ are divided into two ones. i.e., when $\frac{A}{a} - \frac{2n-1}{L_x} = 0$,

$$A_{ni} = \frac{1}{2} C_{A0} \sin \frac{A\pi}{2} \cdot \frac{j\omega\mu L_y}{k_{c,AB}} \left[ L_x + \frac{a}{A\pi} \sin \frac{AL_x\pi}{a} \right] \tag{6.83}$$

and when $\frac{A}{a} - \frac{2n-1}{L_x} \neq 0$,

$$A_{ni} = C_{A0} \sin \frac{A\pi}{2} \cdot \frac{j\omega\mu L_y}{k_{c,AB}} \left[ \begin{array}{l} \dfrac{1}{\left(\dfrac{\pi}{a} - \dfrac{(2n-1)}{L_x}\right)\pi} \cdot \sin\left[\dfrac{1}{2}\left(\dfrac{AL_x\pi}{a} - (2n-1)\pi\right)\right] \\[3ex] + \dfrac{1}{\left(\dfrac{\pi}{a} + \dfrac{(2n-1)}{L_x}\right)\pi} \cdot \sin\left[\dfrac{1}{2}\left(\dfrac{AL_x\pi}{a} + (2n-1)\pi\right)\right] \end{array} \right] \tag{6.84}$$

ii) for the second case that $A = 0$, all the $A_{ni}$ vanish to zero, i.e.,

$$A_{ni} = 0 \tag{6.85}$$

and

iii) for the third case that $A, B \neq 0$, the expressions for $Y_{ni}$ are divided into two ones, i.e., when $\frac{A}{a} - \frac{n}{L_x} = 0$,

$$A_{ni} = C_{AB} \sin \frac{A\pi}{2} \cdot \frac{j\omega\mu Ab}{k_{c,AB}^2 aB} \cos \frac{B\pi}{2} \sin \frac{B\pi L_y}{2b} \left[ L_x + \frac{a}{A\pi} \sin \frac{AL_x\pi}{a} \right] \tag{6.86}$$

and when $\frac{A}{a} - \frac{n}{L_x} \neq 0$,

$$A_{ni} = 2C_{AB} \sin \frac{A\pi}{2}$$

$$\cdot \frac{j\omega\mu Ab}{k_{c,AB}^2 aB} \cos \frac{B\pi}{2} \sin \frac{B\pi L_y}{2b} \left[ \begin{array}{l} \dfrac{1}{\left(\dfrac{A}{a} - \dfrac{(2n-1)}{L_x}\right)\pi} \cdot \sin\left[\dfrac{1}{2}\left(\dfrac{AL_x\pi}{a} - (2n-1)\pi\right)\right] \\[3ex] + \dfrac{1}{\left(\dfrac{A}{a} + \dfrac{(2n-1)}{L_x}\right)\pi} \cdot \sin\left[\dfrac{1}{2}\left(\dfrac{AL_x\pi}{a} + (2n-1)\pi\right)\right] \end{array} \right] \tag{6.87}$$

For TM mode, when $A = 0$ or $B = 0$, $A_{ni} = 0$ and the expressions for $A_{ni}$ for all the modes that are assumed to be cutoff in the text are divided into two ones, i.e., when $\frac{A}{a} - \frac{2n-1}{L_x} = 0$,

$$A_{ni} = D_{AB}\sin\frac{A\pi}{2}\cdot\frac{\alpha_{TM,AB}}{k_{c,AB}^2}\cos\frac{B\pi}{2}\sin\frac{B\pi L_y}{2b}\left[L_x + \frac{a}{A\pi}\sin\frac{AL_x\pi}{a}\right] \tag{6.88}$$

and when $\frac{A}{a} - \frac{2n-1}{L_x} \neq 0$,

$$A_{ni} = 2D_{AB}\sin\frac{A\pi}{2}$$

$$\cdot\frac{\alpha_{TM,AB}}{k_{c,AB}^2}\cos\frac{B\pi}{2}\sin\frac{B\pi L_y}{2b}\left[\begin{array}{l}\dfrac{1}{\left(\dfrac{A}{a}-\dfrac{(2n-1)}{L_x}\right)\pi}\cdot\sin\left[\dfrac{1}{2}\left(\dfrac{AL_x\pi}{a}-(2n-1)\pi\right)\right]\\[3mm]+\dfrac{1}{\left(\dfrac{A}{a}+\dfrac{(2n-1)}{L_x}\right)\pi}\cdot\sin\left[\dfrac{1}{2}\left(\dfrac{AL_x\pi}{a}+(2n-1)\pi\right)\right]\end{array}\right]$$

$$\tag{6.89}$$

Using the expressions for $k_i$, $Y_i$, $e_i$, and $A_{ni}$ in (6.71), we obtain the admittance matrix element $\left[Y^b\right]_{N\times N}$ inside the lossy cavity.

## 6.6    Current source matrix

When the unknown magnetic current $M$ is given as a single term of the entire basis function, the current source is expressed as a single scalar term as in (6.4). On the other hand, when the magnetic current $M$ is given as a set of entire basis functions as in (6.45), the current source matrix element $I_n$ is expressed by

$$I_n = 2\iint_{ap} M_n \cdot H_o^i ds, n = 1, 2, 3, \ldots \tag{6.90}$$

Using (6.1(b)) and (6.45) in (6.90) the current column matrix element is found to be

$$I_n = \frac{4}{(2m-1)\pi}\frac{L_x L_y}{\eta_0}\sin\left[\frac{(2m-1)\pi}{2}\right]E_0, m = 1, 2, \ldots, N \tag{6.91}$$

## 6.7    Resonant cavity lengths for the multi-term expansion of magnetic currents

In Section 6.1, we dealt with the method for solving resonant cavity lengths for the single-term expansion of magnetic current. Here for a more general case in actual calculations by the method of moments, we deal with the case where the unknown magnetic current is expanded in terms of a set of expansion functions as in (6.45). In this case, inserting (6.45) instead of $M$ in the expressions for the $K_i$ in (6.68) and

(6.69) leads to the total electromagnetic fields as the sum of a fundamental mode and all higher order modes which are expressed by

$$
E_t\left(-\sum V_n M_n\right) = \sum_i \sum_{n=1}^{N} V_n \iint_{ap} M_n \cdot [a_z \times e_i(x,y)] ds
$$

$$
\cdot e_i \frac{\sin k_i(d-z) + (1-j)r_{si}\cos k_i(d-z)}{\sin k_i d + (1-j)r_{si}\cos k_i d} \tag{6.92}
$$

$$
= \sum_i \left(\sum_{n=1}^{N} V_n A_{ni}\right) \underline{e_i} \frac{\sin k_i(d-z) + (1-j)r_{si}\cos k_i(d-z)}{\sin k_i d + (1-j)r_{si}\cos k_i d}
$$

and

$$
\underline{H_t}\left(-\sum V_n M_n\right) = -j \sum_i \left(\sum_{n=1}^{N} V_n A_{ni}\right) Y_i(a_z \times e_i(x,y))
$$

$$
\cdot \frac{\cos k_i(d-z) - (1-j)r_{si}\sin k_i(d-z)}{\sin k_i d + (1-j)r_{si}\cos k_i d} \tag{6.93}
$$

where

$$
A_{ni} = \iint_{ap} M_n \cdot [a_z \times e_i(x,y)] ds \tag{6.94}
$$

From this equation (6.94), it is seen that $A_{ni}$ is defined as an inner product between the $n$th basis function $M_n$ and $a_z \times e_i$ and corresponds to the amplitude of the $i$th normalized electric field $e_i(M_n)$ which is excited due to the $M_n$ in the equivalent transmission line.

Note that the two equations can be obtained by substituting $\sum V_n M_n$ for $M$ in (6.70). For the sake of calculating the first resonant length, extracting the expression for the total electromagnetic field of the dominant mode $\underline{e_i}$ due to the all the basis functions leads to the following expressions:

$$
E_{t1} = \sum_{n=1}^{N} V_n A_{n1} e_i(x,y) \frac{\sin k_1(d-z) + (1-j)r_s\cos k_1(d-z)}{\sin k_1 d + (1-j)r_s\cos k_1 d}
$$

$$
= A_{11} V_{eq} e_i(x,y) \frac{\sin k_1(d-z) + (1-j)r_s\cos k_1(d-z)}{\sin k_1 d + (1-j)r_s\cos k_1 d} \tag{6.95}
$$

$$
H_{t1} = -j \sum_{n=1}^{N} V_n A_{n1} Y_1(a_z \times e_i(x,y)) \frac{\cos k_1(d-z) - (1-j)r_s\sin k_1(d-z)}{\sin k_1 d + (1-j)r_s\cos k_1 d}
$$

$$
= -j A_{11} V_{eq} Y_1(a_z \times e_i(x,y)) \frac{\cos k_1(d-z) - (1-j)r_s\sin k_1(d-z)}{\sin k_1 d + (1-j)r_s\cos k_1 d}
$$

$$
\tag{6.96}
$$

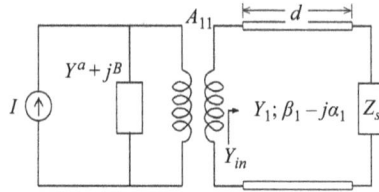

Figure 6.5    *Equivalent circuit for determining the second aperture-cavity resonance for the multi-term expansion case of the magnetic current over the aperture*

where

$$V_{eq} = \sum_{n=1}^{N} \left(\frac{A_{n1}}{A_{11}}\right) V_n \tag{6.97}$$

As seen in (6.95), for the magnitude of the above electric field $|E_{t1}|$ to be maximum, $|V_{eq}|$ has to be maximum. Since the value of $|V_{eq}|$ varies according to the length $d$ of the lossy cavity, as a method for searching for the maximum value of $|V_{eq}|$ and the value of $d$ giving the maximum value of $|V_{eq}|$ as a function of $d$, the iterative "0.618 method" of optimization [12] is used. On the other hand, for the second resonance the basis function for $n = 1$ among the multi terms of magnetic currents $M = \sum_{n=1}^{N} V_n M_n$ dominantly contributes [4] while setting up the fundamental mode in the lossy cavity so the resonant length is determined from the condition that

$$I_m(Y_{11}) = I_m\left(A_{11}^2 Y_{in}\right) + B_{11}' = 0 \tag{6.98}$$

where $Y_{11}$ means the total admittance at the aperture due to only the $V_1 M_1$ under the assumption that all the modes be cutoff except the fundamental mode. So the $I_m(Y_{11})$ means the sum of the input susceptance $\left(A_{11}^2 Y_{in}\right)$ looking into the transformer in the equivalent circuit of Figure 6.5 under the above assumption and the susceptance $B_{11}'$ due to the total stored energy of both higher-order modes due to $-M_1$ in the cavity side and the half-space side due to $M_1$.

## 6.8   Numerical example

As a numerical example, we consider an electromagnetic absorption problem through a small rectangular slot into a rectangular lossy cavity as shown in Figure 6.1 using the above two kinds of resonances and resonance lengths. A plane wave is assumed to be normally incident onto the aperture. The parameters are given in Table 6.1, for which case the numerical results are obtained.

*Table 6.1   Specifications for aperture coupled lossy cavity*

| | |
|---|---|
| Wavelength | $\lambda = 3$ m |
| Cavity dimensions with $d$ adjusted for aperture-cavity resonance | $a = 0.762\lambda$<br>$b = 0.339\lambda$ |
| Aperture dimensions | $L_x = 0.25\lambda$<br>$L_y = 0.02\lambda$ |
| Cavity material | Copper $(\sigma = 5.80 \times 10^7 \text{s/m})$ |

The incident wave is chosen to be

$$\boldsymbol{H}^i = \boldsymbol{a}_x H_0 e^{-jk_0 z} = \boldsymbol{a}_x \frac{E_0}{\eta_0} e^{-jk_0 z} \tag{6.99}$$

where $k_0$ and $\eta_0$ follow the usual meaning. To begin with, we are going to solve the given problem by use of a single expansion function and compare the results with those obtained by use of multi-term expansion of the magnetic current.

i) single-term magnetic current

In this case, we let

$$\boldsymbol{M} = \boldsymbol{a}_x V_0 \cos\left(\frac{\pi x}{L_x}\right) \tag{6.100}$$

Using (6.53) and (6.54) for this single-term magnetic current, the numerical result for $Y^a = G^a + jB^a$ is calculated to be

$$\begin{aligned}
Y^a &= G^a + jB^a \\
&= 9.8418 \times 10^{-7} - j1.6659 \times 10^{-5} \\
&= (0.0082394 - j0.1394670)\frac{L_x L_y}{\eta_0}
\end{aligned} \tag{6.101}$$

On the other hand, the admittance element for the cavity side of the aperture is found to be through the use of (6.73) for the case that $m = n = 1$

$$\begin{aligned}
Y^b &= G^b + jB^b + jB \\
&= A_{01}^2 Y_{in} - j0.1480565\frac{L_x L_y}{\eta_0}
\end{aligned} \tag{6.102}$$

where   $G^b \cong G^a = 0.0082394 \frac{L_x L_y}{\eta_0}$   and   $B^b \cong -(B^a + B) = -B' \cong 0.2875235$. Recall that $\text{Re}\{A_{01}^2 Y_{in}\} = G^b$, $\text{Im}\{A_{01}^2 Y_{in}\} = B^b$ and B' < 0 as expected from the condition under consideration. Here $G^b$ and $B^b$ are determined by $a$, $b$, and the cavity length $d$, whereas B' is almost independent of $d$.

ii) multiple terms of magnetic current

In this case, we expand the equivalent current into the finite number of the entire basis functions as follows:

$$\boldsymbol{M} = \boldsymbol{a}_x \sum_{n=1}^{N} V_n \cos\left\{\frac{(2n-1)\pi x}{L_x}\right\} \tag{6.103}$$

In the actual calculation, $N = 7$ was used. The results are tabulated in Table 6.2, together with those obtained by using $N = 1$.

From Table 6.2, it is seen that for both single-term and multiple-term approximation of the magnetic currents, the results for the resonant lengths and the equivalent voltages agree well with each other. This supports the validity of the single-term approximation of the magnetic current over the aperture.

Figure 6.6 and Figure 6.7 illustrate the dominant-mode electric field distribution inside the cavity respectively at first and second resonances, which have been obtained by use of (6.95) for the respective cavity lengths. From the figures, for both the first and the second resonances, standing wave patterns are seen to be established inside the lossy cavity structure and the maximum electric field magnitude at the center of the lossy cavity is also seen to be larger than that of the

*Table 6.2    Characteristics of two kinds of resonances*

|  | N | First resonance | Second resonance |
|---|---|---|---|
|  | 1 | $0.4937348\lambda_{g1}$ | $0.49999917\lambda_{g1}$ |
|  | 7 | $0.493075\lambda_{g1}$ | $0.49999917\lambda_{g1}$ |
| $V_o$ | 1 | $151.27E_0$ | $0.0034E_0$ |
| $V_{eq}$ | 7 | $(153.03 + j0.213)E_0$ | $0.0034E_0$ |

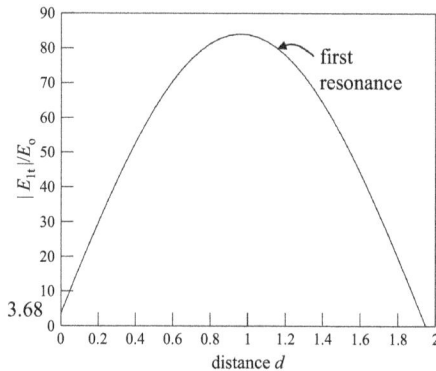

*Figure 6.6    Dominant mode electric field distributions inside a copper cavity at first aperture-cavity resonance at the cavity length of $d_1$*

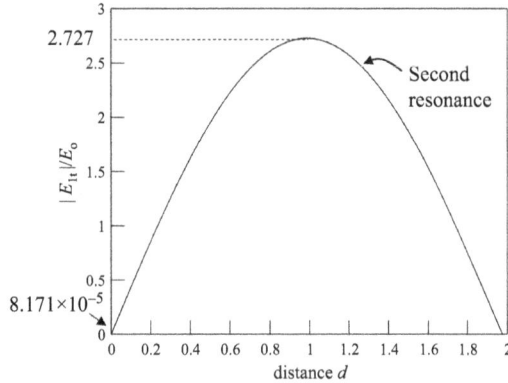

*Figure 6.7   Dominant mode electric field distributions inside a copper cavity at second aperture-cavity resonance at the cavity length of $d_2$*

incident field. In particular, the maximum electric field magnitude for the first resonance is observed to be much larger than that for the second resonance as well as the incident electric field magnitude. This strong standing wave mode setup for the first resonance plays an important role in the maximum absorption phenomena which is called the absorption resonance in the sense that the absorption power at the absorption resonance is comparable to the transmitted power level at the transmission resonance in Chapter 4.

## 6.9   Maximum power absorption condition and absorption cross section

Based upon discussions on the maximum absorption phenomenon through the small aperture into the lossy cavity in Section 6.3, the condition for the maximum absorption is summarized to be

$$G^b = G^a \tag{6.104}$$

and

$$B^b = Im\left(A_{01}^2 Y_{in}\right) = -B' \tag{6.105}$$

Under this condition, all the transmitted power through the aperture is absorbed. In addition, because all the highest order modes except the fundamental mode are assumed to be cutoff inside the lossy cavity, the first expression of (6.39) for the transmitted power through the aperture can be changed to the expression in (6.40). The equality between those two expressions is due to the above impedance matching conditions as given in (6.104) and (6.105).

Similar to the funneling phenomenon in the transmission resonance in Chapters 4 and 5, in the maximum absorption (or absorption resonance) case, incident electromagnetic field power is collected from a much larger area than the geometrical aperture area and is subsequently funneled into the inside region of the lossy cavity and is dissipated into the Joule's heat loss due to the inside wall current of the lossy cavity. Because of this funneling phenomenon, the integration of $\frac{1}{2}\text{Re}(\boldsymbol{E} \times \boldsymbol{H})$ over the aperture can be transformed into the integration of $\frac{1}{2}\text{Re}(\boldsymbol{E}_{t1} \times \boldsymbol{H}_{t1})$ over the cross section of the lossy cavity. Recall that here $\boldsymbol{E}$ and $\boldsymbol{H}$ mean the total electric and magnetic fields at the aperture, whereas $\boldsymbol{E}_{t1}$ and $\boldsymbol{H}_{t1}$ mean the electric and magnetic fields of $TE_{10}$ mode which is assumed to be the only propagating mode in the lossy cavity. By use of (6.91) for $m = 1$ in (6.41), the maximum absorbed power in the lossy cavity is found to be

$$\max P_1 = \frac{I^2}{8G^a} = \frac{\left[\frac{4}{\pi}\left(\frac{L_xL_y}{\eta_0}\right)E_0\right]^2}{8G^a} = \frac{2}{\pi^2}\frac{\left(L_xL_y\right)^2}{G^a\eta_0^2}E_0^2 \quad [W] \tag{6.106}$$

where the subscript 1 in $P_1$ means the first resonance. If we define the absorption cross section of the lossy cavity as the total electromagnetic power absorbed by the lossy cavity, normalized by the incident plane wave flux, the absorption cross-section is expressed to be

$$\tau = \frac{\max P_1}{P_{inc}} = \frac{4}{\pi^2}\frac{\left(L_xL_y\right)^2}{G^a\eta_0} \quad [\text{m}^2], \tag{6.107}$$

where the incident plane wave flux means the incident power density $P_{inc}\left(=\frac{1}{2}\frac{E_0^2}{\eta}\left[\frac{W}{m^2}\right]\right)$. As a numerical example, if we use the value of $G^a$ for the radiation conductance looking into the left half-space which we are considering $G^a = 9.8418 \times 10^{-7} = 0.0082394 \frac{L_xL_y}{\eta_0}$ in the above equation (6.107) for $\tau$, we find the absorption cross section approximately to be

$$\tau = \frac{4}{\pi^2}\frac{\left(L_xL_y\right)^2}{0.0082394 \times \frac{L_xL_y}{\eta_0}\eta_0} = 0.772653 \times \frac{\lambda^2}{\pi} \cong \frac{3\lambda^2}{4\pi} \tag{6.108}$$

It is interesting to note that the absorption cross section which has been obtained for the small coupling aperture backed by a lossy cavity is calculated to be approximately the same as the transmission cross section $\left(\frac{3\lambda^2}{4\pi}\right)$ for a small coupling aperture. The equality between the two cross sections requires that the input admittance looking into the lossy cavity be equal to the input admittance looking into the transmission cavity whose inside wall is made of a perfect conductor. This is a natural result of consideration of the same impedance matching conditions for both transmission and absorption resonances.

For this reason, the condition for the absorption resonance in which the absorbed power in the lossy cavity after transmission of incident field power through the small aperture becomes maximum is approximately the same as that for the transmission resonance in Chapter 4. The deviation of the numerical result

$\left(0.772653 \times \frac{\lambda^2}{\pi}\right)$ for the absorption cross section from the expression of $\frac{3\lambda^2}{4\pi}$ [m$^2$] is thought to be mainly due to the use of the small coupling aperture whose gain in the left half-space of the incident side is 1.5. If we use a more general expression $\frac{2G\lambda^2}{4\pi}$ for the transmission cross section which is valid for the coupling aperture whose directivity is arbitrarily larger than 1.5 for the ideal Hertzian dipole, the numerical result is found to be $0.7687 \times \frac{\lambda^2}{\pi}$ for the small rectangular slot case ($a \times b = 0.25\lambda \times 0.02\lambda$) under consideration whose gain is calculated to be 1.5374. From this, it is seen that the more accurate gain we use, the more accurate result for the absorption cross section we get. For reference for the half wavelength narrow slot of coupling aperture, since $= 1.64$, the absorption cross section amounts to $0.802 \times \frac{\lambda^2}{\pi}$ [m$^2$]. Note that the values of the absorption cross section have been calculated by mere use of the impedance matching condition based on the radiation conductance looking into the left half-space without taking into account the input admittance looking into the lossy cavity.

Next, let us consider the effect of the input admittance by looking into the lossy cavity on the aperture cross section. If the inside wall is made of a good conductor such as copper, the input conductance $G^b (= 2.1 \times 10^{-8})$ is much smaller than the radiation conductance $G^a (= 9.8418 \times 10^{-7})$. So the absorption cross section of the coupling aperture backed by such a lossy cavity is too small for the impedance matching condition to be satisfied. In this case, for the maximum absorption inside the lossy cavity to occur, we investigate what material can be used as inside wall material of the lossy cavity to make $G^a = G^b$ by varying the conductivity (resistivity) of the cavity wall material. Recall that once the geometrical parameters are fixed, the radiation conductance $G^a$ is also fixed and the only variable that can be varied by varying the material constant $\sigma$ of the cavity wall is $Y^b (= G^b + jB^b)$. It is noteworthy that the resonance lengths $d_1$ and $d_2$ remain almost unchanged even when the material constant $\sigma$ is significantly varied. Under this situation, we have investigated what material, within what range of the conductivity values, is suitable for the good matching condition of $G^a = G^b$.

Some data for surface resistance of the end wall $R_s$, lossy cavity length $d_1$ of the first resonance, the conductance $G^b$ looking into the lossy cavity, and absorption cross section $\tau$ are listed for various materials such as copper, bronze, nichrome, iron, graphite, barium, antimony, lutetium, dysprosium, etc. in Table 6.3.

From Table 6.3, it is seen that for $\sigma = 2.7 \times 10^4$ [s/m] the conductance $G^b$ can be made to be almost the same as the radiation conductance $G^a$. Besides for this case of $\sigma = 2.7 \times 10^4$ [s/m], the absorption cross section is seen to be maximized to $0.7815\lambda^2/\pi$[m$^2$], which almost corresponds to the value $0.772653\lambda^2/\pi$[m$^2$] obtained by use of the expression of $\frac{2G\lambda^2}{4\pi}$ [m$^2$]. A little discrepancy between these two values is thought to be mainly due to the intrinsically approximate perturbation method which has been used in calculation procedure of internal field and loss inside the cavity. From the numerical data in Table 6.3, it is also seen that, for the absorption cross section to exceed at least $0.70\lambda^2/\pi$[m$^2$] for good matching, the

*Table 6.3    Variation of $G^b$ and ACS according to material constants ($\sigma$)*

| | $\sigma$ (conductivity)[s/m] | $R_s[\Omega]$ | $d_1$[m] | $G^b[\text{s} \cdot \text{m}^2]$ | Absorption cross section (ACS) $\tau$ [m$^2$] |
|---|---|---|---|---|---|
| Copper | $5.8 \times 10^7$ | 0.0026 | 1.9605 | $2.1 \times 10^{-8}$ | $0.0649\lambda^2/\pi$ |
| Bronze | $10^7$ | 0.0063 | 1.9605 | $5.1 \times 10^{-8}$ | $0.1475\lambda^2/\pi$ |
| Nichrome | $10^6$ | 0.0199 | 1.9605 | $1.62 \times 10^{-7}$ | $0.3804\lambda^2/\pi$ |
| Iron | $10^5$ | 0.0628 | 1.9605 | $5.11 \times 10^{-7}$ | $0.7050\lambda^2/\pi$ |
| Graphite | $7 \times 10^4$ | 0.0751 | 1.9605 | $6.1 \times 10^{-7}$ | $0.7401\lambda^2/\pi$ |
| Barium | $3 \times 10^4$ | 0.1147 | 1.9605 | $9.31 \times 10^{-7}$ | $0.7812\lambda^2/\pi$ |
| Antimony | $2.7 \times 10^4$ | 0.1209 | 1.9605 | $9.84 \times 10^{-7}$ | $0.7815\lambda^2/\pi$ |
| Lutetium | $2 \times 10^4$ | 0.1405 | 1.9605 | $1.12 \times 10^{-6}$ | $0.7765\lambda^2/\pi$ |
| Dysprosium | $10^4$ | 0.1987 | 1.9605 | $1.62 \times 10^{-5}$ | $0.7327\lambda^2/\pi$ |
| Graphite | $10^3$ | 0.6283 | 1.9605 | $5 \times 10^{-6}$ | $0.4093\lambda^2/\pi$ |

conductivity $\sigma$ of the wall material of the lossy cavity should lie in the range of $7 \times 10^3 < \sigma < 10^5$ [s/m]. The results tabulated in Table 6.3 have been obtained by using $N = 7$ though a single term($N = 1$) half-cosine function also gives quite good results compared with those obtained by using $N = 7$. For reference, in the above calculations, a total of about 5,000 high-order modes, from $TE_{01}$ to $TE_{50,50}$ and from $TM_{11}$ to $TM_{50,50}$ have been included as in [4].

## 6.10    Transmission cross section and its relation to the absorption cross section

Through the study on the 3D (three-dimensional) electromagnetic coupling problem through a small aperture system [13] or through an aperture-to-cavity-to aperture system [13], the transmission cross sections as a measure of the maximum transmission under the transmission resonance condition are the same expression of $\frac{3\lambda^2}{4\pi}$ [m$^2$] for both cases as discussed in Chapters 4 and 5. Note that the above aperture-to-cavity-to aperture system means the case where the input aperture and the output aperture are the same to each other. If not, the transmission cross section of the unsymmetrical system is mainly determined by the directivity of the input aperture of the incident side as discussed in Chapter 4.

Recall that the scattering cross section of the ideal Hertzian electric dipole can be found to be $\frac{3\lambda^2}{2\pi}$ [m$^2$] by shorting the load resistance only to retain the radiation resistance of its equivalent circuit model under the assumption of the complex conjugate matching between the load and radiation impedances. This scattering cross section [14] of a shorted Hertzian dipole as a scatterer of incident electromagnetic energy is thus four times as great as its maximum effective aperture of $\frac{3\lambda^2}{8\pi}$ [m$^2$] as an absorber of energy [15].

For reference, the backscattering (radar) cross section of the ideal small scatterer at resonance is well known to be $\frac{9\lambda^2}{4\pi}$ [m$^2$] under the assumption of omnidirectional reradiation [16]. If we assume that the small resonant scatterer radiates a Hertzian dipole whose $\theta$ dependence is given by $\sin^2\theta$, the above backscattering (radar) cross-section is modified to $\frac{3\lambda^2}{2\pi}$ [m$^2$], which is compatible with the above result for the scattering cross section obtained by use of the equivalent circuit model. It is interesting to note also that the expression for the scattering cross section is the same as that of electric dipole in the quantum mechanical scattering theory [17].

For the above two cases of small coupling aperture in both TRA and TRC structures, the coupling hole can be taken as a magnetic Hertzian dipole whose directivity $D$ is 1.5. So we assume the general expression to be $\frac{2D\lambda^2}{4\pi}$ [m$^2$] for $D > 1.5$ where the area of the coupling aperture is extended to that of larger areas than that of the Hertzian dipole type and check the validity. To this end, we have investigated numerical transmission cross sections for various arbitrarily shaped apertures such as complementary split ring resonator (CSRR) [18]. Through the comparison of the numerical result with the close expression, the close expression has been numerically validated as in Chapter 5. A recent study on a general relation between the transmission cross section and the directivity also supports the validity of the expression of $\frac{2G\lambda^2}{4\pi}$ [m$^2$], which has been proved from the coupled mode theory [19]. Through this study, the transmission cross section of an infinitely long resonant slit in a thick conducting screen with an isotropic radiation pattern was found to be $\frac{\lambda}{\pi}$ [m] as two-dimensional slit problem. In addition, it has been also found that in the case of anisotropic radiation which can be obtained for example by introducing surface corrugation or by modifying the entrance to the slits, the transmission cross-section can be made to go beyond the limit of $\frac{\lambda}{\pi}$ for the isotropic case, amounting to $\frac{G\lambda}{\pi}$ [m] [19].

Though the discussion on the relation between the transmission cross section and the directivity was limited to the two-dimensional long slit problem, in the three-dimensional resonant aperture case too, the directivity $D$ is thought to be included as in the expression of $\frac{G\lambda^2}{2\pi}$ [m$^2$], where $\frac{\lambda^2}{2\pi}$ [m$^2$] is a well-known expression [19] for the three-dimensional isotropic resonant aperture. In this case too, the directivity $G$ can be increased by introducing surface corrugation or by modifying the entrances to the apertures or by changing the shape or size of the apertures.

Regarding the absorption resonance problem, the aperture cross section is very close to the value of $\frac{2G\lambda^2}{4\pi}$, which corresponds to the expression for the transmission cross section of the resonant aperture. Recall that the structure for the absorption resonance problem has been chosen to be made of the small coupling aperture backed by a lossy cavity, which has been analyzed by use of the perturbation approach. Despite the intrinsic error source of the perturbation approach, the absorption cross section is almost the same as the transmission cross section. This can be clearly explained in terms of the impedance matching concept at the coupling aperture. That is to say, for transmission, this can be achieved when the radiation impedance of one side at the aperture is complex-conjugately matched to the input impedance into the other side whereas, for absorption resonance problem, this requires a specific lossy material for the inside wall of the cavity such that the input impedance looking into the aperture may be complex-conjugately matched to the aperture radiation impedance. Note that the above impedance matching condition for the transmission cross section is valid for both TRA structure in Chapter 4 and TRC structure in Chapter 5.

In the sense that a transmission or absorption resonance occurs under the same impedance matching condition only excepting that the input impedance looking into the opposite side to the incident side in the transmission resonance problem is replaced with the input impedance looking into the lossy cavity through the coupling aperture in the absorption resonance problem, the expression for the absorption cross section is expected to be the same as that for the transmission cross section.

Further, research results for optimizing aperture-enhanced photodetectors revealed that the expression for the maximum absorption in the case of two-dimensional narrow slit filled with a lossy dielectric material and the bottom of the slit closed off with a perfect conductor is given by the same expression of $\frac{D\lambda}{\pi}$ [m] [19] for the two-dimensional transmission resonance problem. This also strengthens the equality between the absorption cross section and the transmission cross section. Lately, two structures were considered to investigate the ability of resonant slits to concentrate efficiently electromagnetic energy into the volume of absorbing material placed inside or directly behind the slit for a wide range of applications [20] including high-speed, low-capacitance photodetectors with deep sub-wavelength active regions. Through the investigation, the structure of the slit filled with absorbing material is much more favored from the viewpoint of enhancing the absorption cross section, the expression for the absorption cross section for this case still remains to be investigated further.

Note that the above expression of $\frac{2G\lambda^2}{4\pi}$ [m$^2$] corresponds to doubling the general effective aperture [21] of antennas appearing in the transmission formula of Friis. In summary, the expression for the absorption cross section for the small aperture backed by a lossy cavity under consideration is essentially the same as that for the transmission cross section in the transmission resonance problem from the viewpoint of impedance matching, which has been widely investigated in connection with the extraordinary optical transmission (EOT) problem. Recall that in particular two-dimensional resonant slit problem in thick conducting screens has been widely studied for that purpose [22–25].

## 6.11    Absorption frequency selective surface

As discussed, the similarity between the transmission cross section (TCS) of the transmission resonant aperture (TRA) structures and the absorption cross section (ACS) of the absorption resonant structure composed of a coupling small aperture backed by a lossy cavity in the previous Section 6.10 has been pointed out. In more detail, the TCS of the TRC structure has been observed to be almost the same as the ACS of the above absorption resonant structure.

### 6.11.1   Structure of the absorbing frequency selective surface

Based upon the above discussion on the absorption cavity, a kind of absorption frequency selective surface (AFSS) is considered by composing a two-dimensional array of the absorption resonant structure of an aperture backed by a lossy cavity and next eliminating all the cavity walls of the lossy cavity except lossy end walls. So, the resultant structure is made of the small aperture array over an absorbing lossy ground, which is essentially different from the conventional AFSS [26,27] which is composed of the lossy periodic patch array over the perfect conducting ground.

Accordingly, the performance of the present AFSS can be designed by adjusting the three main degrees of freedom such as the size and shape of the coupling apertures on the upper perfectly conducting plane, the conductivity of the lower absorbing (lossy) ground plane, and the separation between the two planes. Figure 6.8 shows the three-dimensional view of the absorbing FSS structure under consideration, where unit cell length along the $x$ and $y$ axes, $T_x$ and $T_y$ and the distance $d$ between the upper FSS plane and the lower lossy ground plane are chosen to be the same as geometrical parameters such as cross section and longitudinal length of the previous absorption cavity.

Here the separation $d$ between the upper FSS plane and the lower lossy ground plane is 21.7434559 [mm]. The conductivity $\sigma$ of the lossy ground plane and the coupling aperture size are chosen to be the same as those for the absorption cavity in Section 6.1.

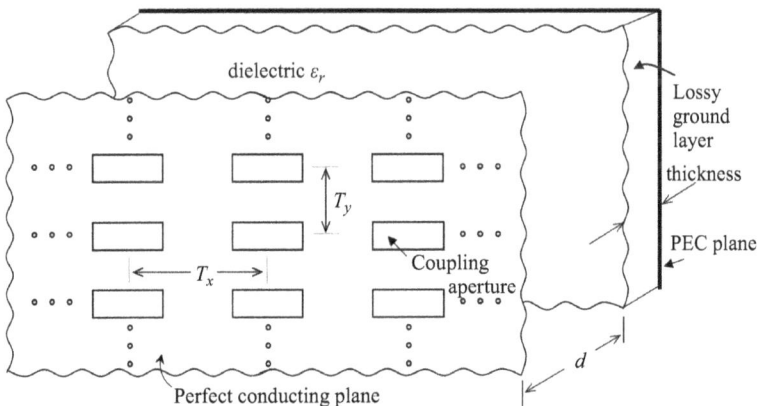

*Figure 6.8   Three-dimensional view of the absorbing FSS structure*

## 6.11.2    Analysis method

We consider an analysis method for the present absorbing FSS structure which is applicable to the case of an arbitrary incidence angle and polarization for an arbitrary shaped coupling aperture. As is depicted [28] in Figure 6.9, the incident angles $\theta$ and $\phi$ are such that $\theta$ is the angle between the $z$-axis and the incident unit vector $\underline{n}$ and $\phi$ between the $x$-axis and the incident unit vector's projection on the $x$–$y$ plane, the segment OA. The origin O is not marked to avoid the crowdedness in the figure.

On the other hand, the polarization angle $\gamma$ measured in the clockwise direction for the incident unit vector is the angle between the arbitrarily polarized incident electric field vector $\boldsymbol{E}_{inc}$ and the segment OA' in the $x$–$y$ plane which is perpendicular to the segment OA. For example, when $\gamma = 0$ and, the incident electric field vector is parallel to the $x$–$y$ plane, and so is the incident magnetic field, when $\gamma = \pm\frac{\pi}{2}$.

Based upon the above discussion, an arbitrarily polarized incident electric field upon the periodic aperture plane under consideration can be decomposed into two kinds of mode electric fields. One is the TE mode of the electric field defined for $\gamma = 0$ in Figure 6.9, which is given by

$$\boldsymbol{E}^i_{TE}(x,y,z) = \left(\boldsymbol{a}_x \sin\phi - \boldsymbol{a}_y \cos\phi\right)E_0 \cos\gamma \; e^{jk^i_T \rho} \; e^{jk_z z} \tag{6.109}$$

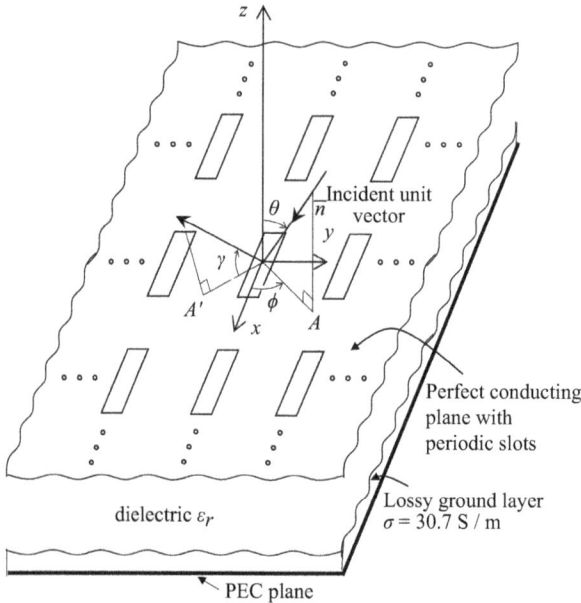

*Figure 6.9    Incident angle $(\theta, \phi)$ and the polarization angle $\gamma$ on the absorbing FSS plane*

and the other is the TM mode of the electric field defined for $\gamma = 90°$, which is given by

$$\boldsymbol{E}^i_{TM}(x,y,z) = \left(-\boldsymbol{a}_x\cos\theta\cos\phi - \boldsymbol{a}_y\cos\theta\sin\phi\right)E_0\sin\gamma\, e^{jk^i_T\rho}\, e^{jk_z z}, \qquad (6.110)$$

where $\rho(x,y)$ means the position vector on $x$–$y$ plane, $\boldsymbol{k}^i_T = \boldsymbol{a}_x k^i_x + \boldsymbol{a}_y k^i_y$, $k^i_x = k_0\sin\theta\cos\phi$, and $k^i_y = k_0\sin\theta\sin\phi$.

## 6.11.3   Field representations for each region and formulation procedure

Here we express the incident fields of arbitrarily incident angle and polarization and the scattered fields due to the periodic apertures, at $z = 0$ and the fields for $-d \leq z \leq 0$ in terms of Froquet mode functions. To begin with, the transverse $x$ and $y$ components (to the $z$-direction) of the incident electromagnetic fields as shown in Figure 6.9 are expressed as a combination of TE and TM waves as follows:

$$E^i_x = \left(E'^i k^i_x - E'''^i k^i_y\right)\frac{e^{jk_0 \cdot \underline{r}}}{\sqrt{\left(k^i_T\right)^2}} \qquad (6.111a)$$

$$E^i_y = \left(E'^i k^i_y + E'''^i k^i_x\right)\frac{e^{jk_0 \cdot \underline{r}}}{\sqrt{\left(k^i_T\right)^2}} \qquad (6.111b)$$

$$H^i_x = \left(E'^i \widehat{Y}'_0 k^i_y + E'''^i \widehat{Y}''_0 k^i_x\right)\frac{e^{jk_0 \cdot \underline{r}}}{\sqrt{\left(k^i_T\right)^2}} \qquad (6.111c)$$

$$H^i_y = \left(-E'^i \widehat{Y}'_0 k^i_x + E'''^i \widehat{Y}''_0 k^i_y\right)\frac{e^{jk_0 \cdot \underline{r}}}{\sqrt{\left(k^i_T\right)^2}} \qquad (6.111d)$$

where all the superscripts $i$ mean the incident field, single and double prime mean the TM and TE waves, respectively, and $k^i_x$, $k^i_y$, and $k^i_z$ mean each cartesian component of the propagation vector $\underline{k}_0$ of the incident wave as

$$k^i_x = k_0\sin\theta\cos\phi$$

$$k^i_y = k_0\sin\theta\sin\phi$$

$$k^i_z = k_0\cos\theta$$

and $\left(k^i_T\right)^2 = \left(k^i_x\right)^2 + \left(k^i_y\right)^2 = k_0^2\sin^2\theta$, $\widehat{Y}'_0 = \widehat{Y}''_0 = \sqrt{\dfrac{\varepsilon_0}{\mu_0}}$

The transverse $x$ and $y$ components of the scattered fields for the upper half-space region ($z > 0$) are also expressed as a combination of TE (double primed quantity) wave and TM (single primed quantity) wave as

$$E_x^s = \sum_p \sum_q \left( V'^s_{pq} k_{xp} + V''^s_{pq} k_{yp} \right) \frac{e^{j\mathbf{k}_T \cdot \boldsymbol{\rho}} \cdot e^{-jk_z z}}{\sqrt{k_T^2}\sqrt{T_x T_y}} \tag{6.112a}$$

$$E_y^s = \sum_p \sum_q \left( V'^s_{pq} k_{yq} - V''^s_{pq} k_{xp} \right) \frac{e^{j\mathbf{k}_T \cdot \boldsymbol{\rho}} \cdot e^{-jk_z z}}{\sqrt{k_T^2}\sqrt{T_x T_y}} \tag{6.112b}$$

$$H_x^s = \sum_p \sum_q \left( -V'^s_{pq} \widehat{Y}'_{pq} k_{yq} + V''^s_{pq} \widehat{Y}''_{pq} k_{xp} \right) \frac{e^{j\mathbf{k}_T \cdot \boldsymbol{\rho}} \cdot e^{-jk_z z}}{\sqrt{k_T^2}\sqrt{T_x T_y}} \tag{6.112c}$$

$$H_y^s = \sum_p \sum_q \left( V'^s_{pq} \widehat{Y}'_{pq} k_{xp} + V''^s_{pq} \widehat{Y}''_{pq} k_{yq} \right) \frac{e^{j\mathbf{k}_T \cdot \boldsymbol{\rho}} \cdot e^{-jk_z z}}{\sqrt{k_T^2}\sqrt{T_x T_y}} \tag{6.112d}$$

where $k_{xp} = k_x^i + \frac{2\pi p}{T_x}$, $k_{yq} = k_y^i + \frac{2\pi q}{T_y}$, $\mathbf{k}_T = \mathbf{a}_x k_{xp} + \mathbf{a}_y k_{yq}$, $k_z = \sqrt{k_0^2 - k_T^2}$, $\widehat{Y}'_{pq} = \frac{\omega \varepsilon_0}{k_z}$, and $\widehat{Y}''_{pq} = \frac{k_z}{\omega \mu_0}$. $T_x$ and $T_y$ mean the periodicities along $x$ and $y$ axes, respectively.

Note that signs of the transverse electric and magnetic fields are chosen such that the direction of the Poynting vector may direct toward $+z$ direction as expected from the scattered power flow.

Also inside the dielectric substrate region for $-d \leq z \leq 0$ between the conducting plane perforated periodically with apertures at $z = 0$ and the lossy ground plane at $= -d$, the transverse $x$ and $y$ components (to $z$) are expressed as follows:

$$E_x^c = \sum_p \sum_q \left\{ \left( V'^+_{pq} k_{xp} + V''^+_{pq} k_{yq} \right) e^{-jk_{ez} z} + \left( V'^-_{pq} k_{xp} - V''^-_{pq} k_{yq} \right) e^{jk_{ez} z} \right\} \frac{e^{j\mathbf{k}_T \cdot \boldsymbol{\rho}}}{\sqrt{k_T^2}\sqrt{T_x T_y}}$$

$$\tag{6.113a}$$

$$E_y^c = \sum_p \sum_q \left\{ \left( V'^+_{pq} k_{yp} - V''^+_{pq} k_{xp} \right) e^{-jk_{ez} z} + \left( V'^-_{pq} k_{yp} + V''^-_{pq} k_{xp} \right) e^{jk_{ez} z} \right\} \frac{e^{j\mathbf{k}_T \cdot \boldsymbol{\rho}}}{\sqrt{k_T^2}\sqrt{T_x T_y}}$$

$$\tag{6.113b}$$

$$H_x^c = \sum_p \sum_q \left\{ \left( -V'^+_{pq} Y'_{pq} k_{yq} + V''^+_{pq} Y''_{pq} k_{xp} \right) e^{-jk_{ez} z} \right.$$

$$\left. + \left( V'^-_{pq} Y'_{pq} k_{yq} + V''^-_{pq} Y''_{pq} k_{xp} \right) e^{jk_{ez} z} \right\} \frac{e^{j\mathbf{k}_T \cdot \boldsymbol{\rho}}}{\sqrt{k_T^2}\sqrt{T_x T_y}} \tag{6.113c}$$

$$H_y^c = \sum_p \sum_q \left\{ \left( V'^+_{pq} Y'_{pq} k_{xp} + V''^+_{pq} Y''_{pq} k_{yq} \right) e^{-jk_{ez} z} \right.$$

$$\left. + \left( -V'^-_{pq} Y'_{pq} k_{xp} + V''^-_{pq} Y''_{pq} k_{yq} \right) e^{jk_{ez} z} \right\} \frac{e^{j\mathbf{k}_T \cdot \boldsymbol{\rho}}}{\sqrt{k_T^2}\sqrt{T_x T_y}} \tag{6.113d}$$

where $k_{\bar{e}z} = \sqrt{k_0^2 \varepsilon_r - k_{xp}^2 - k_{yq}^2}$, $\widehat{Y}'_{pq} = \frac{\omega \varepsilon_0 \varepsilon_r}{k_{\bar{e}z}}$, and $\widehat{Y}''_{pq} = \frac{k_{\bar{e}z}}{\omega \mu_0}$ and the superscript $c$ which means the cavity region for $-d \leq z \leq 0$ has been attached to emphasize that the strong resonant field is a prerequisite for the maximum absorption by the lossy ground plane. The superscripts $+$ and $-$ mean the propagation toward $+z$ and $-z$ directions, respectively. The boundary condition on the tangential electromagnetic fields at $z = 0$ requires that

$$E_t^i + E_t^s = E_t^c, \text{ at } z = 0 \tag{6.114a}$$

$$H_t^i + H_t^s = H_t^c, \text{ over the aperture} \tag{6.114b}$$

Substitution of tangential electric components, $x$ and $y$ components in (6.111), (6.112), and (6.113) into (6.114a) yields the following expression, respectively, for TM and TE waves:

$$E''^i \sqrt{T_x T_y} + V'_{oo}R = V'_{oo} + + V'_{oo}- \tag{6.115a}$$

$$-E''^i \sqrt{T_x T_y} + V''^R_{oo} = V''^+_{oo} - V''^-_{oo} \tag{6.115b}$$

for zeroth order Floquet mode and

$$V'^R_{pq} = V'_{pq} + + V'_{pq}- \tag{6.116a}$$

$$V''^R_{pq} = V''^+_{pq} - V''^-_{pq} \tag{6.116b}$$

for higher-order Floquet modes.

Similarly, the substitution of tangential magnetic field components, in (6.111), (6.112), and (6.113) into (6.114b) yields the following expression for the continuity of the $x$-component tangential magnetic fields over the aperture:

$$\sum_p \sum_q \left\{ \left( -V'^+_{pq} + V'^-_{pq} \right) Y'_{pq} k_{yq} + \left( V''^+_{pq} + V''^-_{pq} \right) Y''_{pq} k_{xp} \right\} \frac{e^{jk_T \cdot \rho}}{\sqrt{k_T^2} \sqrt{T_x T_y}}$$

$$= \sum_p \sum_q \left( -V'^R_{pq} \widehat{Y}'_{pq} k_{yq} + V''^R_{pq} \widehat{Y}''_{pq} k_{xp} \right) \frac{e^{jk_T \cdot \rho}}{\sqrt{k_T^2} \sqrt{T_x T_y}} \tag{6.117}$$

$$+ \left( E'^i \widehat{Y}'_0 k_y^i + E''^i \widehat{Y}''_0 k_x^i \right) \frac{e^{jk_T^i \cdot \rho}}{\sqrt{\left(k_T^i\right)^2}}$$

In this equation, collecting the zeroth order Floquet terms and using the expressions for $V'^R_{oo}$ and $V''^R_{oo}$ from (6.115) leads to the zeroth-order mode equation for $p = 0$ and $q = 0$ as

$$\left(-V'^{+}_{oo} + V'^{-}_{oo}\right) Y'_{oo} k^i_y + \left(V''^{+}_{oo} + V''^{-}_{oo}\right) Y''_{oo} k^i_x$$

$$= \left\{-\left(V'^{+}_{oo} + V'^{-}_{oo}\right) \widehat{Y}'_{oo} k^i_y + \left(V''^{+}_{oo} - V''^{-}_{oo}\right) \widehat{Y}'_{oo} k^i_x\right\} \tag{6.118}$$

$$+ 2\left(E''^{i} \widehat{Y}'_{o} k^i_y + E'''^{i} \widehat{Y}''_{o} k^i_x\right) \sqrt{T_x T_y}$$

Similarly for the higher-order modes for $p \neq 0$ and $\neq 0$, using (6.116), we obtain

$$\sum_p \sum_q \left\{\left(-V'^{+}_{pq} + V'^{-}_{pq}\right) Y'_{pq} k_{yq} + \left(V''^{+}_{pq} + V''^{-}_{pq}\right) Y''_{pq} k_{xp}\right\} \frac{e^{j\frac{2\pi p}{T_x}x} \cdot e^{j\frac{2\pi q}{T_y}y}}{\sqrt{k_T^2}}$$

$$= \sum_p \sum_q \left\{-\left(V'^{+}_{pq} + V'^{-}_{pq}\right) \widehat{Y}'_{pq} k_{yq} + \left(V''^{+}_{pq} - V''^{-}_{pq}\right) \widehat{Y}''_{pq} k_{xp}\right\} \frac{e^{j\frac{2\pi p}{T_x}x} \cdot e^{j\frac{2\pi q}{T_y}y}}{\sqrt{k_T^2}}$$

$$\tag{6.119}$$

Combining (6.118) for zeroth order ($p = 0$ and $q = 0$) and (6.119) for higher order ($p \neq 0$ and $q \neq 0$) leads to a desired equation as follows:

$$\sum_p \sum_q \left\{\left(-V'^{+}_{pq} + V'^{-}_{pq}\right) Y'_{pq} k_{yq} + \left(V'^{+}_{pq} + V'^{-}_{pq}\right) \widehat{Y}'_{pq} k_{yq}\right.$$

$$\left. + \left(V''^{+}_{pq} + V''^{-}_{pq}\right) Y''_{pq} k_{xp} - \left(V''^{+}_{pq} - V''^{-}_{pq}\right) \widehat{Y}''_{pq} k_{xp}\right\} \cdot \frac{e^{j\frac{2\pi p}{T_x}x} \cdot e^{j\frac{2\pi q}{T_y}y}}{\sqrt{k_T^2}\sqrt{T_x T_y}}$$

$$= 2\left(E'^{i} \widehat{Y}'_{o} k^i_y + E''^{i} \widehat{Y}''_{o} k^i_x\right) \frac{1}{\sqrt{(k_T^i)^2}}$$

$$\tag{6.120}$$

Repeating the above procedure to the continuity condition of the $y$-component tangential magnetic fields over apertures, we obtain the remaining desired equation as

$$\sum_p \sum_q \left\{\left(V'^{+}_{pq} - V'^{-}_{pq}\right) Y'_{pq} k_{xp} - \left(V'^{+}_{pq} + V'^{-}_{pq}\right) \widehat{Y}'_{pq} k_{xp}\right.$$

$$\left. + \left(V''^{+}_{pq} + V''^{-}_{pq}\right) Y''_{pq} k_{yq} - \left(V''^{+}_{pq} - V''^{-}_{pq}\right) \widehat{Y}''_{pq} k_{yq}\right\} \cdot \frac{e^{j\frac{2\pi p}{T_x}x} e^{j\frac{2\pi q}{T_y}y}}{\sqrt{k_T^2}\sqrt{T_x T_y}}$$

$$= 2\left(-E'^{i} \widehat{Y}'_{o} k^i_x + E''^{i} \widehat{Y}''_{o} k^i_y\right) \frac{1}{\sqrt{(k_T^i)^2}}$$

$$\tag{6.121}$$

If we put the equivalent magnetic current density $M$ over the aperture, i.e., on the side of $z = 0^+$, its relation to the electric field $E$ over the aperture is given by

$$E \times n(= -a_z) = M \tag{6.122}$$

Expressing $M$ as

$$M(x,y) = \{M_x(x,y)a_x + M_y(x,y)a_y\}e^{jk_T^i \cdot \rho} \tag{6.123}$$

and using (6.113a) and (6.113b) based on the continuity condition of the tangential electric fields across the aperture, one obtains the following two expressions:

$$\sum_p \sum_q \left\{ \left(V_{pq}^{'+}k_{xp} + V_{pq}^{''+}k_{yq}\right) + \left(V_{pq}^{'-}k_{xp} - V_{pq}^{''-}k_{yq}\right)\right\} \frac{e^{jk_T \cdot \rho}}{\sqrt{k_T^2}\sqrt{T_x T_y}}$$
$$= -M_y(x,y)e^{jk_T^i \cdot \rho} \tag{6.124a}$$

$$\sum_p \sum_q \left\{ \left(V_{pq}^{'+}k_{yq} - V_{pq}^{''+}k_{xp}\right) + \left(V_{pq}^{'-}k_{yq} + V_{pq}^{''-}k_{xp}\right)\right\} \frac{e^{jk_T \cdot \rho}}{\sqrt{k_T^2}\sqrt{T_x T_y}}$$
$$= M_x(x,y)e^{jk_T^i \cdot \rho} \tag{6.124b}$$

where the two relations

$$E_x = -M_y(x,y)e^{jk_T^i \cdot \rho} \tag{6.125a}$$

$$E_y = M_x(x,y)e^{jk_T^i \cdot \rho} \tag{6.125b}$$

have been used.

Multiplying both sides by $e^{-j\frac{2\pi p'}{T_x}x}e^{-j\frac{2\pi q'}{T_y}y}$ in the two equations (6.124a) and (6.124b) and integrating over one cell area of $T_x T_y$ lead to the following desired two equations:

$$\left\{ \left(V_{pq}^{'+}k_{xp} + V_{pq}^{''+}k_{yq}\right) + \left(V_{pq}^{'-}k_{xp} - V_{pq}^{''-}k_{yq}\right)\right\} \frac{1}{\sqrt{k_T^2}\sqrt{T_x T_y}} = -\tilde{M}_y \tag{6.126a}$$

$$\left\{ \left(V_{pq}^{'+}k_{yq} - V_{pq}^{''+}k_{xp}\right) + \left(V_{pq}^{'-}k_{yq} + V_{pq}^{''-}k_{xp}\right)\right\} \frac{1}{\sqrt{k_T^2}\sqrt{T_x T_y}} = \tilde{M}_x \tag{6.126b}$$

where $\tilde{M}_T = \frac{1}{T_x T_y}\int_{cell}M_T(x,y)e^{-j\frac{2\pi p}{T_x}x}e^{-j\frac{2\pi q}{T_y}y}dxdy$, here the subscript $T$ in $M_T$ means $x$ or $y$ and the integration range of cell means $T_x T_y$.

Finally, we consider the desired equation relevant to the boundary condition of the lossy ground at $z = -d$. Here it would be convenient to define the ratio of $\frac{V_{pq}^{'+}}{V_{pq}^{'-}}$

and $\frac{V''^{+}_{pq}}{V''^{-}_{pq}}$ at $z = 0^{-}$ in terms of the reflection coefficients $\Gamma'_{pq}$ and $-\Gamma''_{pq}$ at $z = -d$ as follows:

$$\frac{V'^{+}_{pq}}{V'^{-}_{pq}} = \frac{Y'_{pq} - Y'_{pq.l}}{Y'_{pq} + Y'_{pq.l}}e^{-j2k_{ez}d} = \frac{Z'_{pq.l}Y'_{pq} - 1}{Z'_{pq.l}Y'_{pq} + 1}e^{-j2k_{ez}d} = \Gamma'_{pq}e^{-j2k_{ez}d} \qquad (6.127a)$$

and

$$\frac{V''^{+}_{pq}}{V''^{-}_{pq}} = -\frac{Y''_{pq} - Y''_{pq.l}}{Y''_{pq} + Y''_{pq.l}}e^{-j2k_{ez}d} = \frac{Z''_{pq.l}Y''_{pq} - 1}{Z''_{pq.l}Y''_{pq} + 1}e^{-j2k_{ez}d} = \Gamma''_{pq}e^{-j2k_{ez}d}, \qquad (6.127b)$$

where $Y'_{pq}\left(Y''_{pq}\right)$ are the characteristic admittance of the medium 2 between the FSS plane at $z = 0$ and lossy ground plane at $z = -d$ for the TM(TE) mode case as mentioned above and $Y'_{pq.l}\left(Y''_{pq.l}\right)$ means the characteristic admittance of the lossy ground medium for the TM(TE) mode, which is given by

$$Y'_{pq.l} = \frac{(\omega\varepsilon_0 - j\sigma)}{k_z} \qquad (6.128a)$$

$$Y''_{pq.l} = \frac{k_z}{\omega\mu_0} \qquad (6.128b)$$

Here $k_z = \sqrt{k_0^2\left(1 - \frac{j\sigma}{\omega\varepsilon_0}\right) - k_{xp}^2 - k_{yq}^2}$, $Z'_{pq.l}\left(Z''_{pq.l}\right)$ means the characteristic impedance of the lossy ground medium for TM(TE) mode as a reciprocal of $Y'_{pq.l}\left(Y''_{pq.l}\right)$, and the thickness of the lossy ground is assumed to be infinite. Recall that the material constants of the dielectric slab between the FSS plane and lossy ground plane are $(\varepsilon_0\varepsilon_r, \mu_0)$ and those for the loss ground are $(\varepsilon_0, \mu_0, \sigma)$. All the material constants follow the usual meaning. Using (6.127) in (6.126), after some algebra, one obtains the expressions for $V'^{-}_{pq}$, $V'^{+}_{pq}$, $V''^{-}_{pq}$, and $V''^{+}_{pq}$ in terms of $M_x$ and $M_y$ as follows:

$$V'^{-}_{pq} = \frac{\left(\tilde{M}_x k_{yq} - \tilde{M}_y k_{xp}\right)}{\left(1 + \Gamma'_{pq}e^{-j2k_{ez}d}\right)} \cdot \frac{\sqrt{T_x T_y}}{\sqrt{k_T^2}} \qquad (6.129a)$$

$$V'^{+}_{pq} = \frac{\left(\tilde{M}_x k_{yq} - \tilde{M}_y k_{xp}\right)\Gamma'_{pq}e^{-j2k_{ez}d}}{\left(1 + \Gamma'_{pq}e^{-j2k_{ez}d}\right)} \cdot \frac{\sqrt{T_x T_y}}{\sqrt{k_T^2}} \qquad (6.129b)$$

$$V''^{-}_{pq} = \frac{\left(\tilde{M}_x k_{xp} + \tilde{M}_y k_{yq}\right)}{\left(1 + \Gamma''_{pq}e^{-j2k_{ez}d}\right)} \cdot \frac{\sqrt{T_x T_y}}{\sqrt{k_T^2}} \qquad (6.129c)$$

$$V''^{+}_{pq} = \frac{-\left(\tilde{M}_x k_{xp} + \tilde{M}_y k_{yq}\right)\Gamma''_{pq}e^{-j2k_{ez}d}}{\left(1 + \Gamma''_{pq}e^{-j2k_{ez}d}\right)} \cdot \frac{\sqrt{T_x T_y}}{\sqrt{k_T^2}} \qquad (6.129d)$$

Inserting (6.129a)–(6.129d) into (6.120) and (6.121) and using the modal admittance for TM and TE waves looking into the dielectric region from $z = 0^-$ plane as given below:

$$\vec{Y}'_{pq} = Y'_{pq} \frac{1 - \Gamma'_{pq}}{1 + \Gamma'_{pq}} \tag{6.130a}$$

$$\vec{Y}''_{pq} = Y''_{pq} \frac{1 - \Gamma''_{pq}}{1 + \Gamma''_{pq}} \tag{6.130b}$$

leads to the following expressions for the desired linear equations:

$$\sum_p \sum_q \left\{ G^{xx}_{pq} \tilde{M}_x + G^{xy}_{pq} \tilde{M}_y \right\} e^{j\frac{2\pi p}{T_x}x} e^{j\frac{2\pi q}{T_y}y} = S_x \tag{6.131a}$$

$$\sum_p \sum_q \left\{ G^{yx}_{pq} \tilde{M}_x + G^{yy}_{pq} \tilde{M}_y \right\} e^{j\frac{2\pi p}{T_x}x} e^{j\frac{2\pi q}{T_y}y} = S_y \tag{6.131b}$$

where $G^{xx}_{pq} = \dfrac{\left( k^2_{xp} \overset{\leftrightarrow}{Y}''_{pq} + k^2_{yq} \overset{\leftrightarrow}{Y}'_{pq} \right)}{k^2_T}$

$G^{xy}_{pq} = G^{yx}_{pq} = \dfrac{k_{xp} k_{yq} \left( \overset{\leftrightarrow}{Y}''_{pq} - \overset{\leftrightarrow}{Y}'_{pq} \right)}{k^2_T}$

$S_x = \dfrac{2\left( E'^i \widehat{Y_0'} k^i_y + E'''^i \widehat{Y_0} k^i_x \right)}{\sqrt{k^2_T}}$

$G^{yx}_{pq} = \dfrac{k_{xp} k_{yq} \left( \overset{\leftrightarrow}{Y}''_{pq} - \overset{\leftrightarrow}{Y}'_{pq} \right)}{k^2_T}$

$G^{yy}_{pq} = \dfrac{\left( k^2_{yq} \overset{\leftrightarrow}{Y}''_{pq} + k^2_{xp} \overset{\leftrightarrow}{Y}'_{pq} \right)}{k^2_T}$

$S_y = \dfrac{2\left( -E'^i \widehat{Y_0'} k^i_x + E'''^i \widehat{Y_0''} k^i_y \right)}{\sqrt{k^2_T}}$

In (6.130), $Y'_{pq}$ and $Y''_{pq}$ designates the modal admittances used for transverse electromagnetic fields for TM and TE modes in (6.113a)–(6.113d).

In the above expression, $\vec{Y}'_{pq}$ and $\vec{Y}''_{pq}$ mean the sum of modal admittances $\widehat{Y}'_{pq} + \vec{Y}'_{pq}$ and $\widehat{Y}''_{pq} + \vec{Y}''_{pq}$, respectively, for TM and TE modes. Recall that $\widehat{Y}'_{pq} \left( \widehat{Y}''_{pq} \right)$ and $\vec{Y}'_{pq} (\vec{Y}''_{pq})$ mean the modal admittance looking into the upper region from $z = 0^+$ and into the lower dielectric-filled region from $z = 0^-$ for TM (TE) mode case.

### 6.11.4    Choice of basis function $\overline{M_T}$ and its Fourier transform $\widetilde{M_T}$

Let us consider a solving method for the linear equations in (6.131a) and (6.131b). To this end, the aperture chosen to be rectangular for analysis convenience is divided into rectangular cells whose each size is $\Delta x \Delta y$. As basis functions for the equivalent surface magnetic current, a rooftop function is selected. That is, two adjacent rectangular cells, sharing a common border perpendicular to the $x(y)$-direction, for example, will form an $x(y)$-directed current cell as shown in Figure 6.10.

For the rectangular aperture with $m \times n$ rectangular cells, the number of $x$-directed current cells is $M = (m-1)n$ and that of $y$-directed current cells $N = m(n-1)$ as seen in Figure 6.10. The magnetic current density $\boldsymbol{M_T}$ as unknown is expressed as two vector components each of which can be given as sum [29] of multiplication of the rooftop function $\Lambda_n$ and unit pulse function $\Pi_n$ as follows:

$$
\overline{\boldsymbol{M}}_T = \boldsymbol{M}_x + \boldsymbol{M}_y
$$

$$
= \boldsymbol{a}_x \sum_{m=0}^{M_1-2N_1-1} \sum_{n=0}^{} \dot{M}_{xmn} \Lambda_m(x) \Pi_n(y) + \boldsymbol{a}_y \sum_{m=0}^{M_1-1N_1-2} \sum_{n=0}^{} \dot{M}_{ymn} \Pi_m(x) \Lambda_n(y) \tag{6.132}
$$

where

$$
\Lambda_n(x) = \begin{cases} \dfrac{1}{\Delta x}(x - m\Delta x), & \text{for } m\Delta x \le x \le (m+1)\Delta x \\[2ex] \dfrac{-1}{\Delta x}\{x - (m+2)\Delta x\}, & \text{for } (m+1)\Delta x \le x \le (m+2)\Delta x \end{cases}
$$

$$
\Pi_n(y) = 1, \quad \text{for } n\Delta y \le y \le (n+1)\Delta y
$$

Using the above expression for $\overline{\boldsymbol{M}}_T$ in (6.132) for $\widetilde{M_T}$ defined in (6.126) and inserting the expression for $\widetilde{M_T}$ into (6.131), and arranging the resultant equation

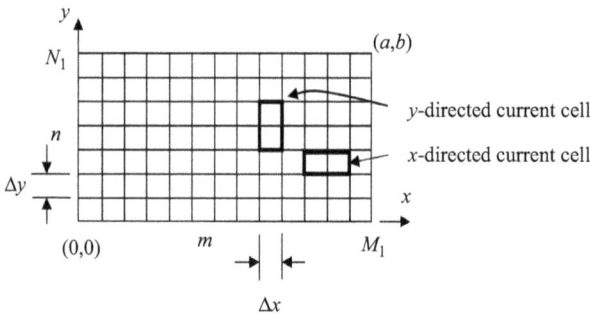

*Figure 6.10    Segmentation of the rectangular aperture $(a \times b)$ and $x(y)$-directed current cell*

such that it may be facilitated to get the desired linear equation whose unknowns are $\{\dot{M}_{xmn}, \dot{M}_{ymn}\}$, one obtains

$$
\sum_{m=0}^{M_1-2N_1-1}\sum_{n=0} \dot{M}_{xmn}\left(\sum_p\sum_q G_{pq}^{xx}\widetilde{\Lambda}_{mp}\widetilde{\Pi}_{nq}e^{j\frac{2\pi p}{T_x}x}e^{j\frac{2\pi q}{T_y}y}\right)
$$
$$
+\sum_{m=0}^{M_1-1N_1-2}\sum_{n=0}\dot{M}_{ymn}\left(\sum_p\sum_q G_{pq}^{xy}\widetilde{\Pi}_{mp}\widetilde{\Lambda}_{nq}e^{j\frac{2\pi p}{T_x}x}e^{j\frac{2\pi q}{T_y}y}\right)=S_x
$$

(6.133a)

$$
\sum_{m=0}^{M_1-2N_1-1}\sum_{n=0} \dot{M}_{xmn}\left(\sum_p\sum_q G_{pq}^{yx}\widetilde{\Lambda}_{mp}\widetilde{\Pi}_{nq}e^{j\frac{2\pi p}{T_x}x}e^{j\frac{2\pi q}{T_y}y}\right)
$$
$$
+\sum_{m=0}^{M_1-1N_1-2}\sum_{n=0}\dot{M}_{ymn}\left(\sum_p\sum_q G_{pq}^{yy}\widetilde{\Pi}_{mp}\widetilde{\Lambda}_{nq}e^{j\frac{2\pi p}{T_x}x}e^{j\frac{2\pi q}{T_y}y}\right)=S_y
$$

(6.133b)

where $\widetilde{\Lambda}_{mp}=\frac{1}{T_x}\int_{ap}\Lambda_m(x)e^{-j\frac{2\pi p}{T_x}x}dx$

$$\widetilde{\Pi}_{nq}=\frac{1}{T_y}\int_{ap}\Pi_n(y)e^{-j\frac{2\pi q}{T_y}y}dy$$

$$\widetilde{\Lambda}_{nq}=\frac{1}{T_y}\int_{ap}\Lambda_n(y)e^{-j\frac{2\pi q}{T_y}y}dy$$

$$\widetilde{\Pi}_{mp}=\frac{1}{T_x}\int_{ap}\Pi_m(x)e^{-j\frac{2\pi p}{T_x}x}dx$$

As a final step, performing the Galerkin's testing procedure in (6.133a) and (6.133b), one can obtain the desired two coupled linear equations as follows:

$$
\sum_{m=0}^{M_1-2N_1-1}\sum_{n=0}\dot{M}_{xmn}\left\{\sum_p\sum_q\left(G_{pq}^{xx}\widetilde{\Lambda}_{mp}\widetilde{\Pi}_{nq}\widetilde{\Lambda}^*_{sp}\widetilde{\Pi}^*_{tq}\right)\right\}
$$
$$
+\sum_{m=0}^{M_1-1N_1-2}\sum_{n=0}\dot{M}_{ymn}\left\{\sum_p\sum_q\left(G_{pq}^{xy}\widetilde{\Pi}_{mp}\widetilde{\Lambda}_{nq}\widetilde{\Lambda}^*_{sp}\widetilde{\Pi}^*_{tq}\right)\right\}=S_x\widetilde{\Lambda}^*_{s0}\widetilde{\Pi}^*_{t0}
$$

(6.134a)

$$
\sum_{m=0}^{M_1-2N_1-1}\sum_{n=0}\dot{M}_{xmn}\left\{\sum_p\sum_q\left(G_{pq}^{yx}\widetilde{\Lambda}_{mp}\widetilde{\Pi}_{nq}\widetilde{\Pi}^*_{sp}\widetilde{\Lambda}^*_{tq}\right)\right\}
$$
$$
+\sum_{m=0}^{M_1-1N_1-2}\sum_{n=0}\dot{M}_{ymn}\left\{\sum_p\sum_q\left(G_{pq}^{yy}\widetilde{\Pi}_{mp}\widetilde{\Lambda}_{nq}\widetilde{\Pi}^*_{sp}\widetilde{\Lambda}^*_{tq}\right)\right\}=S_y\widetilde{\Pi}^*_{s0}\widetilde{\Lambda}^*_{t0}
$$

(6.134b)

Casting this into the convenient matrix form, it is written as

$$
\begin{bmatrix} U_{xx,mnst} & U_{xy,mnst} \\ U_{yx,mnst} & U_{yy,mnst} \end{bmatrix} \begin{bmatrix} \dot{M}_{xmn} \\ \dot{M}_{ymn} \end{bmatrix} = \begin{bmatrix} S_x \widetilde{\Lambda}^*_{s0} \widetilde{\Pi}^*_{t0} \\ S_y \widetilde{\Pi}^*_{s0} \widetilde{\Lambda}^*_{t0} \end{bmatrix}
\tag{6.135}
$$

where

$$
U_{xx,mnst} = \sum_p \sum_q \left( G^{xx}_{pq} \widetilde{\Lambda}_{mp} \widetilde{\Pi}_{nq} \widetilde{\Lambda}^*_{sp} \widetilde{\Pi}^*_{tq} \right)
$$

$$
U_{xy,mnst} = \sum_p \sum_q \left( G^{xy}_{pq} \widetilde{\Pi}_{mp} \widetilde{\Lambda}_{nq} \widetilde{\Lambda}^*_{sp} \widetilde{\Pi}^*_{tq} \right)
$$

$$
U_{yx,mnst} = \sum_p \sum_q \left( G^{yx}_{pq} \widetilde{\Lambda}_{mp} \widetilde{\Pi}_{nq} \widetilde{\Pi}^*_{sp} \widetilde{\Lambda}^*_{tq} \right)
$$

$$
U_{yy,mnst} = \sum_p \sum_q \left( G^{yy}_{pq} \widetilde{\Pi}_{mp} \widetilde{\Lambda}_{nq} \widetilde{\Pi}^*_{sp} \widetilde{\Lambda}^*_{tq} \right)
$$

The subscript, *mnst*, for example, means the *y*-component magnetic field at the subinterval corresponding to two adjacent cells where one rooftop function is defined at $(s,t)$ th subinterval due to the *x*-component magnetic current at the subinterval order set of $(m,n)$. Here $(m,n)$ is the subcell order set for basis expansion and $(s,t)$ is the subcell order set for testing. As discussed above, the subcell comprises *x*- and *y*-directed unknown magnetic surface current densities. Note that the subcell size of the rooftop function for the *x*-directed magnetic current density is $(2\Delta x)(\Delta y)$, whereas the subcell size of the rooftop function for the *y*-directed magnetic current density is $(\Delta x)(2\Delta y)$.

Let us consider one example of the matrix composition in which case the number of the single segment cell along the *x*-direction is $M_1 = M$ and the number of the single segment cell along the *y*-direction is $N_1 = N$ as shown in Figure 6.10. In this case, the matrix representation for the desired linear equation is given as

$$
\begin{bmatrix} [U_{xx}] & [U_{xy}] \\ [U_{yx}] & [U_{yy}] \end{bmatrix} \begin{bmatrix} [M_x] \\ [M_y] \end{bmatrix} = \begin{bmatrix} [S'_x] \\ [S'_y] \end{bmatrix},
\tag{6.136}
$$

where the submatrices are given by

$$
[U_{xx}] = \begin{bmatrix}
U_{xx,0000} & U_{xx,0100} \cdots\cdots & U_{xx,(M-2)(N-1)00} \\
U_{xx,0001} & U_{xx,0101} \cdots\cdots & U_{xx,(M-2)(N-1)01} \\
\vdots & \vdots & \vdots \\
\vdots & \vdots & \vdots \\
U_{xx,00(M-2)(N-1)} & U_{xx,01(M-2)(N-1)} \cdots\cdots & U_{xx,(M-2)(N-1)(M-2)(N-1)}
\end{bmatrix}
$$

$$
[U_{xy}] = \begin{bmatrix}
U_{xy,0000} & U_{xy,0100}\cdots\cdots & U_{xy,(M-1)(N-2)00} \\
U_{xy,0001} & U_{xy,0101}\cdots\cdots & U_{xy,(M-1)(N-2)01} \\
\vdots & \vdots & \vdots \\
\vdots & \vdots & \vdots \\
U_{xy,00(M-2)(N-1)} & U_{xy,01(M-2)(N-1)}\cdots\cdots & U_{xy,(M-1)(N-2)(M-2)(N-1)}
\end{bmatrix}
$$

$$
[U_{yx}] = \begin{bmatrix}
U_{yx,0000} & U_{yx,0100}\cdots\cdots & U_{yx,(M-2)(N-1)00} \\
U_{yx,0001} & U_{yx,0101}\cdots\cdots & U_{yx,(M-2)(N-1)01} \\
\vdots & \vdots & \vdots \\
\vdots & \vdots & \vdots \\
U_{yx,00(M-1)(N-2)} & U_{yx,01(M-1)(N-2)}\cdots\cdots & U_{yx,(M-2)(N-1)(M-1)(N-2)}
\end{bmatrix}
$$

$$
[U_{yy}] = \begin{bmatrix}
U_{yy,0000} & U_{yy,0100}\cdots\cdots & U_{yy,(M-1)(N-2)00} \\
U_{yy,0010} & U_{yy,0110}\cdots\cdots & U_{yy,(M-1)(N-2)01} \\
\vdots & \vdots & \vdots \\
\vdots & \vdots & \vdots \\
U_{yy,00(M-1)(N-2)} & U_{yy,01(M-1)(N-2)}\cdots\cdots & U_{yy,(M-1)(N-2)(M-1)(N-2)}
\end{bmatrix}
$$

$$
[M_x] = \begin{bmatrix}
M_{x00} \\
M_{x01} \\
M_{x02} \\
\vdots \\
M_{x0(N-1)} \\
M_{x10} \\
M_{x11} \\
\vdots \\
M_{x1(N-1)} \\
M_{x20} \\
\vdots \\
M_{x2(N-1)} \\
\vdots \\
M_{x(M-2)(N-1)}
\end{bmatrix}
\qquad
[M_y] = \begin{bmatrix}
M_{y00} \\
M_{y01} \\
M_{y02} \\
\vdots \\
M_{y0(N-2)} \\
M_{y10} \\
M_{y11} \\
\vdots \\
M_{y1(N-2)} \\
M_{y20} \\
\vdots \\
M_{y2(N-2)} \\
\vdots \\
M_{y(M-1)(N-2)}
\end{bmatrix}
$$

and

$$[S'_x] = \begin{bmatrix} S_x\tilde{\Lambda}^*_{00}\tilde{\Pi}^*_{00} \\ S_x\tilde{\Lambda}^*_{00}\tilde{\Pi}^*_{10} \\ S_x\tilde{\Lambda}^*_{00}\tilde{\Pi}^*_{20} \\ \vdots \\ S_x\tilde{\Lambda}^*_{00}\tilde{\Pi}^*_{(N-1)0} \\ S_x\tilde{\Lambda}^*_{10}\tilde{\Pi}^*_{00} \\ \vdots \\ S_x\tilde{\Lambda}^*_{10}\tilde{\Pi}^*_{(N-1)0} \\ \vdots \\ \vdots \\ S_x\tilde{\Lambda}^*_{(M-2)0}\tilde{\Pi}^*_{(N-1)0} \end{bmatrix} \quad [S'_y] = \begin{bmatrix} S_y\tilde{\Lambda}^*_{00}\tilde{\Pi}^*_{00} \\ S_y\tilde{\Lambda}^*_{00}\tilde{\Pi}^*_{10} \\ S_y\tilde{\Lambda}^*_{00}\tilde{\Pi}^*_{20} \\ \vdots \\ S_y\tilde{\Lambda}^*_{00}\tilde{\Pi}^*_{(N-2)0} \\ S_y\tilde{\Lambda}^*_{10}\tilde{\Pi}^*_{00} \\ \vdots \\ S_y\tilde{\Lambda}^*_{10}\tilde{\Pi}^*_{(N-2)0} \\ \vdots \\ \vdots \\ S_y\tilde{\Lambda}^*_{(M-1)0}\tilde{\Pi}^*_{(N-2)0} \end{bmatrix}$$

Note that in the cell numbering for $\{M_x\}$, for example, with the first cell order of $(0,0)$ at the bottom left corner of the rectangular aperture under consideration, the order is numbered in the vertical ($y$) direction to the $(0, N-1)$ and then goes to $(1,0)$ which corresponds to the bottom of just the right side column and repeat the same procedure of ordering to the last $M$th column in Figure 6.10.

As mentioned already, two adjacent rectangular cells, sharing a common border perpendicular to $x(y)$-direction will form an $x(y)$-directed rooftop magnetic current cell. So an automatic overlapping of magnetic current (rooftop) cells is obtained in this manner. As a result for rectangular aperture with $M$(in $x$-direction) and $N$(in $y$-direction) cells, the number of $x$-directed rooftop functions (magnetic current cells) is $(M-1)N$ and that of $y$-directed magnetic current cells $M(N-1)$. So for the left end column corresponding to the real left edge of the aperture under consideration, the $x$-directed magnetic current (rooftop) cell is first numbered from $(0,0)$ to $(0, N-1)$ and the next procedure of numbering from $(1,0)$ to $(1, N-1)$ for the next column. The same way of numbering is repeated till the right end column is numbered from $(M-2,0)$ to $(M-2, N-1)$.

The reason why the first number for the last cell order of $(M-2, N-1)$ for the $x$-directed magnetic current cell is given as $(M-2)$ is that the numbering starts with zeroth counting (not first counting) and two adjacent rectangular cells, sharing a common border perpendicular to $x$-direction will form an $x$-directed rooftop magnetic current cell, and two adjacent rooftop cells for $\{M_x\}$ are defined to be overlapped over one rectangular cell.

Also for the $y$-directed magnetic current densities, similar to above, the two adjacent rectangular cells, sharing a common border perpendicular to the $y$-direction will form a $y$-directed rooftop magnetic current cell and two adjacent rooftop cells for $\{M_y\}$ are defined to be overlapped over one rectangular cell. So the $y$-directed magnetic current rooftop cells are first numbered from $(0,0)$ to $(0, N-2)$ for the left end column and next from $(1,0)$ to $(1, N-2)$ for just the next right column, and the same way of numbering is repeated till the right end column is numbered from $(M - 1, 0)$ to $(M - 1, N - 2)$.

So for the rectangular aperture, if segmented into $M$ (along $x$-direction) $\times N$ ( along $y$-direction) cells, the total number of the magnetic current cells (corresponding to the rooftop functions) [29] is $(M - 1)N + M(N - 1)$ whereas the total number of the magnetic charge cells is $MN$. The pair of subscripts in $\{M_x\}$ and $\{M_y\}$ has been numbered based on this method. The pair of first subscripts of $\Lambda^*$ and $\Pi^*$ also follow the same numbering method as above.

## 6.11.5    Numerical results and discussions on absorbing FSS structure

One of the structures of the radar absorbing materials (RAM) [27,30] which have been widely studied for their various applications [31] in stealth technology, communication antennas, and anechoic chambers, as well as electromagnetic interference (EMI) reduction and electromagnetic compatibility (EMC) problem solver, is an absorbing frequency selective surface (FSS). As is well known, it is composed of a lossy frequency selective surface in the vicinity of the perfect electric conducting (PEC) plane. In previous research, the primary goal was to achieve the characteristics of the Salisbury [32] and Taumann [33] screens by employing a slimmer design. Here we deal with the various physical characteristics of the absorbing FSS structure whose analysis method has been considered in the foregoing section. The present structure is different from the conventional one in that the former comprises a combination of perfect conducting FSS and lossy ground structures whereas the latter comprises a combination of lossy FSS surface and perfect conducting ground.

In addition, the present structure can be taken as the case where a periodic lossless FSS is used as a replacement for the perfect conducting ground in the original Salisbury structure and the incident wave impinges upon the FSS side. Of course, the lossy ground in the present structure is assumed to be thick enough to block the incident electromagnetic energy. It seems to be rare to deal with the present structure in the literature about this subject. What is more, the cavity is formed between the lossless FSS and lossy ground plane of this structure and a strong electromagnetic field can be established. For this reason, this structure has been used for a basic architecture of photovoltaic devices [34] where the FSS patterned region is used as an anode and the bottom of the lossy ground is used as a cathode.

As will be seen later, the use of a properly combined structure of the FSS, for example, rectangular loop type aperture and the lossy ground plane allows us to design a novel compact RAM family similar to the metamaterial-inspired design [35,36]. In addition, for broadening bandwidth characteristics of the Jaumann type absorber, the employment of the FSS structure whose element is a planar multi-loop type [37,38] of apertures is thought to be a good solution. As a unit cell of aperture type of FSS, two cases of rectangular aperture and rectangular loop type of aperture are considered as shown in Figure 6.11. First we investigate near-perfect

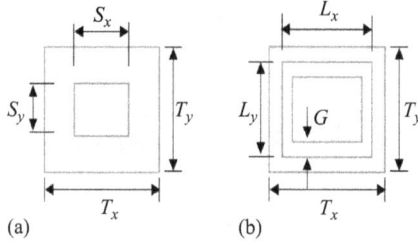

*Figure 6.11   Two unit cell geometries under consideration. (a) Rectangular aperture. (b) Rectangular loop type of aperture*

absorption phenomena which are observed in the rectangular aperture array backed by a lossy (resistive) ground.

The geometrical parameters and material constants are as follows: the slot size is $S_x \times S_y = 10$ mm $\times$ 4 mm, the periodicities $T_x$ and $T_y$ along the $x$ and $y$ directions are $T_x = T_y = 20$ mm. The separation $d$ between the FSS plane and the lossy ground plane is $d = 22$ mm. The relative dielectric constant $\varepsilon_r$ of the dielectric substrate material between the two planes is $\varepsilon_r = 1.5$ and the conductivity $\sigma$ of the lossy ground is $\sigma = 30.7$ s/m. For the numerical example normal incidence case in which $\theta = 0$ and $\phi = 0$ is considered, i.e., the incident electric field vector is chosen to be perpendicular to the long side edge ($x$-axis) of the aperture structure. Through the use of the above MoM approach and the matrix inversion in (6.136), the unknown elements of the column matrices $\{M_x\}$ and $\{M_y\}$ can be obtained. From knowledge of the elements of $\{M_x\}$ and $\{M_y\}$, $V'^{-}_{pq}, V'^{+}_{pq}, V''^{-}_{pq}$, and $V''^{+}_{pq}$ can be found by use of (6.129a)–(6.129d). Next imposing the boundary condition of tangential electric fields across the aperture by use of (6.115a) and (6.115b) for the zeroth-order Floquet mode, we can get the electric field reflection coefficient by use of (6.112a) and (6.112b).

Figure 6.12 shows the absorption characteristics versus frequencies for the geometrical parameter and material constants given above and the comparison between the present MoM results and those obtained by commercially available software such as HFSS and CST MWS. A comparison shows good agreements.

Note that, to use commercial software such as HFSS and CST MWS for calculating the absorption characteristics for example as in Figure 6.12, we need to specify the thickness of the lossy ground plane. To this end, we have only to choose the thickness of the ground plane to be at least greater than one and half times the skin depth because penetrated electromagnetic fields into the lossy conducting medium exponentially decay to zero. Here taking this matter into account, the thickness of the lossy conducting ground is chosen to be 2 mm at the operating frequency of 10 GHz.

To assess the effect of the conductivity $\sigma$ on the absorption characteristics of the structure under consideration in Figure 6.9, we have investigated the reflection characteristics for the case that $T_x = T_y = 15$ mm (square lattice) and $S_x = S_y = 10$ mm (square aperture) under the normal incidence of 10 GHz plane

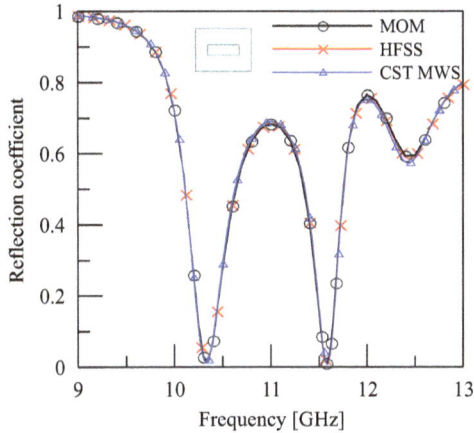

*Figure 6.12*    *Absorption characteristics versus frequencies obtained by the present MoM and comparison with those obtained by use of commercial software such as HFSS and CST MWS*

wave whose electric field is parallel to the $y$-axis. Under such a condition, the geometrical parameter $d$ and material constant $\sigma$ for near-zero reflection is searched while varying the parameters $d$ and $\sigma$. As a result, the dielectric thickness $d$ and the conductivity of the lossy ground are chosen to be $d = 8.57$ mm and $\sigma = 24.86$ s/m, respectively. For reference, the reflection coefficients for various values of $\sigma$ have been investigated and illustrated in Figure 6.13.

In Figure 6.13, the curve for $\sigma = 0$ corresponds to the reflection characteristics of the transmission FSS loaded with dielectric slab only without the lossy ground plane, which shows total transmission, i.e., near-zero reflection at a frequency of 13.95 GHz. When the above transmission FSS is backed by a lossy ground plane and the operating frequency becomes deviated from the above transmission resonance frequency of 13.95 GHz, the resonant cavity can be formed between the FSS plane and the lossy ground plane and so a strong mode field is established. As a result, the near-zero reflection due to perfect absorption by the lossy ground layer can be achieved as shown in Figure 6.13. It is seen that for all three cases of non-zero $\sigma (= 5, 24.86, 60 \left[\frac{S}{m}\right])$ the curves of reflection characteristics versus frequencies go through two minimum points, near 10 GHz and 16 GHz. The two minimum points are also seen to move somewhat to higher frequencies as the conductivity $\sigma$ of the lossy ground layer increases.

To clarify the effects of the aperture size $S_x \times S_y$, periodicities $T_x$ and $T_y$, dielectric thickness $d$ on the reflection (or absorption) characteristics, we have investigated the reflection characteristics versus frequencies with each geometrical or material constant as a parameter while keeping the conductivity $\sigma = 24.86$ S/m of the lossy ground layer fixed. In all these cases the two minimum points of the reflection curves are observed.

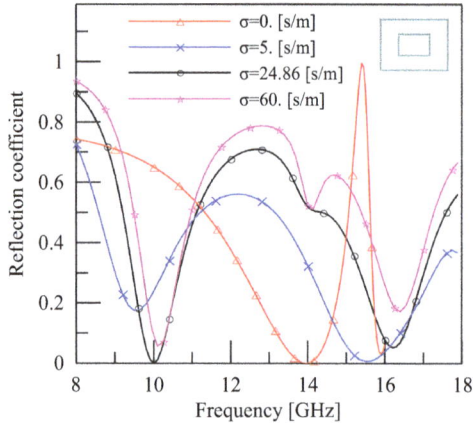

*Figure 6.13    Calculated reflection coefficients versus frequencies with conductivity σ of the lossy ground as a parameter for the square array of the square aperture*

Figure 6.14(a) shows a reflection coefficient versus frequencies which has been obtained with the periodicity $T_x = T_y$ a parameter while fixing the aperture size $S_x = S_y = 10$ mm as a constant. It is interesting to note that the reflection minimum (dip) frequency of the lower frequency side of 10 GHz remains almost the same whereas the reflection minimum frequency of the higher frequency side around 16 GHz significantly changes as the periodicity is varied. This means that the reflection minimum (dip) of the lower frequency side depends entirely on the size of the aperture, in particular, the resonant length of $S_x$ corresponding to the half-wavelength dipole. On the other hand, the reflection minimum of the higher frequency side is mainly determined by the periodicity of the array. It is worthwhile to mention that this dependence of the absorption peaks (reflection minimum) on two ingredients, aperture element resonance and periodicity, is similar to that of the transmission resonance in the previous extraordinary optical transmission (EOT) problem to the same two ingredients [39,40].

Figure 6.14(b) shows the effect of the dielectric thickness corresponding to the separation between the periodic aperture (FSS) plane and the lossy ground plane on the reflection (or absorption) characteristics versus frequencies. As $d$ increases, both the two reflection minimum frequencies are observed to move to the lower frequency side as expected from the fact that the resonant cavity is formed between the FSS and lossy ground plane. To examine the effect of the aperture resonance on the reflection (or absorption) characteristics, we have investigated the reflection characteristics with aperture width $S_x$ and aperture length $S_y$ as parameters while keeping $T_x = T_y = 15$ mm, $d = 8.57$ mm, $\varepsilon_r = 2$, and $\sigma = 24.86$ S/m.

If we obtain the reflection characteristics with the aperture length $S_x$ as a parameter while fixing the aperture width $S_y$ as a constant, both the reflection minimum frequencies of lower and higher sides are seen to vary together as shown

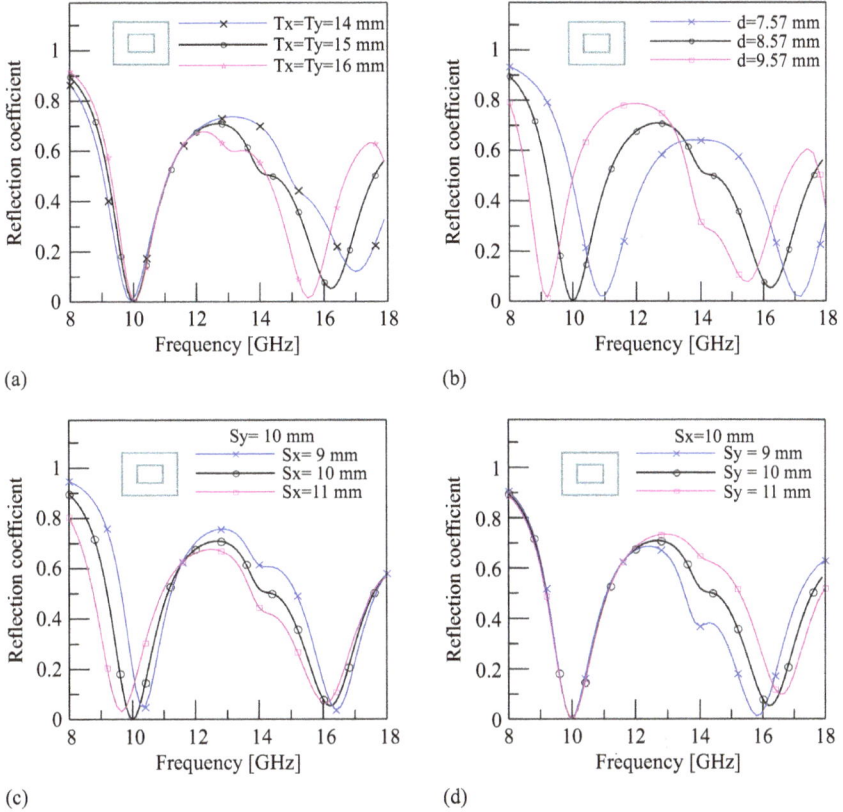

*Figure 6.14    Reflection coefficients versus frequencies obtained (a) with a periodicity $T_x = T_y$ as a parameter, (b) with a dielectric thickness d as a parameter, (c) with an aperture width $S_y$ as a parameter, and (d) with an aperture length $S_x$ as a parameter*

in Figure 6.14(c). In contrast to this, if we obtain the reflection characteristics with the aperture width $S_y$ as a parameter for the aperture length $S_x$ as a constant, the reflection minimum frequency of the lower side remains the same, whereas the reflection minimum frequency of the higher side is seen to vary significantly as shown in Figure 6.14(d). This observation is thought to be consistent with the previous one that, between two reflection minimum frequencies, relevant to the aperture resonance and periodicity, the reflection minimum frequency of the lower frequency side relevant to the aperture resonance remains almost the same though periodicity is varied as seen in Figure 6.14(a). Note that the aperture resonance frequency is mainly determined by the aperture length $S_x$.

For the present structure in Figure 6.8 to be used as radar absorbing material (RAM), the thickness of the lossy conducting ground layer $d_r$ should be kept as thin as possible. So we should examine the thickness effect of the lossy conducting

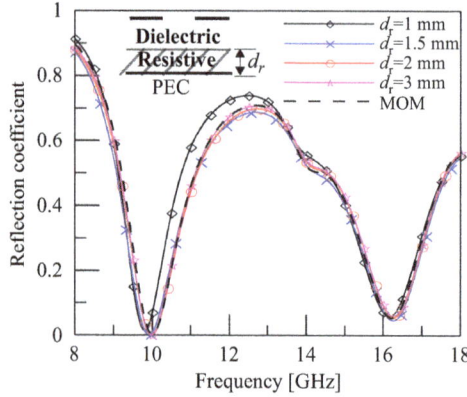

*Figure 6.15    Reflection coefficient versus frequencies with a thickness of the lossy ground layer as a parameter*

ground layer on the present structure. For that purpose, we have investigated the reflection characteristics while varying the thickness of the lossy ground layer backed by the perfect conducting plane as shown in Figure 6.8.

Figure 6.15 shows the reflection coefficient characteristics for the case of 24.86 S/m with the thickness of the lossy ground layer as a parameter. The reflection characteristics curves are seen to converge to the curve for the thickness $d = 3$ mm. The results for $d = 1, 1.5, 2,$ and 3 mm have been obtained by use of the Ansoft HFSS SW and those for $d = 3$ mm by use of the present MoM. A comparison shows good agreement. For reference, the skin depth at 10 GHz is found to be about 1    mm. As shown in the Figure, the reflection characteristics do not change for $d \geq 1.5$ mm (=one and half times the skin depth).

Next, we have investigated the reflection characteristics for the case that the foregoing rectangular aperture is replaced by a rectangular loop type of aperture in the periodic aperture array structure combined with the lossy ground conducting layer under consideration. For the periodicity $T_x = T_y = 15$ mm, the unit cell size $L_x = L_y = 10$ mm of the rectangular loop, the slot width $G = 1$ mm of the loop, the thickness $d$ of the dielectric ($\varepsilon_r = 2$) and the conductivity $\sigma$ of the lossy ground layer for near-zero reflection at 10 GHz has been obtained to be $d = 1.76$ mm and $\sigma = 27.89$S/m. This means that if we use the rectangular loop type of aperture as a unit element of the array, a very thin absorbing structure can be achieved in comparison with the above case of the rectangular aperture.

For reference for three values of the conductivity $\sigma$ of the lossy ground layer, $\sigma = 0, 5, 27.89,$ and 60 [S/m], we have examined the reflection characteristics versus frequencies with main interest centering on the frequency dependence of the reflection minimum location on the conductivity of the lossy ground layer. Figure 6.16 shows reflection coefficient characteristics whose minimum points move to higher frequencies as the conductivity of the lossy ground layer increases from 0 to 60 [S/m]. It is seen that when $\sigma = 0$, the reflection minimum occurs at

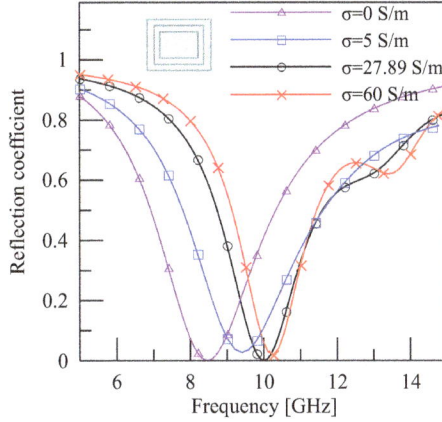

*Figure 6.16   Reflection coefficient versus frequencies with the conductivity σ of the lossy ground layer as a parameter*

8.5 GHz. When $\sigma = 0$, the present structure corresponds to the transmission type of FSS loaded with a dielectric slab. So the reflection minimum at 8.5 GHz corresponds to the transmission maximum. But when $\sigma = 5, 27.89, 60$ [S/m], the reflection minimum corresponds to the absorption maximum, though the absorption maximum (or reflection minimum) frequencies are slightly increased with an increase of $\sigma$. In these absorptions, maximum cases for $\sigma = 5, 27.89, 60$ [S/m], the resonant cavity is formed between the FSS plane whose element of the periodic apertures is of the rectangular loop type and the lossy ground plane whereas for $\sigma = 0$ the resonant cavity is not. The transmission problem for $\sigma = 0$ and the absorption problem for $\sigma = 5, 27.89$, and 60 [S/m] can be viewed as an impedance matching problem. So the frequency characteristics for both cases are expected to follow a similar curve to each other except discrepancy between the impedance matching point corresponding to the transmission and absorption resonance frequencies as shown in Figure 6.16.

The reflection minimum frequencies are seen to move upward gradually as $\sigma$ increases. This is because, as $\sigma$ increases, the inductance becomes decreased due to a decrease of the field penetration distance in the lossy ground layer while the capacitance of the small rectangular loop type of aperture remains constant.

Next, we have investigated the effects of the periodicities $T_x = T_y$, dielectric thickness $d$, square loop size $L_x = L_y$, and loop gap $G$ on the reflection characteristics. Figure 6.17(a) shows the reflection characteristics obtained with the periodicities as a parameter. It is seen that the curve shape and the frequency of the reflection minimum remain almost the same irrespective of the periodicities $T_x = T_y = 14, 15, 16$ [mm]. This is similar to the case of the reflection minimum location due to the single aperture resonance as shown in Figure 6.14(a).

Figure 6.17(b) shows the variation of the reflection coefficient due to the variation of the dielectric thickness $d$. As the dielectric thickness $d$ varies, while the

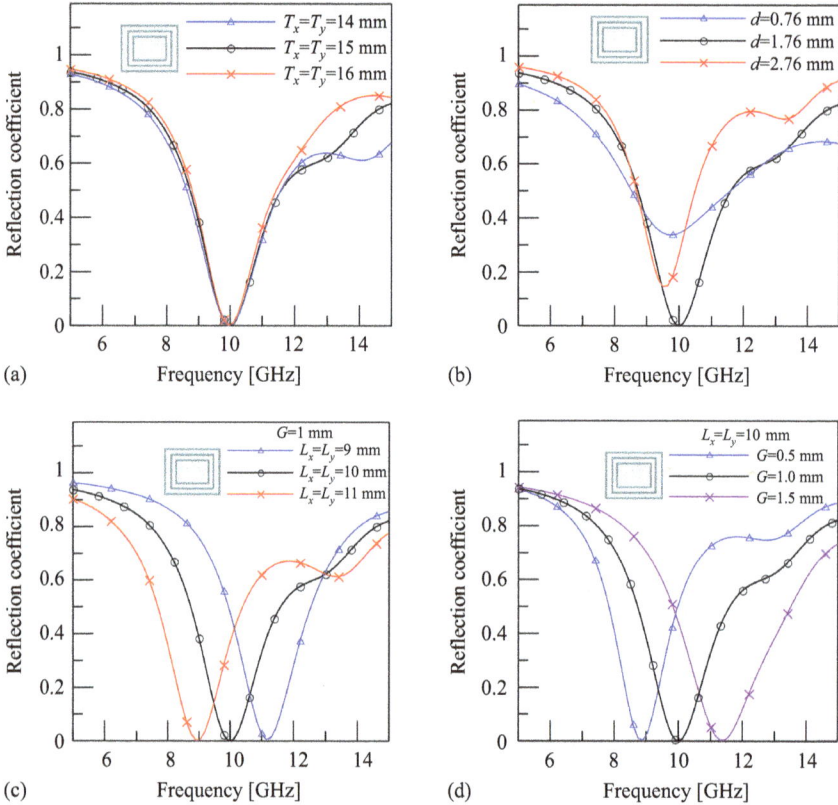

*Figure 6.17*    *Reflection coefficient characteristics versus frequencies (a) obtained
with periodicities $T_x = T_y$ as a parameter, (b) obtained with
dielectric thickness d as a parameter, (c) obtained with the loop size
$L_x = L_y$ as a parameter, and (d) obtained with the loop gap G as a
parameter*

reflection minimum frequency varies only slightly, the reflection characteristics of
the curve shape are seen to vary significantly.

Figure 6.17(c) shows the effect of the loop size $L_x = L_y$ on the reflection
coefficient. As expected, the reflection minimum frequency decreases as the loop
size $L_x = L_y$ increases.

Figure 6.17(d) shows the effect of the loop gap size $G$ on the reflection characteristics. The reflection minimum frequency is observed to be lowered as the loop
gap size $G$ decreases. This is because the total capacitance cross the gap around the
loop per unit cell $(T_x \times T_y)$ increases as the loop gap size decreases.

This increased capacitance is combined with the inductance corresponding to
the approximate input admittance toward the lossy ground, to constitute the lossy

resonant cavity. As a result, near-zero reflection $(R \cong 0)$ or almost perfect absorption $(A = 1 - R \cong 1)$ can be achieved even for a very thin structure.

Here the very thin structure means that the distance $d$ between the aperture types of the FSS plane and the lossy ground plane is very small in comparison with the wavelength $\lambda$ inside the dielectric. In more detail, the thickness $d(= 1.76$ mm) here is much smaller than that for the previous rectangular aperture case and corresponds to approximately $0.08\lambda$ in terms of the wavelength $\lambda$ in the dielectric.

As far as the total thickness is concerned for application to the radar absorbing material (RAM), the thickness $d(= 1.76$ mm) here is comparable to those of the previous novel compact RAM family [26,30,35].

We have investigated the reflection coefficient versus frequencies also for the oblique incidence case. Figure 6.18(a) shows a reflection coefficient versus frequencies for $\theta = 0°, 15°, 30°$, and $45°$ for the TE case where the incident electric field is parallel to the $y$-axis. Reflection characteristics are seen to remain almost the same even if the incident angle varies from $0°$ to $45°$ while keeping $\emptyset$ to be $0°$. Figure 6.18(b) shows a reflection coefficient versus frequencies for various values of the incidence angle $\theta = 0°, 15°, 30°, 45°$ for TM case where the incident magnetic field is parallel to the $y$-axis. In this case too, reflection characteristics remain almost the same even if the incident angle $\theta$ varies from $0°$ to $45°$ while keeping $\phi$ to be $0°$, similar to the TE case. The reason for the insensitivity of the incidence angle to the reflectivity curve for both TE and TM waves is that the loop gap $G$ is so small that the transmission characteristics do not vary appreciably for the incident angle variation from $0°$ to $45°$.

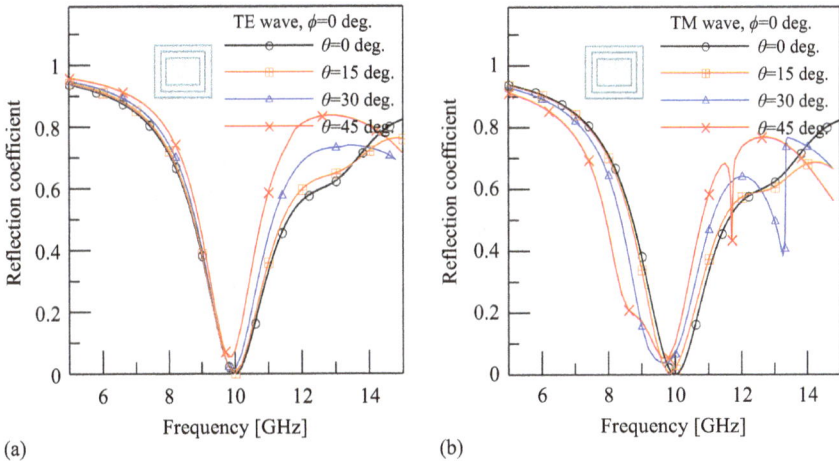

(a)                                                 (b)

*Figure 6.18    Reflection coefficient versus frequencies in case of oblique incidence for $T_x = T_y = 15$ mm, $L_x = L_y = 10$ mm, $G = 1$ mm, $\varepsilon_r = 2$, $d = 1.76$ mm, $\sigma = 27.89$S/m with incident angle as a parameter (a) for TE case, (b) for TM case*

For broadening bandwidth, the multi-loop type [37] of aperture is thought to be a good solution. This multi-loop type would provide a compact solution in comparison with the Jaumann absorber of the broadband type which is composed of multiple lossy layers over a conducting ground plane. So, the loop type of aperture along with its multiloop type can be employed for synthesizing both thin and wideband absorbers that strongly outperform the conventional Salisbury [32] and Jaumann [33] configurations.

As a future research direction, it is necessary to explore the relationship between the optimal absorption structures for photovoltaics [41] and photodetectors [42] and the absorption structure being considered for radar absorbing materials (RAM). In addition, a promising line of research is to clarify how the structure of extraordinary optical transmission (EOT) [43] is related to that of extraordinary optical absorption (EOA) under consideration and further how to transform the former into the latter structure. This research is very helpful to have a unified physics on design for various devices such as filters, photovoltaic devices [41], photodetector devices [42], etc.

Another promising avenue is to revisit Wood's anomaly phenomena [44–46], in the areas of guided-mode resonance [47–49] and abnormal optical absorption (AOA) [50] as the main cause for Wood's anomaly. Here the guided mode resonance means the transverse resonance observed when the incident wave is phase-matched to a leaky waveguide mode [48]. In contrast, abnormal optical absorption means a kind of absorption resonance that may occur due to the field enhancement within the channel such as deep waveguide groove, i.e., cavity mode in the real conducting medium [50] and very shallow groove types in the real metal medium [51]. It is worthwhile to mention that chronologically the topic of the guided mode resonance, starting to be investigated in connection with the leaky wave-related Wood anomaly phenomena [45], was later extended to the filter design area of integrated optics. It is interesting to note that the above abnormal optical absorption observed in the periodic groove of very shallow depth is analogous to the ultra-thin electromagnetic absorber structures [26,35].

For the design of a thin type of microwave absorber, the capacitive screen is favored [52] since the input impedance looking into the ground plane through thin dielectric is reactive. This is thought to be analogous to the abnormal optical absorption [51] observed in the electrostatic regime. What is more, this structure corresponds to the one-layer thin meta-surface absorber structure [53]. It would be interesting to consider the analytic properties of the reflection coefficient of a single-layer thin type of absorber structure as a general parameter [54] for the ultimate thickness-to-bandwidth ratio of radar absorbers.

## Appendix A.    Proof for the identity of (6.49)

Here we are going to prove the identity that is given by

$$\iint_{s'}\iint_{s}\boldsymbol{M}_m\left(\underline{r}_t\right)\cdot\nabla\left\{\nabla\cdot\left[\boldsymbol{M}_n\left(\underline{r}'_t\right)\varphi\left(\underline{r},\underline{r}'_t\right)\right]\right\}dsds'$$

$$=-\iint_{s'}\iint_{s}\left(\nabla\cdot\boldsymbol{M}_m\left(\underline{r}_t\right)\right)\left(\nabla\cdot\boldsymbol{M}_n\left(\underline{r}'_t\right)\right)\varphi\left(\underline{r},\underline{r}'_t\right)dsds' \tag{A.1}$$

In this context, we have dealt with the entire basis function type which is a function of only $x$ variable under the assumption of narrow slit. Here we deal with a more general case that the basis function is a function of two tangential components of $x$ and $y$. Then the basis function $\boldsymbol{M}_m\left(\underline{r}_t\right)$ is given by

$$\boldsymbol{M}_m\left(\underline{r}_t\right)=\boldsymbol{a}_xM_m^x(x,y)+\boldsymbol{a}_yM_m^y(x,y) \tag{A.2}$$

So $\nabla\cdot\left[\boldsymbol{M}_n\left(\underline{r}'_t\right)\varphi\left(\underline{r},\underline{r}'_t\right)\right]$

$$=M_n^x(x',y')\frac{\partial\varphi}{\partial x}+M_n^y(x',y')\frac{\partial\varphi}{\partial y}=N \tag{A.3}$$

where the vector identity $\nabla\cdot(V\boldsymbol{F})=V\nabla\cdot\boldsymbol{F}+\boldsymbol{F}\cdot\nabla V$ has been used for the case that $\boldsymbol{F}=\boldsymbol{M}_n\left(\underline{r}'_t\right)$ and $V=\varphi\left(\underline{r},\underline{r}'_t\right)$.

Inserting (A.3) into the left side of (A.1) leads to

$$\iint_{s'}\iint_{s}\boldsymbol{M}_m\left(\underline{r}_t\right)\cdot\nabla\left\{\nabla\cdot\left[\boldsymbol{M}_n\left(\underline{r}'_t\right)\varphi\left(\underline{r},\underline{r}'_t\right)\right]\right\}dsds'$$

$$=\iint_{s'}\iint_{s}\boldsymbol{M}_m\left(\underline{r}_t\right)\cdot\underline{\nabla}Ndsds'$$

$$=\iint_{s'}\iint_{s}\left(M_m^x(x,y)\frac{\partial N}{\partial x}+M_m^y(x,y)\frac{\partial N}{\partial y}\right)dsds'$$

$$=\iint_{s'}\iint_{s}\left\{M_m^x(x,y)\frac{\partial}{\partial x}\left[\frac{\partial}{\partial x}\left(M_n^x(x',y')\varphi\left(\underline{r},\underline{r}'_t\right)\right)\right]\right.$$

$$+M_m^x(x,y)\frac{\partial}{\partial x}\left[\frac{\partial}{\partial y}\left(M_n^y(x',y')\varphi\left(\underline{r},\underline{r}'_t\right)\right)\right] \tag{A.4}$$

$$+M_m^y(x,y)\frac{\partial}{\partial y}\left[\frac{\partial}{\partial x}\left(M_n^x(x',y')\varphi\left(\underline{r},\underline{r}'_t\right)\right)\right]$$

$$\left.+M_m^y(x,y)\frac{\partial}{\partial y}\left[\frac{\partial}{\partial y}\left(M_n^y(x',y')\varphi\left(\underline{r},\underline{r}'_t\right)\right)\right]\right\}dsds'$$

Let us consider the first term of the right side of (A.4). By use of the method of integration by parts and the boundary condition on the equivalent magnetic current

$\underline{M}$, the expression can be changed as follows:

$$\iint_{S'}\iint_S M_m^x(x,y)\frac{\partial}{\partial x}\left[\frac{\partial}{\partial x}\left(M_n^x(x',y')\varphi\left(\underline{r},\underline{r}'_t\right)\right)\right]\mathrm{d}s\mathrm{d}s'$$

$$=\iint_{\Delta s'}\iint_{\Delta s} M_m^x(x,y)\frac{\partial}{\partial x}\left[M_n^x(x',y')\frac{\partial}{\partial x}\varphi\left(\underline{r},\underline{r}'_t\right)\right]\mathrm{d}s\mathrm{d}s'$$

$$=\iint_{\Delta s'}\int_{\Delta y}\left[M_m^x(x,y)M_n^x(x',y')\frac{\partial\varphi\left(\underline{r},\underline{r}'_t\right)}{\partial x}\right]\Bigg|_{\Delta x} \tag{A.5}$$

$$-\int_{\Delta x}\frac{\partial M_m^x(x,y)}{\partial x}M_n^x(x',y')\frac{\partial}{\partial x}\varphi\left(\underline{r},\underline{r}'_t\right)\mathrm{d}x\Big]\mathrm{d}y$$

$$=-\iint_{\Delta s'}\iint_{\Delta s}\frac{\partial M_m^x(x,y)}{\partial x}M_n^x(x',y')\frac{\partial}{\partial x}\phi\left(\underline{r},\underline{r}'_t\right)\mathrm{d}s\mathrm{d}s',$$

where the $\Delta S(\Delta S')$ means the aperture region $\Delta x\Delta y(\Delta x'\Delta y')$ in terms of the observation (source) coordinate and the boundary condition along the aperture periphery $M_m^x(x,y)|_{x=\pm\frac{L_x}{2}}=0$ has been imposed for the small rectangular aperture case under consideration as in Figure 1.

We proceed to follow the method of integration by parts after using the equation

$$\frac{\partial}{\partial x}\varphi\left(\underline{r},\underline{r}'_t\right)=-\frac{\partial}{\partial x'}\varphi\left(\underline{r},\underline{r}'_t\right) \tag{A.6}$$

and changing the integration order, i.e., integrating the last integrand in (A.5) for the source (primed) coordinate first as follows:

$$-\iint_{\Delta s}\iint_{\Delta s'}\frac{\partial M_m^x(x,y)}{\partial x}M_n^x(x',y')\frac{\partial}{\partial x}\varphi\left(\underline{r},\underline{r}'_t\right)\mathrm{d}x\mathrm{d}y\mathrm{d}x'\mathrm{d}y'$$

$$=\iint_{\Delta s}\iint_{\Delta s'}\frac{\partial M_m^x(x,y)}{\partial x}M_n^x(x',y')\frac{\partial}{\partial x'}\varphi\left(\underline{r},\underline{r}'_t\right)\mathrm{d}x\mathrm{d}y\mathrm{d}x'\mathrm{d}y'$$

$$=\iint_{\Delta s}\int_{\Delta y'}\frac{\partial M_m^x(x,y)}{\partial x}\int_{\Delta x'}M_n^x(x',y')\frac{\partial}{\partial x'}\varphi\left(\underline{r},\underline{r}'_t\right)\mathrm{d}x'\mathrm{d}y'\mathrm{d}x\mathrm{d}y$$

$$=\iint_{\Delta x\Delta y}\int_{\Delta y'}\frac{\partial M_m^x(x,y)}{\partial x}\left[M_n^x(x',y')\varphi\left(\underline{r},\underline{r}'_t\right)\Big|_{-\frac{L_x}{2}}^{\frac{L_x}{2}}-\int_{\Delta x'}\frac{\partial M_n^x(x',y')}{\partial x'}\varphi\left(\underline{r},\underline{r}'_t\right)\mathrm{d}x'\right]\mathrm{d}y'\mathrm{d}x\mathrm{d}y$$

$$=-\iint_{\Delta x\Delta y}\int_{\Delta y'}\int_{\Delta x'}\frac{\partial M_m^x(x,y)\partial M_n^x(x',y')}{\partial x}\frac{}{\partial x'}\varphi\left(\underline{r},\underline{r}'_t\right)\mathrm{d}x'\mathrm{d}y'\mathrm{d}x\mathrm{d}y$$

$$\tag{A.7}$$

where $M_n^x(x',y')|_{x'=\pm\frac{L_x}{2}}=0$ has been used.

Similar to this method, the remaining three terms in (A.4) can be obtained to be

$$\iint_{\Delta s'}\iint_{\Delta s}M_m^x(x,y)\frac{\partial}{\partial x}\left[\frac{\partial}{\partial y}\left(M_n^x(x',y')\varphi\left(\underline{r},\underline{r}'_t\right)\right)\right]dsds'$$

$$=-\iint_{\Delta s}\iint_{\Delta s'}\left[\frac{\partial}{\partial x}M_m^x(x,y)\right]\left[\frac{\partial}{\partial y'}M_n^y(x',y')\right]\varphi\left(\underline{r},\underline{r}'_t\right)ds'ds$$

(A.8)

$$\iint_{\Delta s'}\iint_{\Delta s}M_m^y(x,y)\frac{\partial}{\partial y}\left[\frac{\partial}{\partial x}\left(M_n^y(x',y')\varphi\left(\underline{r},\underline{r}'_t\right)\right)\right]dsds'$$

$$=-\iint_{\Delta s}\iint_{\Delta s'}\left[\frac{\partial}{\partial y}M_m^y(x,y)\right]\left[\frac{\partial}{\partial x'}M_n^x(x',y')\right]\varphi\left(\underline{r},\underline{r}'_t\right)ds'ds$$

(A.9)

$$\iint_{\Delta s'}\iint_{\Delta s}M_m^y(x,y)\frac{\partial}{\partial y}\left[\frac{\partial}{\partial y}\left(M_n^y(x',y')\varphi\left(\underline{r},\underline{r}'_t\right)\right)\right]dsds'$$

$$=-\iint_{\Delta s}\iint_{\Delta s'}\left[\frac{\partial}{\partial y}M_m^y(x,y)\right]\left[\frac{\partial}{\partial y'}M_n^y(x',y')\right]\varphi\left(\underline{r},\underline{r}'_t\right)ds'ds$$

(A.10)

Summing the above four terms (A.7)–(A.10), the following equality is proved to hold

$$\iint_{\Delta s'}\iint_{\Delta s}\boldsymbol{M}_m\left(\underline{r}_t\right)\cdot\nabla\left\{\nabla\cdot\left[\boldsymbol{M}_n\left(\underline{r}'_t\right)\varphi\left(\underline{r},\underline{r}'_t\right)\right]\right\}dsds'$$

$$=-\iint_{\Delta s'}\iint_{\Delta s}\left[\frac{\partial}{\partial x}M_m^x(x,y)+\frac{\partial}{\partial y}M_m^y(x,y)\right]\left[\frac{\partial}{\partial x'}M_n^x(x',y')+\frac{\partial}{\partial y'}M_n^y(x',y')\right]\varphi\left(\underline{r},\underline{r}'_t\right)dsds'$$

$$=-\iint_{\Delta s'}\iint_{\Delta s}\left[\nabla\cdot\boldsymbol{M}_m\left(\underline{r}_t\right)\right]\left[\nabla'\cdot\boldsymbol{M}_n\left(\underline{r}'_t\right)\right]\varphi\left(\underline{r},\underline{r}'_t\right)dsds'$$

(A.11)

In summary, it can be said that the singularity problem in the integration procedure can be overcome by employing Galerkin's scheme of cosinusoidal entire basis function expansion.

## Appendix B.    Derivation of the unknown constant $k_i$ in (6.70)

Rewriting the expression for the electric field of the $i$th mode inside the lossy cavity in (6.58), it is given as

$$\underline{E}_t(-\boldsymbol{M})=\sum_i k_i\boldsymbol{e}_i\frac{\sin k_i(d-z)+(1-j)r_{si}\cos k_i(d-z)}{\sin k_id+(1-j)r_{si}\cos k_id}$$

(B.1)

Resorting to equivalence principle at $z=0^+$ of the cavity side, we get

$$-\boldsymbol{M}=\sum_i k_i\boldsymbol{e}_i\times\boldsymbol{a}_z$$

(B.2)

Performing the dot product of the both sides of (B.2) and $\mathbf{a}_z \times \mathbf{e}_i$ and integrating the resultant over the rectangular cross section of the lossy cavity leads to the following expression:

$$\iint_{ap} \mathbf{M} \cdot (\mathbf{a}_z \times \mathbf{e}_i) is = \iint_{ap} \sum_i k_i (\mathbf{a}_z \times \mathbf{e}_i) \cdot (\mathbf{a}_z \times \mathbf{e}_j) ds$$

$$= \iint_{ap} \sum_i k_i (\mathbf{e}_i \cdot \mathbf{e}_j) ds = \sum_i k_i \iint_{ap} (\mathbf{e}_i \cdot \mathbf{e}_j) ds \qquad (B.3)$$

$$= \sum_i k_i \delta_{ij} = k_j,$$

where the mode power orthogonality has been used. In addition, when the magnetic current $\mathbf{M}$ is given by multi-terms of entire basis functions as in (6.70), the constant $k_j$ is expressed as

$$k_j = \iint_{ap} \mathbf{M} \cdot (\mathbf{a}_z \times \mathbf{e}_j) ds$$

$$= \sum_{n=1}^{N} V_n \iint_{ap} \mathbf{M}_n \cdot [\mathbf{a}_z \times \mathbf{e}_i(x,y)] ds \qquad (B.4)$$

$$= V_1 A_{1j} + V_2 A_{2j} + V_3 A_{3j} + \cdots + V_N A_{Nj}$$

where

$$A_{ni} = \iint_{ap} \mathbf{M}_n \cdot [\mathbf{a}_z \times \mathbf{e}_i(x,y)] ds \qquad (B.5)$$

# References

[1]  Ebbesen T.W., Lezec H.J., Ghaemi H.F., Thio T., Wolff P.A. 'Extraordinary optical transmission through sub-wavelength hole arrays.' *Nature.* 1998, vol. 391, pp. 667–669

[2]  Chern R.L. Chen Y.T., Lin H.Y. 'Anomalous optical absorption in metallic grating with subwavelength slits.' *Opt. Express*, 2010, vol. 18(19), pp. 19510–19521

[3]  White J. S., Veronis G., Yu Z., Barnard E.S., Chandran A., Fan S., Brongersma, M.L. 'Extraordinary optical absorption through subwavelength slits.' *Opt. Lett.* 2009, vol. 34(5), pp. 686–688

[4]  Liang C.H., Cheng D. K. 'Electromagnetic fields coupled into a cavity with a slot-aperture under resonant condition.' *IEEE Trans. Antennas Propagat.* 1982, vol. 30(4), pp. 664–672

[5]  Harrington R.F. 'Resonant behavior of a small aperture backed by a conducting body.' *IEEE Trans. Antennas Propagat.* vol. 30(2), pp. 205–212

[6]  Harrington R.F. Mautz J.R. 'A generalized network formulation for aperture problems.' *IEEE Trans. Antennas Propagat.* 1976, vol. 24, pp. 870–872

[7] Harrington R.F. Mautz J.R., 'Electromagnetic transmission through an aperture in a conducting plane.' *AEU*, 1977, vol. 31, pp. 81–87

[8] Collin, R.E., *Field Theory of Guided Waves.* New York: McGraw-Hill; 1960, p. 192

[9] Pozar D.M. *Microwave Wave Engineering.* New York: Wiley; 1998, p. 125

[10] Harrington R. F. *Time-Harmonic Electromagnetic Fields.* McGraw-Hill; 1961

[11] Teodoridis S. T., Gardiol F. 'The reflection in open-ended rectangular waveguides.' *Mikrowellen Mag.* 1985, vol. 11(3), pp. 234–238

[12] Cheng D.K., Liang C.H. 'Optimum directivity of elliptic loop antennas.' *Electron. Lett.* 1981, vol. 17(20), pp. 736–738

[13] Leviatan Y., Harrington R.F., Mautz J.R. 'Electromagnetic transmission through apertures in a cavity in a thick conductor.' *IEEE Trans. Antennas Propagat.* 1982, vol. 30(6), pp. 1153–1165

[14] Kraus J.D. *Antennas.* New York: McGraw-Hill; 1988, pp. 27–36

[15] Collin R. E. *Antennas and Radiowave Propagation.* New York: McGraw-Hill; 1985

[16] Harrington R.F. 'Small resonant scatterer and their use for field measurements.' *IRE Trans. Microwave Theory Tech.* 1962, vol. 10, pp. 165–174

[17] Lin-Liu Y.R., Ikezi H., Ohkawa T. 'Radiation damping and resonance scattering.' *Am. J. Phys.* 1988, vol. 56(4): p. 373

[18] Caloz C., Itoh T., *Electromagnetic Metamaterials: Transmission Line Theory and Microwave Applications.* Hoboken, NJ: Wiley-Interscience, 2006

[19] Verslegers L., Yu Z., Catrysse P.B., Fan, S. 'Temporal coupled-mode theory for resonant apertures.' *J. Opt. Soc. Am. B*, 2010, vol. 27(10), pp. 1947–1956

[20] White J.S., Veronis G., Yu Z., *et al.* 'Extraordinary optical absorption through subwavelength slits.' *Opt. Lett.* 2009, vol. 34(5), pp. 686–688

[21] Tai C.T. 'On the definition of the effective aperture of antennas.' *IRE Trans. Antennas Propagat.* 1961, vol. 9, pp. 224–225

[22] Harrington R.F., Auckland D.T. 'Electromagnetic transmission through narrow slots in thick conducting screens.' *IEEE Trans. Antennas Propagat.* 1980, vol. 28(5), pp. 616–622

[23] Porto J.A., Garcia-Vidal F.J., Pendry J.B. 'Transmission resonances on metallic gratings with very narrow slits.' *Phys. Rev. Lett.* 1999, vol. 83, pp. 2845–2848

[24] Takakura Y. 'Optical resonance in a narrow slit in a thick metallic screen.' *Phys. Rev. Lett.* 2001, vol. 86, pp. 5601–5603

[25] Yang F., Sambles J.R. 'Resonant transmission of microwaves through a narrow metallic slit.' *Phys. Rev. Lett.* 2002, vol. 89, p. 063901

[26] Costa F., Monorchio A., Manara G. 'Analysis and design of ultra thin electromagnetic absorbers comprising resistively loaded high impedance surfaces.' *IEEE Trans. Antennas Propagat.* 2010, vol. 58(5), pp. 1551–1558

[27] Li S., Gao J., Cao X., Li W., Zhang Z., Zhang D. 'Wideband, thin, and polarization-insensitive perfect absorber based the double octagonal rings

metamaterials and lumped resistances.' *J. Appl. Phys.* 2014, vol. 116, p. 043710

[28]   Uchida K., Noda T., Matsunaga T. 'Electromagnetic wave scattering by an infinite plane metallic grating in case of oblique incidence and arbitrary polarization.' *IEEE Trans. Antennas Propagat.* 1988,vol. 36(3), pp. 415–422

[29]   Mosig J.R., Gardiol F.E. 'General integral equation formulation for microstrip antennas and scatters.' *Proc. IEEE*, 1985, vol. 132(7), pt. H, pp. 424–432

[30]   Sun L., Cheng H., Zhou Y.J., Wang J. 'Broadband metamaterial absorber based on coupling resistive frequency selective surface.' *Opt. Express.* 2012, vol. 20(4), pp. 4675–4680

[31]   Singh D., Kumar A., Meena S., Agarwala V., 'Analysis of frequency selective surfaces for radar absorbing materials.' *Prog. Electromagn. Res. B.* 2012, vol. 38, pp. 297–314

[32]   Fante R.L., McCormact M.T. 'Reflection properties of the Salisbury screen.' *IEEE Trans. Antennas Propagat.* 1988, vol. 36(7), pp. 1443–1454

[33]   du Toit L.J. 'The design of Jaumann absorber.' *IEEE Trans. Antennas Propagat. Mag.* 1994, vol. 36(6), pp. 17–25

[34]   Lindquist N.C., Luman W.A., Oh S-H., Holmes R.J. 'Plasmonic nanocavity arrays for enhanced efficiency in organic photovoltaic cells.' *Appl. Phys. Lett.* 2008, vol. 93, p. 123308

[35]   Engheta N. 'Thin absorbing screen using metamaterial surfaces.' *IEEE Antennas Propag. Soc. Int. Symp.* 2002, vol. 2, pp. 392–395

[36]   Zadeh A.K., Karlsson A. 'Capacitive circuit method for fast and efficient design of wideband radar absorbers.' *IEEE Trans. Antennas Propagat.* 2009, vol. 57(8), pp. 2307–2314

[37]   Munk B.A. *Frequency Selective Surface-Theory and Design.* New York: Wiley; 2000

[38]   Wu T.K. *Frequency Selective Surface and Grid Array.* New York: Wiley; 1995

[39]   Medina F., Mesa F., Marques R. 'Extraordinary transmission through arrays of electrically small holes from a circuit theory perspective.' *IEEE Trans. MTT.* 2008, vol. 56(12), pp. 3108–3120

[40]   Garcia de Abajo F.J., Saenz J.J., Campillo I., Dolado, J.S. 'Site and lattice resonances in metallic hole arrays.' *Opt. Express.* 2006, vol. 14(1), pp. 7–18

[41]   Seassal C., Park Y., Fave A., Drouard E., Fourmond E., Kaminski A., Lemiti M., Letartre X., Victorovitch P. 'Photonic crystal assisted ultra-thin silicon photovoltaic solar cell.' *Proc. Of SPIE.* vol. 7002, pp. 700207-1–700207-8

[42]   Unlu M.S., Strite S. 'Resonant cavity enhanced photonic devices.' *J. Appl. Phys.* 1995, vol. 78(2), pp. 607–639

[43]   Medina F., Mesa F., Skigin D.C. 'Extraordinary transmission through arrays of slits: a circuit theory model.' *IEEE Trans. Microw. Theory Tech.* 2010, vol. 58(1), pp. 105–115

[44]   Wood R.W. 'On a remarkable case of uneven distribution of light in a diffraction grating spectrum.' *Philos. Mag.* 1902, vol. 4, pp. 396–402

[45]  Hessel A., Oliner A.A. 'A new theory of Wood's anomalies on optical gratings.' *Appl. Opt.* 1965, vol. 4(10), pp. 1275–1297

[46]  Hagglund J., Sellberg F. 'Reflection, absorption, and emission of light by opaque optical gratings.' *J. Opt. Soc. Am.* 1966, vol. 56(8), pp. 1031–1040

[47]  Rosenblatt D., Sharon A., Friesem A.A. 'Resonant grating waveguide structures.' *IEEE J. Quant. Elect.* 1997, vol. 33, pp. 2038–2059

[48]  Ding Y., Magnusson R. 'Use of nondegenerate resonant leaky modes to fashion diverse optical spectra.' *Opt. Express.* 2004, vol. 12(9), pp. 1885–1891

[49]  Brundrett D.L., Glytsis E.N., Gaylord T.K., Bendickson J.M. 'Effects of modulation strength in guided-mode resonant subwavelength grating at normal incidence.' *J. Opt. Soc. Am.* 2000, vol. 17(7), pp. 1221–1230

[50]  Wirgin A., Lopez T. 'Can surface-enhanced Raman scattering be caused by a waveguide resonances.' *Opt. Commun.* 1984, vol. 48(6), pp. 416–420

[51]  LePerchec J., Quemerais P., Barbara A., Lopez-Rios T. 'Why metallic surfaces with grooves a few nanometers deep and wide may strongly absorb visible light.' *Phys. Rev. Lett.* 2008, vol. 100(6), pp. 066408

[52]  Lee J.H., Lee B.S. 'Design of thin microwave absorbers using capacitive screen.' *Int. Symp. Antenna Technol.* 2015, pp. 199–201

[53]  Mosallaei H., Sarabandi K. 'A one-layer ultra-thin meta-surface absorber.' *Proc. IEEE Int. Symp. Antennas Propag.* 2005, pp. 615–618

[54]  Rozanov K.N. 'Ultimate thickness to bandwidth ratio of radar absorbers.' *IEEE Trans. Antennas Propagat.* 2000, vol. 48(8), pp. 1230–1234

*Chapter 7*

# Scattering resonance

## Synopsis

The scattering resonance is an example of Wood anomalies that manifest themselves as rapid variations in the intensities of the diffracted beams over a narrow range of frequencies or observation angles. They were first discovered by Wood in 1902 in experiments on optical reflection grating and were termed "anomalies" because the effects could not be explained by ordinary grating theory. Since then, these effects constituted a topic of many experimental and theoretical investigations. Significant progress in the physical understanding of the topic did not occur until 1965 when Hessel and Oliner reported their work on it. They pointed out that there are basically two types of anomalous effects, the Rayleigh type due to the emergence of a new spectral order at grazing angle and a resonance type which is related to the leaky wave supportable by the grating.

Between the two types, the Rayleigh type has been well understood, whereas the resonance type has been investigated in connection with the ongoing research topics such as metasurface relevant to Bragg and off-Bragg blazing phenomena in reflection gratings, guided mode resonance filter (GMRF) in the diffractive optics area, and optical transmission problem in a nanostructured slab.

In this chapter, we deal with Bragg and off-Bragg blazing phenomena in the reflection grating as examples of resonance type of Wood anomalies. Here in a broad sense, the resonance type of anomalies are taken as those that are attributed to the coupling between the incident wave and the leaky wave sustained by the reflection grating. In addition, two types of Wood anomalies are discussed, Lorentz and Fano types. It should be noted that Wood's first reported anomaly in 1902 now falls under the field of Fano-type plasmonics. The Bragg blazing to be dealt with here corresponds to the retroreflection phenomena in the metasurface area. By designing the transmission grating whose complex propagation constant is chosen to be the same as that for the reflection grating, we make the transmission angle of the transmission grating symmetrical to the reflection angle of the reflection grating for the grating plane, which is analogous to the above retroreflection phenomena. In addition, the transmission grating structure is modified to implement a kind of transmission filter of guided mode resonance filter (GMRF) type for normal incidence.

Finally, various engineering problems associated with homogeneous leaky wave solutions such as the scan blindness problem in the phased array problem,

Smith–Purcell radiation problem, and a common feature of the working principle between dichroic surface and grating coupler are briefly touched upon.

For simplicity of analysis, we consider a periodic conducting strip array structure over grounded dielectric as a reflection grating and later deal with the transmission grating which comprises double layers of periodic conducting strip of the same geometrical parameters as those of the above reflection grating with dielectric slab interposed between them.

The discussion will be restricted to the Bragg and off-Bragg blazing phenomena relevant to leaky waves for both transverse magnetic (TM) and transverse electric (TE) polarization cases because the Rayleigh type of Wood anomaly has been well understood.

## 7.1 Scattering resonance by a reflection grating for TM case

When Wood's anomalies [1] on optical reflection grating were reported in 1902, they could not be explained through ordinary grating theory. Since that time the phenomena of the Wood's anomalies constituted a topic of many researches. Among them, the representative is Hessel and Oliner's work [2] through which the relevancy of the leaky wave to the Wood's anomalies is emphasized. They reported their works on that topic for the slow and fast wave structures [2–4]. In the case of the fast wave structure which is composed of the parallel-plate waveguide structure with periodic narrow slits in its upper plate, the transverse resonance method for the equivalent circuit representation is used to obtain the complex root for the leaky wave supportable along the structure. On the other hand for the slow wave structure as a theoretical model for general reflection grating, the periodically modulated reactive surface model is used to investigate diffraction properties of various types of Wood's anomalies according to the number of possible propagating diffracted order modes under the plane wave incidence at arbitrary angle.

In this chapter, we focus our attention on the case of two propagating modes in which only one higher diffracted mode plus the specularly reflected mode are assumed to be propagating. The reason for this assumption is that it facilitates the analysis of the diffraction properties and dispersion characteristics of the Bragg and off-Bragg phenomena, which have been taken as a kind of Wood's anomalies. Here the Bragg-angle blazing (or Bragg blazing) means that all the incident power is scattered back into the direction of incidence which equals the first-order Bragg backscattering angle. On the other hand, off-Bragg blazing means that its scattered wave propagates in a direction different from the Bragg backscattering angle.

It deserves mentioning that the approach for analyzing the Wood's anomaly under analysis as well as the previous approaches [2–4] has a drawback in that these approaches have not dealt with the absorption [5] properties of Wood's anomalies. Note that, historically the resonance type of Wood anomalies was investigated first in the reflection grating [2–4] and later extended to the study [6,7] on the transmission grating [6,7] including the extraordinary optical transmission phenomena. The meaning of the resonance type of Wood anomalies will be clarified as discussions proceed.

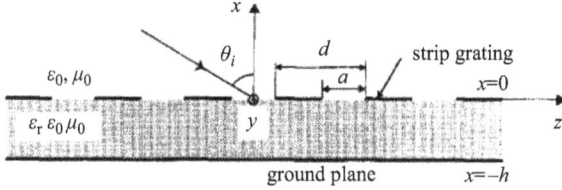

*Figure 7.1    Geometry and notations*

We deal with electromagnetic scattering phenomena such as Bragg and off-Bragg blazing as a kind of scattering resonance problem which is observed in the reflection grating whose structure is shown in Figure 7.1 and discuss the relation between the scattering resonance problem and the leaky wave (homogeneous) problem.

The reason for the choice of the periodic conducting strip grating is first to deal with a practically realizable structure in comparison with the theoretical model of the periodically modulated reactance surface (PMRS) which was used as an approximate model for various reflection gratings and second to deal with the scattering resonance problem numerically rigorously without resorting to the approximate transverse resonance method using the equivalent circuit representation as in [2].

For that purpose, the numerical methods for both the reflection grating and the leaky-wave antenna problems of the geometry under consideration are briefly outlined. The dielectric in the region for $0 < x < -h$ is assumed lossless with parameters $\varepsilon_r\varepsilon_0$ and $\mu_0$. The perfectly conducting strips of width $d$-$a$ and negligible thickness are arranged with a period $d$. Time dependence $e^{j\omega t}$ is assumed.

## 7.1.1    Reflection grating problem

Let us consider the TM polarization case [8] where the magnetic field is parallel to the strip axis. In this case, the incident plane wave at incident angle $\theta_i$ has the following fields

$$\boldsymbol{H}^i = \boldsymbol{y}_0 H_0 e^{-j\beta_0 z} e^{jk_x x} \tag{7.1}$$

$$\boldsymbol{E}^i = \boldsymbol{z}_0 \frac{\gamma_0}{\omega\varepsilon_0} H_0 e^{-j\beta_0 z} e^{jk_x x} + \boldsymbol{x}_0 \frac{\beta_0}{\omega\varepsilon_0} H_0 e^{-j\beta_0 z} e^{jk_x x} \tag{7.2}$$

where

$$\beta_0 = k_0 \sin\theta_i$$

$$k_x = k_0 \cos\theta_i$$

$$k_0 = \omega\sqrt{\mu_0\varepsilon_0}$$

The scattered electromagnetic fields in region (1) $x \geq 0$ are expressed as a summation of space harmonic by

$$\boldsymbol{H}^{(1)} = \boldsymbol{y}_0 H_0 \sum_{n=-\infty}^{\infty} A_n e^{-jk_{1x}^n x} e^{-j\beta_n z} \tag{7.3}$$

$$\boldsymbol{E}^{(1)} = \boldsymbol{z}_0 H_0 \sum_{n=-\infty}^{\infty} \frac{-k_{1x}^n}{\omega \varepsilon_0} A_n e^{-jk_{1x}^n x} e^{-j\beta_n z}$$

$$+ \boldsymbol{z}_0 H_0 \sum_{n=-\infty}^{\infty} \frac{\beta_n}{\omega \varepsilon_0} A_n e^{-jk_{1x}^n x} e^{-j\beta_n z} \tag{7.4}$$

where

$$\beta_n = \beta_0 + \frac{2n\pi}{d}$$

$$k_{1x} = \sqrt{k_0^2 - \beta_n^2}$$

and $A_n$ are unknown amplitudes to be determined. In region (2), $-h \leq z \leq 0$, the electromagnetic fields satisfying the boundary condition that the tangential electric field should vanish at $z = -h$ are given by

$$\boldsymbol{H}^{(2)} = \boldsymbol{y}_0 H_0 \sum_{n=-\infty}^{\infty} B_n \left( e^{jk_{2x}^n x} + e^{-2jk_{2x}^n h} e^{-jk_{2x}^n x} \right) e^{-j\beta_n z} \tag{7.5}$$

$$\boldsymbol{E}^{(2)} = \boldsymbol{x}_0 H_0 \sum_{n=-\infty}^{\infty} \frac{\beta_n}{\omega \varepsilon_0 \varepsilon_r} B_n \left( e^{jk_{2x}^n x} + e^{-j2k_{2x}^n h} e^{-jk_{2x}^n x} \right) e^{-j\beta_n z}$$

$$+ \boldsymbol{z}_0 H_0 \sum_{n=-\infty}^{\infty} \frac{k_{2x}^n}{\omega \varepsilon_0 \varepsilon_r} B_n \left( e^{jk_{2x}^n x} - e^{-j2k_{2x}^n h} \cdot e^{-jk_{2x}^n x} \right) e^{-j\beta_n z} \tag{7.6}$$

where $k_{2x}^n = \sqrt{\varepsilon_r k_0^2 - \beta_n^2}$ and $B_n$ are unknown amplitudes.

At $x = 0$ interface, the tangential electric field must be continuous for all $z$. From (7.2), (7.4), and (7.6) at $x = 0$ we obtain

$$A_n = \delta_n - \frac{k_{2x}^n}{k_{1x}^n} \frac{1}{\varepsilon_r} B_n \left( 1 - e^{-j2k_{2x}^n h} \right) \tag{7.7}$$

where $\delta(n) = 1$ for $n = 1$ and $\delta(n) = 0$ for $n \neq 0$.

The unknown equivalent surface magnetic current density $M_x (= -E_z^{(2)})$ over the shorted slot in the region (2) can be expanded into a product of a series of Chebyshev polynomials $T_i$ (of the first kind) and a function satisfying the edge condition as follows:

$$\boldsymbol{M} = \boldsymbol{y}_0 e^{-j\beta_0 z} \sum_{l=0}^{\infty} f_l \frac{T_i(2z/a)}{\sqrt{1 - (2z/a)^2}}, \quad -\frac{a}{2} < z < \frac{a}{2} \tag{7.8}$$

where $f_l$ are unknown coefficients.

Using the relationship between the surface magnetic current density $M$ and the electric field $E^{(2)}$ over the shorted slot in region (2), that is, $M = E^{(2)} \times (-x_0)$ in (7.6) and (7.8), we obtain

$$
H_0 \frac{\eta_0}{k_0 \varepsilon_r} \sum_{n=-\infty}^{\infty} k_{2x}^n B_n \left(1 - e^{-j2k_{2x}^n h}\right) e^{-j\left(\frac{2\pi n}{d}\right)z}
$$

$$
= \begin{cases} -\displaystyle\sum_{l=0}^{\infty} f_l \frac{T_l\left(\dfrac{2z}{a}\right)}{\sqrt{1 - \left(\dfrac{2z}{a}\right)^2}}, & |z| < \dfrac{a}{2} \\[4mm] 0, \text{ over the strip} \end{cases} \tag{7.9}
$$

where $\eta_0 = \sqrt{\mu_0/\varepsilon_0}$.

We multiply both sides of (7.9) by $e^{j\left(\frac{2\pi n}{d}\right)z}$ and integrate over one period $-\frac{d}{2} < z < \frac{d}{2}$ to obtain

$$
B_n = -\frac{k_0 \varepsilon_r}{H_0 k_{2x}^n \eta_0 d \left(1 - e^{-j2k_{2x}^n h}\right)} \sum f_l H_{ln}^* \tag{7.10}
$$

where

$$
H_{ln} = \int_{-\left(\frac{a}{2}\right)}^{\frac{a}{2}} \frac{T_l(2z/a)}{\sqrt{1 - (2z/a)^2}} e^{-j(2\pi n/d)z} dz
$$

$$
= \begin{cases} \dfrac{\pi a}{2} (j)^l J_l\left(-\dfrac{n\pi a}{d}\right), & n < 0 \\[3mm] \dfrac{\pi a}{2}, & n = 0 \text{ and } l = 0 \\[2mm] 0, & n = 0 \text{ and } l \neq 0 \\[2mm] \dfrac{\pi a}{2} (-j)^l J_l\left(\dfrac{n\pi a}{d}\right), & n > 0 \end{cases}
$$

Here $J_l$ are Bessel functions of order $l$ and the asterisk denotes the complex conjugate. Across the slots, the tangential magnetic field should be continuous. Enforcing this condition on (7.1), (7.3), and (7.5), we get

$$
\sum_{n=-\infty}^{\infty} \left[\delta(n) + A_n - B_n\left(1 + e^{-j2k_{2x}^n h}\right)e^{-j\left(\frac{2\pi n}{d}\right)z}\right] = 0 \tag{7.11}
$$

Substituting (7.7) into (7.11) gives

$$
\sum_{n=-\infty}^{\infty} \left[1 + e^{-j2k_{2x}^n h} + \frac{1}{\varepsilon_r}\left(1 - e^{-j2k_{2x}^n h}\right)\cdot \frac{k_{2x}^n}{k_{1x}^n}\right] \cdot B_n e^{-j\left(\frac{2\pi n}{d}\right)z} = 2\delta(n) \tag{7.12}
$$

We multiply both sides of (7.12) by $\dfrac{T_i(2z/a)}{\sqrt{1-(2z/a)^2}}$ and integrate over the slot region $-\frac{q}{2} < z < \frac{q}{2}$ to obtain

$$
\sum_{n=-\infty}^{\infty} \left[ 1 + e^{-j2k_{2x}^n h} + \frac{1}{\varepsilon_r} \left( 1 - e^{-j2k_{2x}^n h} \right) \cdot \frac{k_{2x}^n}{k_{1x}^n} \right] \cdot B_n H_{in}
$$
$$
= \begin{cases} \pi a, i = 0 \\ 0, i \neq 0 \end{cases}
\tag{7.13}
$$

Substituting (7.10) for $B_n$ in (7.13) gives the following system of simultaneous linear equations:

$$
\sum_{l=0}^{\infty} f_l Y_{il} = I_i, i = 0, 1, 2, \ldots
\tag{7.14}
$$

where

$$
Y_{ie} = -\frac{k_0 \varepsilon_r}{\eta_0 d} \sum_{n=-\infty}^{\infty} \left[ \frac{\cot\left( k_{2x}^n h \right)}{j k_{2x}^n} + \frac{1}{\varepsilon_r} \frac{1}{k_{1x}^n} \right] H_{in}^* H_{in},
$$

$$
I_i = \begin{cases} H_0 \pi a, \ i = 0 \\ 0, \ i \neq 0 \end{cases}
$$

Once $f_l$ are known, $A_n$ and $B_n$ are computed from (7.9) and (7.10). This has been discussion on the scattering problem.

## 7.1.2  Leaky wave problem

The transcendental equation for the leaky wave problem corresponding to the above scattering problem is obtained by replacing $\beta_0(= k_0 \sin \theta_i)$ with the complex propagation constant $(\beta - j\alpha)$ and setting the right side term to zero in (7.14) as follows:

$$
\sum_{l=0}^{\infty} f_l Y_{il} = 0, i = 0, 1, 2, \ldots
\tag{7.15}
$$

where

$$
Y_{il} = -\frac{k_0 \varepsilon_r}{\eta_0 d} \sum_{n=-\infty}^{\infty} \left( \frac{\cot\left( k_{2x}^n h \right)}{j k_{2x}^n} + \frac{1}{\varepsilon_r} \frac{1}{k_{1x}^n} \right) H_{in}^* H_{in},
$$

$$
k_{1x}^n = \sqrt{k_0^2 - \beta_n^2}
$$

$$
k_{2x}^n = \sqrt{\varepsilon_r k_0^2 - \beta_n^2}
$$

$$
\beta_n = \beta - j\alpha + \frac{2n\pi}{d}
$$

and in this leaky-wave problem case, the expression for $A_n$ is obtained by deleting the incident field term $\delta(n)$ in the above formulation procedure for the reflection grating case as

$$A_n = -\frac{k_{2x}^n}{k_{1x}^n}\frac{1}{\varepsilon_r} B_n \left(1 - e^{-j2k_{2x}^n h}\right) \tag{7.16}$$

and $B_n$ is given by (7.10).

Enforcing the condition that the determinant of (7.15) should be zero, we obtain a complex propagation constant $\beta - j\alpha$, from which $f_l, A_n$, and $B_n$ are computed by use of (7.15), (7.16), and (7.10), respectively.

Using the large argument approximation of the Hankel function $H_0^{(2)}$ and the relationship $(M_y = E_z^{(1)}$ in (7.4) over the slot) between the equivalent surface magnetic current and the tangential electric field over the slots in region (1) in the following expression, which gives the magnetic field by integrating the magnetic current density across a finite number $N$ of slots:

$$H_y^{(1)} \cong -j\omega\varepsilon_0 \sum_{n=1}^{N} \int_{(n-1)d-\frac{a}{2}}^{(n-1)d+\frac{a}{2}} M_y \frac{1}{2j} H_0^{(2)}(k_0|\rho - z'z_0|)\mathrm{d}z' \tag{7.17}$$

We can define the far-zone magnetic field (radiation pattern) as follows:

$$H_y^{(1)} \cong \sqrt{\frac{2j}{\pi k_0 \rho}} e^{-jk_0\rho} H_0 \sum_{n=-\infty}^{\infty} A_n k_{1x}^n \cdot \frac{\sin\left[(\beta_n - k_0\sin\varphi)\frac{a}{2}\right]}{\beta_n - k_0\sin\varphi} \frac{1 - e^{-j(\beta_n - k_0\sin\varphi)Nd}}{1 - e^{-j(\beta_n - k_0\sin\varphi)d}} \tag{7.18}$$

where $\rho = \sqrt{x^2 + z^2}$ and the finite number $N$ of slots for the radiation pattern calculation has been determined so that the radiation pattern does not appreciably change even if a larger number of slots than $N$ are taken into account. The reason why the radiation pattern for the infinite structure of the leaky wave case can be defined as above is that, wherever the finite number $N$ of slots is chosen for the radiation pattern calculation, we get the same shape of the radiation pattern, except that the relative level of the radiation pattern changes depending on the position of the first slot of $N$ slots. This is because the amplitudes of the magnetic currents over each slot decay exponentially along the $z$-axis according to the attenuation constant. This definition of the radiation pattern for the infinite leaky-wave antenna structure has conceptual utility as a reference radiation pattern [9], particularly when compared with the radiation pattern of the practically finite leaky-wave structure to evaluate the performance of the finite leaky-wave structure.

## 7.1.3 Bragg and off-Bragg blazing

Here we deal with Bragg and off-Bragg blazing phenomena as examples of scattering resonance in the sense that these two phenomena are attributed to the coupling between the incident wave and the leaky wave sustained by the reflection grating structure.

As mentioned already the Bragg blazing, interchangeably used as a retro-reflection, means the phenomena in which all the incident power is scattered back in the direction of incidence. On the other hand, the off-Bragg blazing means the phenomena, where the incident electromagnetic wave is scattered at angles other than the Bragg angle as mentioned above.

By use of the above analysis method, we investigate Bragg and off-Bragg blazing phenomena which are observed in the leaky-wave supportable grating. For this purpose, we consider first the Bragg blazing phenomena in the periodic strip array on a grounded dielectric for which the previous experimental work [10] on the phenomena is available. Our theoretical results have been obtained for the variation of specularly reflected power at 10.9 GHz for the case where the value of $\theta_i$ corresponds to the Bragg angle in the periodic structure, that is, $\theta_i = \sin^{-1}(\lambda/2d) = 27.28°$, where $\lambda$ is the free-space wavelength. The results are compared with those of the previous experimental work [10] in Figure 7.2 where our results show three positions $(A, B, C)$ of minimum reflected power (sharp dips), whereas the prior experimental results (dashed line) show only two sharp dips.

From the comparison, it is felt that the sharp dip corresponding to position B may be missing in the experimental results because the sharp dip of position B is observed to be much narrower than that of position C. However, except for this discrepancy and some deviations in the positions of sharp dips, two results are observed to be in reasonably good agreement.

Because we are dealing with the TM-polarization case with the magnetic field parallel to the strip axis, only the lowest transverse electromagnetic (TEM) mode can propagate for the range $\frac{h}{\lambda} < \frac{1}{2\sqrt{\varepsilon_r}} (= 0.3119)$, and TEM and $TM_1$ modes can propagate for $0.3119 < \frac{h}{\lambda} < 0.6238$ in the parallel-plate waveguide region under the strip, and so the guided (leaky) wave is supportable by the grating over all the range of $\frac{h}{\lambda}$ in the abscissa of Figure 7.2.

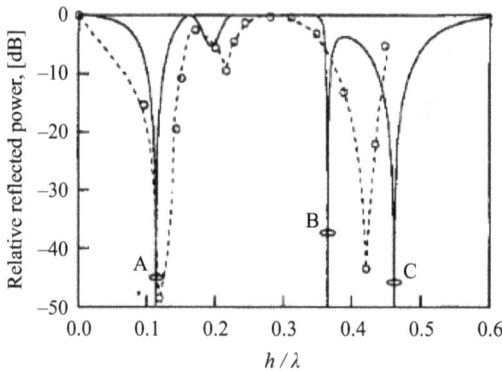

*Figure 7.2    Relative reflected power [dB] versus normalized dielectric thickness (h/λ) for d = 30 mm, a = 0.566d, $\varepsilon_r$ = 2.57, and f = 10.9 GHz. Dashed line, Jose and Nair [10]; Solid line, the present approach.*

To examine the minimum reflection (anomaly) cases A, B, and C more closely, we have calculated the relative powers ($P_0$ and $P_{-1}$) of $n = 0$ and $n = -1$ spectral orders versus angle of incidence $\theta_i$ for the case of $\frac{h}{\lambda} = 0.1128$ (minimum position A case), as examples, which are given graphically in Figure 7.3(a).

In this case, only two space harmonics are propagating and the sum of powers of two space harmonics is calculated to be larger than 99.999% over the range of incidence angle $10° - 50°$.

Let us further examine the anomaly of position A in connection with the leaky wave. For the given parameters in Figure 7.3(a), the complex propagation constant $\beta_0 - j\alpha_0$ is found to be 276.26 [rad/m]-$j$15.90 [Np/m], from which the radiation pattern from the viewpoint of the leaky-wave antenna can be obtained by use of (7.18). For reference, the radiation pattern is given in Figure 7.3(b), where the two maximum beam angles are $-38.6°$ and $17.0°$, respectively. It is interesting to note that the two incident angles, where the relative power of the $n = 0(n = -1)$ mode is maximum (minimum) in Figure 7.3(a) corresponds to $17.0°$ and $38.6°$, respectively.

It should be noted that the two incidence angles correspond to the two angles which are obtained by making two radiation angles of the leaky wave the same sign.

The maximum power conversion into the $n = -1$ spectral order is observed to take place between these two angles under the Bragg condition, in which case the maximum power conversion efficiency is about 99.999%.

In Figure 7.2, it is seen that there are two types [4,5] of dip profiles. One is a Lorentz profile which is symmetrical about the minimum dip point and the other is a Fano profile which is unsymmetrical and goes through an abrupt dip to zero (maximum) locus.

Cases A and C belong to the Lorentz profile whereas the case B belongs to the Fano profile type. Though all the Bragg Blazing phenomena of cases A, B, and C have been observed on the grating where the leaky wave is supportable as mentioned above, there is a fundamental difference between the Lorentzian profile type (A and C) and Fano profile type (B), which will be discussed later in the next section.

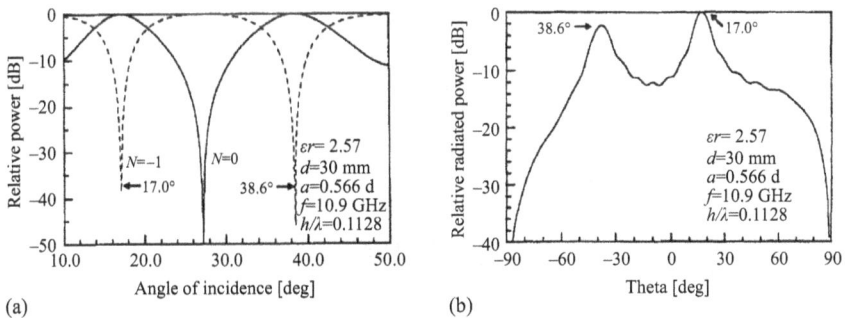

*Figure 7.3*    *Scattering problem and leaky wave problem under the Bragg blazing. (a) Relative power of $P_0$ and $P_{-1}$. (b) Radiation pattern from the viewpoint of leaky-wave.*

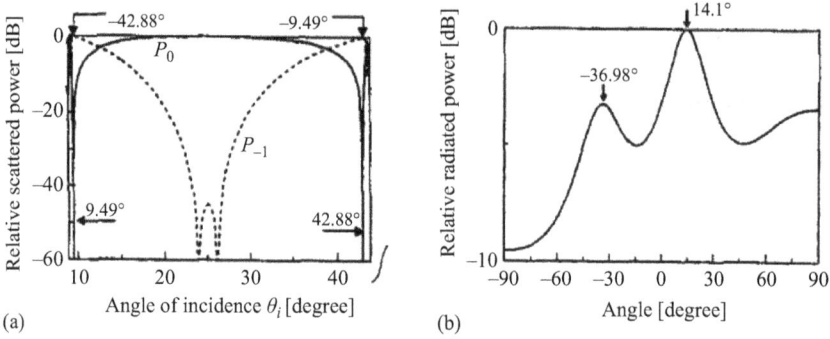

(a)    Angle of incidence $\theta_i$ [degree]

(b)    Angle [degree]

*Figure 7.4    Scattering problem and leaky wave problem under the off-Bragg blazing. (a) Relative scattered power against the angle of incidence for the case of A' point in Figure 7.7 (d = 1.1831λ, a = 0.5d, h = 0.462d, $\varepsilon_r$ = 1.03). (b) Radiation pattern from the viewpoint of leaky wave antenna for the case of A' point in Figure 7.7 (d = 1.1831λ, a = 0.5d, h = 0.462d, $\varepsilon_r$ = 1.03).*

As a typical example of off-Bragg blazing phenomena [11] the results for variations of the relative powers ($P_0$ and $P_{-1}$) of $n = 0$ and $n = -1$ spectral orders for the TM polarization against angle of incidence $\theta_i$ are given graphically in Figure 7.4(a). Here most of the power incident in the direction of $\theta_i = 9.49°$ is seen to be converted into the first-order power scattered at an angle of $\theta_{-1} = -42.88°$ (and conversely).

The reason for selecting different geometrical and material parameters in the current off-Bragg blazing case, as opposed to those used in the previous Bragg blazing case, is to ensure continuity in the forthcoming discussions regarding the loci of Bragg and off-Bragg blazing phenomena within the dispersion curve.

The radiation pattern, calculated from the viewpoint of the leaky wave radiation [12,13], is also given in Figure 7.4(b). Consider two angle pairs in Figures 7.4 (a) and (b); one pair of 9.49° (dip angle in the scattering pattern) and 14.1° (beam angle of the leaky wave radiated pattern), the other of 42.88° and 36.98°. Generally, the dip of sharp minimum and the beam angles of each paired angle like these have been observed to approach closer to each other as the attenuation constant ($= \alpha$ in complex propagation constant $\beta - j\alpha$) becomes smaller. This can be expected from the fact that the plane waves in the scattering problem possess real wave numbers while the wave number of the leaky wave is complex.

## 7.1.4    Consecutive occurrence of Bragg blazing of Lorentzian and Fano type profile; characteristic mode current distribution

We are to discuss the Bragg blazing phenomena in more detail. To this end, theoretical results for the variation of the normalized dielectric thickness $\left(\frac{h}{d}\right)$ for the incidence (Bragg) angle $\theta_i = \sin^{-1}\left(\frac{\lambda}{2d}\right) = 25°$ ($\lambda$: free-space wavelength) for the

TM polarization, which has been obtained by use of the method in Section 7.1.1, are given in Figure 7.5. Here six cases of Bragg blazing phenomena corresponding to points A–F are observed.

In the figure, the points A, C, and E belong to the Lorentzian type profile whereas B, D, and F belong to the Fano type profile which results from interference between two contributions: directly reflected wave from the grounded dielectric (non-resonant mode) and the specularly reflected wave after interaction with the periodic structure (resonant mode).

Note that this consecutive occurrence of Lorentz and Fano types is similar to that in the scattering resonance peaks of the guided-mode resonance problem [14]. To understand further the Bragg blazing cases, the induced current distributions over the strip are investigated. As a representative example among B, D, and F cases for which all the current distributions are very similar to each other, the current distribution corresponding to case B in Figure 7.5 is illustrated in Figure 7.6 (a). On the other hand, the current distributions for A, C, and E cases have been observed to be quite different from those for B, D, and F cases. As a sampled example, the induced current distribution for case A in Figure 7.5 is given in Figure 7.6(b) for comparison.

The magnitude and the phase of the current distribution in Figure 7.6(a) are almost the same as those for the homogeneous (leaky wave) problem of all the cases of B, D, and F. In contrast to the case of Figure 7.6(a), such a conspicuous feature is not observed in Figure 7.6(b). What is more, the magnitude of the current distribution in Figure 7.6(a) is observed to be much larger than that in Figure 7.6(b).

Based on these observations, the Bragg blazing phenomena in Figure 7.5 can be divided into two categories, depending on whether the solution (i.e., current distribution over the strip) for the scattering (inhomogeneous or forced resonance) problem case is almost the same as that for the leaky wave radiation (homogeneous or free resonance) problem case or not: resonance and non-resonance type, respectively. So, while B, D, and F cases belong to the resonance type, A, C, and E cases belong to the

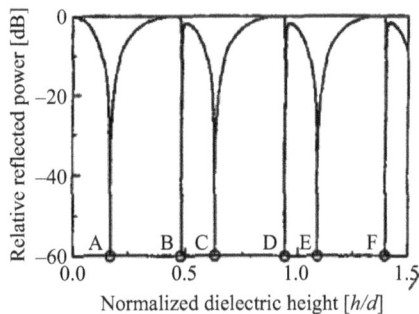

*Figure 7.5   Relative reflected power against normalized dielectric slab height*
*(h/d) $\left( d = 1.1831\lambda, a = 0.5d, \varepsilon_r = 1.03, \theta_i = 25^\circ \right)$*

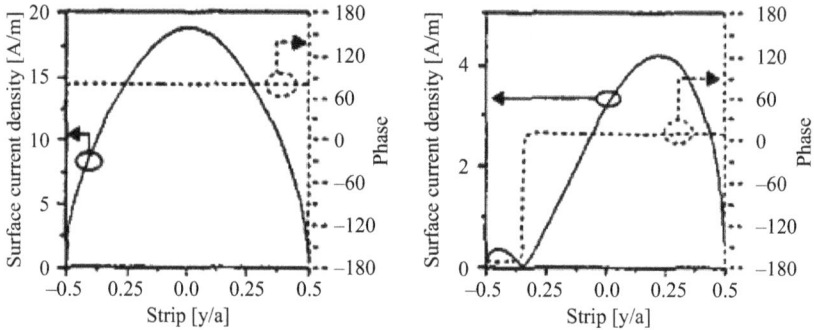

*Figure 7.6    Induced current distribution on the strip. (a) Resonance type corresponding to point B in Figures 7.5 and 7.7. (b) Non-resonance type corresponding to point A in Figures 7.5 and 7.7.*

non-resonance type. Recall that the scattering (leaky wave radiation) problem corresponds to the forced resonance (free resonance) problem [1].

The current distribution over the strip for the resonant type of Bragg blazing phenomena is seen to be very similar to the characteristic mode [15–17] of the grating structure under consideration since the current has a precisely constant phase across the strip although the magnitude of the current varies across the strip.

In the broad sense, the Bragg blazing of the resonance type means that it is solely associated with the complex leaky root that satisfies the transverse resonance condition of the homogeneous leaky wave system, as discussed in previous work [1]. On the other hand, in the strict sense, the coupling point condition [18,19] including the Bragg condition is additionally imposed on the meaning of the resonance type in the broad sense as discussed later.

## 7.1.5    Dispersion characteristics of Bragg and off-Bragg blazing phenomena

To examine the relationship between the Bragg blazing and leaky waves, the dispersion diagram, which has been obtained for $d = 1.1831\lambda, a = 0.5d$ and $\varepsilon_r = 1.03$, is given in Figure 7.7, where locations corresponding to Bragg blazing phenomena (points A–F) in Figure 7.5 are marked.

The first subscript $m$ in $\alpha_{mo}$ and $\beta_{mo}$ denotes the order number relevant to the dielectric substrate height $h$ which is analogous to the mode order number of the unperturbed parallel-plate waveguide and the second subscript $o$ denotes the 0th order among the Floquet modes. From Figure 7.7, all the Bragg blazings of the resonance type (corresponding to points B, D, and F) are observed to occur only when $\beta_{mo}d = \pi$ (and also $2\pi$ like G case to be discussed later), in which cases the two leaky beam angles become symmetrical about the $x$-axis.

Equating this condition to the Bragg condition, we obtain $\beta_{mo} = k_0\sin\theta_i = \pi/d$ (first condition). At the same time, $\alpha_{mo}d$ curve (simply $\alpha_{mo}$ curve) is observed to become minimum, that is, $\alpha_{mo} = $ minimum (second condition) except for real solution cases [18,19]. Here $k_0$ is a wave number in free space. Interestingly the two conditions correspond to the coupling point conditions [19,20] in the usual open periodic structures as discussed later. These two conditions lead to a conclusion that the Bragg blazing phenomena of the resonance type occur at scattering angles very close to the leaky beam angles as described in [1,2]. On the other hand, the Bragg blazing phenomena corresponding to A, C, and E cases in Figure 7.7 occur entirely irrespective of the minimum point of $\alpha$ in the dispersion curve even though the leaky wave is supportable by the grating. That is, these types are observed to occur irrespectively of the coupling point condition. Of course, there are also Bragg blazing phenomena observed when the leaky wave is not supportable as discussed for the TE polarization in [21]. So Bragg blazing phenomena can be divided into three kinds as follows: the first kind is the one which occurs simply due to the periodicity when the leaky wave is not supportable as in [21], the second kind is the one which is observed irrespective of the coupling point condition in the dispersion curve even though the leaky wave is supportable by the grating, and the third is the one which is observed only where the coupling point conditions are met in the dispersion curve as discussed above. It is worth mentioning that the third type is closely related to the working principle of the Guided Mode Resonance Filter (GMRF) discussed later.

Based upon the discussions on the induced current distribution on the conducting strip, it seems legitimate to view the third kind as a resonant type of Bragg blazing in that the induced current distribution for the scattering problem is almost the same as that for the leaky wave problem. Such relation between the current distributions on the strip for scattering and leaky wave problems is not seen for the second case of Bragg blazing, though the second and third cases of Bragg blazing occur when the reflection grating structures support the leaky wave.

Let us further examine the two conditions for the Bragg blazing phenomena of the resonance type. An inspection of Figure 7.7 reveals that the Bragg blazings of

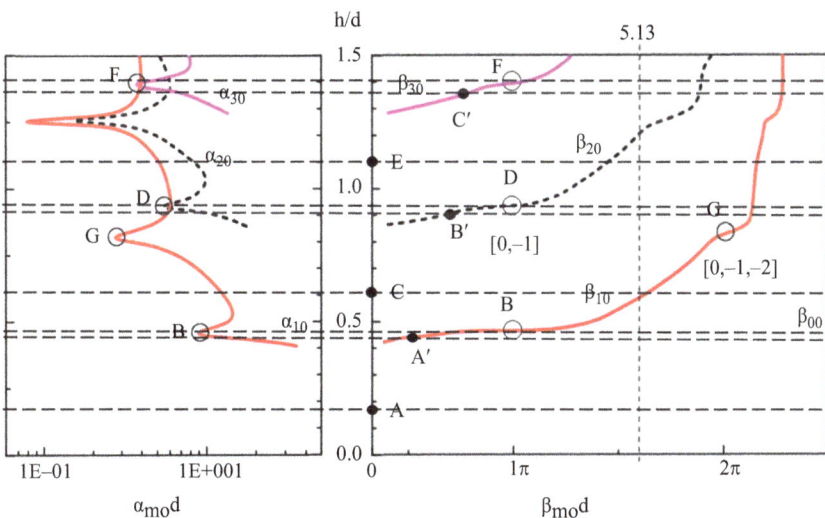

*Figure 7.7   Dispersion diagrams for* $d = 1.1831\lambda, a = 0.5d,$ *and* $\varepsilon_r = 1.03$

the resonance type (B, D, and F as well as G cases) occur only at points corre-sponding to the coupling points [18,19] in the radiation region of the $k - \beta$ dia-gram. In this case, the Floquet modes in the leaky wave problem are coupled in pairs to produce the standing wave pattern along the $z$-axis as follows: $A_0 = A_{-1}$, $A_1 = A_{-2}$, and so on. Here $A_i (i = \pm 1, \pm 2, \ldots)$ are Floquet mode amplitudes in the leaky wave problem case. As this result, the coupling point conditions are formed at $\beta d = \pi$ for two propagating mode problem and at $\beta d = 2\pi$ for three propagating mode problems.

As discussed in Section 7.1.3, in the case of the off-Bragg blazing, as the attenuation constant $\alpha$ in the complex propagation constant $\beta - j\alpha$ becomes smaller, the locations of the sharp dips in scattering pattern and the two main leaky wave beam angles are observed to approach closer to each other. From Figure 7.7, off-Bragg blazing phenomena are observed only at positions corresponding to A′, B′, and C′ points, which seem close to the minimum points of attenuation constant $\alpha$. This is compatible with the above discussions on the relationship between scattering and leaky beam angle and also clearly shows that off-Bragg blazing phenomena occur only when leaky waves are supportable by the grating structure. There have been discussions on TM polarization. The above discussion is valid also for the TE polarization case as will be seen in the following section. Remind that the results in Figure 7.7 have been obtained for the low dielectric ($\varepsilon_r = 1.03$) case. But the dis-cussions above still hold for the case that the dielectric constant increases somewhat except that the relationship between the sharp dips in the scattering pattern and the leaky beam angle tends to be less clearly seen as pointed out in [11].

## 7.2    Scattering resonance by reflection grating for TE case

Here we deal with TE polarization in which the electric field is parallel with the conducting strip axis as shown in Figure 7.8.

In this TE polarization case, the width of the conducting strip is chosen to be "a". Except this, all the geometrical parameters and material constant are the same as those in the previous TM polarization case.

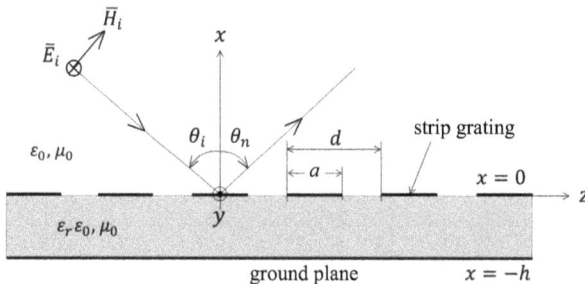

*Figure 7.8    Geometry and notation for TE polarization case*

In this Section, we discuss the numerical analysis method, Bragg blazing (retroreflection) and off-Bragg blazing phenomena, Lorentz profile-like and Fano profile-like type (as a resonant type) of Bragg blazing, current distributions over the conducting strip, and the dispersion characteristics of the Bragg and the off-Bragg blazing in parallel with the discussions corresponding to the previous TM case.

## 7.2.1   Reflection grating problem

In the TE case under consideration, the incident electromagnetic fields, $\boldsymbol{E}^i$ and $\boldsymbol{H}^i$ are given as follows:

$$\boldsymbol{E}^i = \boldsymbol{y}_0 E_0 e^{-j\beta_0 z + jk_x x} \tag{7.19}$$

$$\boldsymbol{H}^i = -\boldsymbol{x}_0 \frac{\beta_0}{\omega\mu_0} E_0 e^{-j\beta_0 z + jk_x x} - \boldsymbol{z}_0 \frac{k_x}{\omega\mu_0} E_0 e^{-j\beta_0 z + jk_x x} \tag{7.20}$$

where $\beta_0$ and $k_x$ have the same meanings as those in the TM polarization case.

In region(1), the scattered electromagnetic field, $\boldsymbol{E}^s_{(1)}$ and $\boldsymbol{H}^s_{(1)}$, due to the induced current over the strip are expressed as a summation of the space harmonics as

$$\boldsymbol{E}^s_{(1)} = \boldsymbol{y}_0 E_0 \sum_{n=-\infty}^{\infty} A_n e^{-j\beta_n z - jk_{1x}^n x} \tag{7.21}$$

$$\boldsymbol{H}^s_{(1)} = -\boldsymbol{x}_0 E_0 \sum_{n=-\infty}^{\infty} \frac{\beta_n}{\omega\mu_0} A_n e^{-j\beta_n z - jk_{1x}^n x} \\
+ \boldsymbol{z}_0 E_0 \sum_{n=-\infty}^{\infty} \frac{k_{1x}^n}{\omega\mu_0} A_n e^{-j\beta_n z - jk_{1x}^n x} \tag{7.22}$$

Here $\beta_n = \beta_0 + \frac{2n\pi}{d}$ and $k_{1x}^n = \sqrt{k_0^2 - \beta_n^2}$, which follow the same definitions as those in the previous TM problem case. $A_n$ is the unknown coefficient of the $n$th space harmonics to be determined. Similarly the scattered electromagnetic fields $\boldsymbol{E}^s_{(2)}$ and $\boldsymbol{H}^s_{(2)}$ in region (2) satisfying the appropriate boundary condition at $x = -h$, i.e., the vanishment of the tangential electric field components, are expressed in terms of the summation of space harmonic as follows:

$$\boldsymbol{E}^s_{(2)} = \boldsymbol{y}_0 E_0 \sum_{n=-\infty}^{\infty} B_n \left( e^{jk_{2x}^n x} - e^{-j2k_{2x}^n h} \cdot e^{-jk_{2x}^n x} \right) e^{-j\beta_n z} \tag{7.23}$$

$$\boldsymbol{H}^s_{(2)} = -\boldsymbol{x}_0 E_0 \sum_{n=-\infty}^{\infty} \frac{\beta_n}{\omega\mu_0} B_n \left( e^{jk_{2x}^n x} - e^{-j2k_{2x}^n h} \cdot e^{-jk_{2x}^n x} \right) e^{-j\beta_n z} \\
- \boldsymbol{z}_0 E_0 \sum_{n=-\infty}^{\infty} \frac{k_{2x}^n}{\omega\mu_0} B_n \left( e^{jk_{2x}^n x} + e^{-j2k_{2x}^n h} \cdot e^{-jk_{2x}^n x} \right) e^{-j\beta_n z} \tag{7.24}$$

where $k_{2x}^n = \sqrt{k_0^2 \varepsilon_r - \beta_n^2}$ and $B_n$ is an unknown coefficient of the space harmonics.

For reference, when all the conducting strips are not present over the dielectric slab, the electric field in region (2) due to the incident field is expressed as

$$E^i_{(2)} = y_0 E_0 \frac{1+R}{1-e^{j2k^0_{2x}h}} \left( e^{-jk^0_{2x}x} - e^{j2k^0_{2x}h}e^{jk^0_{2x}x} \right) \tag{7.25}$$

Here $R$ means the reflection coefficient on the dielectric boundary with no conducting strip array present, which is given by

$$R = \frac{j\cos\theta_i\sin\left(\sqrt{\varepsilon_r-\sin^2\theta_i}k_0h\right) - \sqrt{\varepsilon_r-\sin^2\theta_i}\cos\left(\sqrt{\varepsilon_r-\sin^2\theta_i}k_0h\right)}{j\cos\theta_i\sin\left(\sqrt{\varepsilon_r-\sin^2\theta_i}k_0h\right) + \sqrt{\varepsilon_r-\sin^2\theta_i}\cos\left(\sqrt{\varepsilon_r-\sin^2\theta_i}k_0h\right)} \tag{7.26}$$

Note that the expressions in (7.21) and (7.22), (7.23) and (7.24) mean the scattered electromagnetic fields respectively in region (1) and (2) due to the induced current over the conducting strip, which is used in calculating the radiation pattern from the viewpoint of the leaky wave antenna appearing later.

The unknown surface current density on the conducting strip is expanded into a product of a series of Chebyshev polynomials and a function satisfying the edge conditions [9,22,23], which is expressed to be

$$J_y = y_0 e^{-j\beta_0 z} \sum_{l=0}^{\infty} f_l \frac{T_l(2z/a)}{\sqrt{1-(2z/a)^2}}, |z| \leq \frac{a}{2} \tag{7.27}$$

Here $T_l$ is the Chebyshev polynomial of the first kind and $f_l$ is an unknown expansion coefficient.

From (7.21) and (7.23), imposing the boundary condition that the tangential component of the electric field must be continuous across the boundary at $x = 0$, we get

$$A_n = \left(1 - e^{-j2k^n_{2x}h}\right)B_n \tag{7.28}$$

Next imposing the discontinuity condition of the tangential magnetic field due to the induced current on the conducting strip in (7.22), (7.24), and (7.27) and using (7.28), we get

$$\sum_{n=-\infty}^{\infty} A_n \frac{1}{\omega\mu_0} \left(jk^n_{2x}\cot k^n_{2x}h - k^n_{1x}\right)e^{-j\beta_n z}$$
$$= \begin{cases} e^{-j\beta_0 z} \cdot \sum_{l=0}^{\infty} f_l \dfrac{T_l(2z/a)}{\sqrt{1-(2z/a)^2}}, & |z| \leq \dfrac{a}{2} \\[2mm] 0, \dfrac{a}{2} \leq |z| \leq \dfrac{d}{2} \end{cases} \tag{7.29}$$

Multiplying both sides of (7.29) by $e^{j\frac{2n\pi}{d}z}$ and integrating the resultant terms over one period, we get the following expression:

$$A_n = \frac{k_0\eta_0}{E_0\left(jk_{2x}^n\cot k_{2x}^n h - k_{1x}^n\right)d}\sum_{l=0}^{\infty}f_l H_{ln}^* \tag{7.30}$$

where

$$H_{ln} = \int_{-\frac{a}{2}}^{\frac{a}{2}} \frac{T_l\left(\frac{2z}{a}\right)}{\sqrt{1-\left(\frac{2z}{a}\right)^2}} \cdot e^{-j\frac{2n\pi}{d}z}\,dz$$

$$= \begin{cases} \dfrac{\pi a}{2}(j)^l J_l\left(-\dfrac{n\pi a}{d}\right), n < 0 \\[2mm] \dfrac{\pi a}{2}, n = 0, l = 0 \\[2mm] 0, n = 0, l \neq 0 \\[2mm] \dfrac{\pi a}{2}(-j)^l J_l\left(\dfrac{n\pi a}{d}\right), n > 0 \end{cases} \tag{7.31}$$

Here * means complex conjugate, $J_l$ is the Bessel function of the $l$th order, and $\eta_0\left(=\sqrt{\frac{\mu_0}{\varepsilon_0}}\right)$ is the intrinsic impedance of the free space.

The total electric field $E_{(1)}$ in region (1), which is given by the sum of the incident and reflected waves with no periodic strip present along with the scattered waves due to the induced currents over the conducting strip, is expressed to be

$$E_{(1)} = y_0\left(E_0 e^{-j\beta_0 z + jk_x x} + E_0 R e^{-j\beta_0 z - jk_x x} + E_0\sum_{n=-\infty}^{\infty}A_n e^{-j\beta_n z - jk_{1x}^n x}\right) \tag{7.32}$$

Here $R$ is a reflection coefficient on the dielectric slab at $x = h$ with no conducting strip present as in (7.26). Imposing the boundary condition that the tangential electric field must vanish on the conducting strip, we obtain

$$E_0\left(1 + R + \sum_{n=-\infty}^{\infty}A_n e^{-j\frac{2n\pi}{d}z}\right) = 0, |z| \leq \frac{a}{2} \tag{7.33}$$

Next multiplying both sides of this equation by $\dfrac{T_i(2z/a)}{\sqrt{1-(2z/a)^2}}$ and integrating over the conducting strip width leads to

$$\sum_{l=0}^{\infty}\frac{2k_0\eta_0}{(1+R)\pi da}\sum_{n=-\infty}^{\infty}\frac{H_{ln}^* H_{in}}{jk_{2x}^n\cot k_{2x}^n h - k_{1x}^n}f_l = -E_0\delta_{0i}, \tag{7.34}$$

where $\delta_{0i} = 1$ for $i = 0$ and $\delta_{0i} = 0$ for $i \neq 0$.

Casting the resultant equation into the matrix form, it is given as

$$
\begin{bmatrix} & Z_{il} & \end{bmatrix} [f_l] = \begin{bmatrix} E_0 \\ 0 \\ 0 \\ \vdots \end{bmatrix}
\tag{7.35}
$$

Here the matrix element $Z_{il}$ is given as

$$
Z_{il} = \frac{2k_0\eta_0}{(1+R)\pi da} \sum_{n=-\infty}^{\infty} \frac{H_{ln}^* H_{in}}{jk_{2x}^n \cot k_{2x}^n h - k_{1x}^n}
\tag{7.36}
$$

The unknown coefficients $f_l$ are determined by solving (7.35). From the results for $f_l$, the unknown coefficients $A_n$ and $B_n$ in (7.28) and (7.30) are calculated. From the use of these results, the electromagnetic fields in regions (1) and (2) are specified.

## 7.2.2   Leaky wave problem

Similar to the previous TM polarization case, the transcendental equation for the leaky wave problem also for the present TE polarization case is obtained by replacing $\beta_0 (= k_0 \sin \theta_i)$ with the complex propagation constant $(\beta - j\alpha)$ and setting the right side tern to zero in (7.35) as follows:

$$
\sum_{l=0}^{\infty} Z_{il} f_l^L = 0, i = 1, 2, \ldots
\tag{7.37}
$$

where

$$
Z_{il} = \frac{2k_0\eta_0}{(1+R)\pi da} \sum_{n=-\infty}^{\infty} \frac{H_{ln}^* H_{in}}{jk_{2x}^n \cot k_{2x}^n h - k_{1x}^n}
$$
$$
k_{1x}^n = \sqrt{k_0^2 - \beta_n^2}
$$
$$
k_{2x}^n = \sqrt{\varepsilon_r k_0^2 - \beta_n^2}
\tag{7.38}
$$
$$
\beta_n = \beta - j\alpha + \frac{2n\pi}{d}
$$

It is to be noted that the superscript $L$ in $f_l^L$ for the leaky wave problem is attached to differentiate from the unknown coefficient $f_l$ for the scattering problem and in the calculation procedure for the leaky wave problem, $\beta_n = \beta - j\alpha + \frac{2n\pi}{d}$ is used instead of $\beta_n = \beta_0 + \frac{2n\pi}{d}$ which has been used for the scattering problem.

Enforcing the condition that the determinant of (7.37) should be zero, we obtain a complex propagation constant $\beta - j\alpha$ for $n = 0$. From this result, $f_l^L, A_n,$ and $B_n$ are computed by use of (7.37), (7.30), and (7.28).

Now we discuss how to calculate the radiation pattern from the viewpoint of the leaky wave antenna. Once $A_n$ is known, the electric field in region (1) can be obtained from (7.21). That is, since the electric field $E_y$ over the slot between the conducting strip is given by

$$E_y = \sum_{n=-\infty}^{\infty} A_n e^{-j\beta_n z} \tag{7.39}$$

The equivalent magnetic current density on the side of region (1), i.e., $x = 0^+$, is expressed, according to the equivalence principle $\boldsymbol{M} = \boldsymbol{E} \times \boldsymbol{n}$, as

$$\boldsymbol{M}(z) = -\boldsymbol{z}_0 \sum_{n=-\infty}^{\infty} A_n e^{-j\beta_n z} \tag{7.40}$$

Here $\boldsymbol{n}$ is a unit vector over the slot directing toward region (1).

The electric vector potential $\boldsymbol{F}$ due to the equivalent magnetic current density $\boldsymbol{M}$ is, after taking into account the image effect,

$$\boldsymbol{F} = \frac{\varepsilon_0}{j2} \int H_0^{(2)}(k_0|\rho - \rho'|)\boldsymbol{M}(z')\mathrm{d}z' \tag{7.41}$$

Inserting the expression for $M(z')$ in (7.40) into the (7.41) and making use of the large argument approximation for the Hankel function, we get the expression for far-field electric vector potential $F_z$ as

$$F_z = -\sum_{n=-\infty}^{\infty} A_n \sqrt{\frac{1}{j2\pi k_0\rho}} e^{-jk_0\rho} \cdot \sum_{k=1}^{N_s} \int_{(k-1)d-s/2}^{(k-1)d+s/2} e^{-j(\beta_n - k_0\sin\varphi)z'} \mathrm{d}z' \tag{7.42}$$

In this derivation procedure, the Fraunhofer approximation or the parallel ray approximation has been used [9].

Using the expression for the sum of the geometric series and taking the curl operation on the electric vector potential, i.e., $\boldsymbol{E} = -\nabla \times \boldsymbol{F}$ leads to the following expression for the far field electric field [9]:

$$E_y \cong jk_0\cos\varphi \cdot \sqrt{\frac{2}{j\pi k_0\rho}} e^{-jk_0\rho} \cdot \sum_{n=-\infty}^{\infty} A_n \frac{\sin(\kappa_n s/2)}{\kappa_n} e^{-j\kappa_n d/2} \cdot \frac{1 - e^{-j\kappa_n dN_s}}{1 - e^{-j\kappa_n d}} \tag{7.43}$$

where $\kappa_n = \beta_n - k_0\sin\varphi$, $s = d - a$ and $N_s$ is the total slot number determined so that the radiation pattern does not change appreciably even if a larger number of slots than this are taken into account. The above expression in (7.43) is used to calculate the radiation pattern from the viewpoint of the leaky wave (homogeneous) problem.

## 7.2.3 Bragg and off-Bragg blazing

By use of the above analysis method, we investigate Bragg and off-Bragg blazing phenomena which are observed in the leaky-wave supportable grating structure for the present TE polarization case. For this purpose, we consider first the Bragg blazing phenomena in the periodic conducting strip array on a grounded dielectric for which the previous experimental work [10] on the phenomena is available as in the above TM case.

Our theoretical results have been obtained for the variation of the specularly reflected power at 10.84 GHz due to the variation of the normalized dielectric thickness [ $h/\lambda$ ] and are compared with those of previous experimental work [10] in Figure 7.9.

The incident angle $\theta_i$ has been chosen to correspond to the Bragg angle in the periodic structure in Figure 7.9, that is, $\theta_i = \sin^{-1}(\lambda/2d) = 27.45°$.

Theoretical results show four positions (A, B, C, D) of minimum reflected power (sharp dips), while one sharp dip corresponding to the B position is missing in the experimental results. However, except for this discrepancy and some deviations in the positions of sharp dips, two results are observed to be in good agreement.

Because we are dealing with the TE polarization case with the electric field parallel to the strip axis, we can see that for small values of $h$ like the case A, in the parallel-plate waveguide region under the conducting strip, the lowest $TE_{10}$ mode is cutoff and the guiding structure cannot be formed between two adjacent periodic cells, so the guided leaky mode is not supportable by the grating. Therefore, as pointed out in the previous experimental work [10], the geometry under consideration is expected to behave like a corrugated structure.

Similar to the TM case in Figure 7.2, it is seen that there are two types of dip profiles, Lorentz profile-like and Fano profile-like types, as discussed in Section 7.1.3. Cases A and C belong to the Lorentz profile-like type, whereas cases B and D belong to the Fano profile-like type in that the two cases B and D show an unsymmetrical curve whose one-half side goes through an abrupt dip to zero (maximum) curve. Let us investigate the sharp dip for Bragg blazing of position A. To examine this Bragg (retroreflection) blazing anomaly more closely, we have calculated the relative powers, $P_0$ and $P_{-1}$ of $n = 0$ and $n = -1$ spectral orders

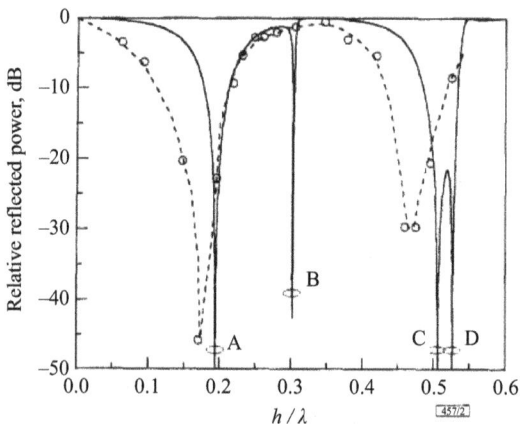

*Figure 7.9    Relative reflected power (dB) against normalized dielectric thickness*

against angle of incidence $\theta_i$ for the case of $\frac{h}{\lambda} = 0.1944$ (minimum position A case), as an example, which are given graphically in Figure 7.10.

In this case too only the two space harmonics are propagating as assumed when we deal with Bragg or off-Bragg blazing problems in the foregoing TM case, and the sum of powers of two space harmonics is calculated to be almost 100% (larger than 99.999%) over the range of incidence angle, 10°–50°.

This type of Bragg blazing anomaly is clearly neither the Rayleigh type nor the resonance type which is closely related to the presence of guided leaky waves supportable by the grating [3]. This anomaly is very similar to the type reported by Tseng *et al.* [24] in that first this anomaly is much broader in appearance than the Rayleigh type and the resonance type (compare this anomaly with the resonance type anomaly corresponding to cases B and D in Figure 7.9), and second this anomaly permits, under appropriate conditions (for suitably adjusted values of parameters such as $a/d$ and $h/\lambda$, etc., as in the case of minimum position A in Figure 7.9), most of the incident power to be converted into $n = -1$ spectral order. The maximum power conversion efficiency amounts to $\sim$99.999% at the center of the anomaly at the Bragg angle of 27.45°. In summary, as mentioned above, the guiding structure cannot be formed between two adjacent cells in the present geometry, which is also the case with rectangular groove geometry, so, both geometries have been found to show almost the same scattering patterns when illuminated by the Gaussian beam wave as discussed in [25]. For the same reason, the relative power pattern of $P_0$ in the present structure for case A is very similar from the viewpoint of broad characteristics in appearance as shown in Figure 7.10 to that in the comb grating [24] under the condition that the guiding structure is not formed along the grating.

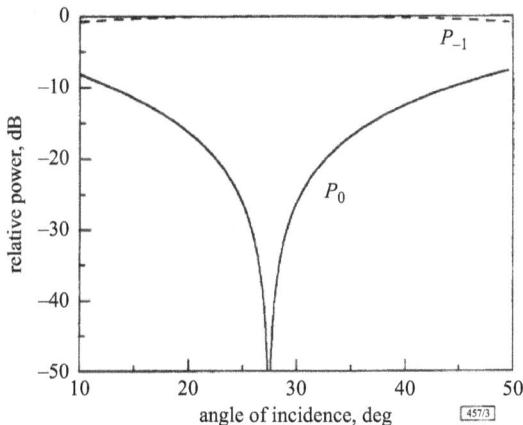

*Figure 7.10*    *Relative power of $P_0$ and $P_{-1}$ against the angle of incidence for the case of $d = 30$ mm, $a = 0.566d$, $f = 10.566$ d GHz, and $\varepsilon_r = 2.57$ (corresponding to case A in Figure 7.9)*

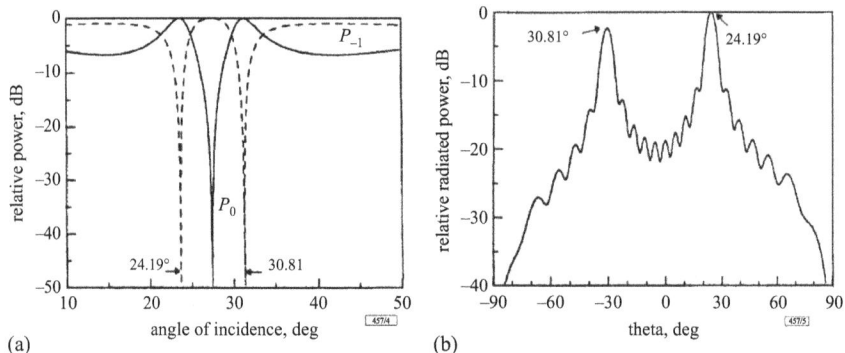

*Figure 7.11    Scattering problem and leaky wave problem under the Bragg blazing for the case $\varepsilon_r = 2.57$, $d = 30$ mm, $a = 0,566d$, $f = 10.840$ GHz, $\frac{h}{\lambda} = 0.5061$ (corresponding to the case A, Figure 7.9). (a) Relative power of $P_0$ and $P_{-1}$. (b) Radiation pattern from viewpoint of leaky wave antenna.*

Similar to the TM polarization case, in this TE polarization case, off-Bragg blazing has been observed only if the guided leaky wave is supportable by the grating under consideration, while Bragg blazing may occur simply due to periodicity even when the leaky wave is not supportable in the grating.

From the above discussions, the observation that off-Bragg blazing is possible only if the guided (leaky) wave is supportable by the present geometry is thought to be compatible with the fact that the off-Bragg blazing in the rectangular groove geometry has not so far been found for the TE polarization case [26,27].

Note that the Bragg blazings corresponding to the three dips of B, C, and D cases in Figure 7.9 occur when the leaky wave is supportable by the grating structure, unlike case A. Among the B, C, and D cases, B and D cases go through an abrupt dip-to-maximum curve whereas the C case does not. To examine the Bragg blazing phenomenon of the C case, we have calculated the relative powers of $n = 0$ and $-1$ spectral orders from the viewpoint of the reflection grating problem for the case that $\frac{h}{\lambda} = 0.5061$ as an example of Lorentz profile type. Figure 7.11(a) shows the relative power pattern of $P_0$ and $P_{-1}$ for case C. In this case too, only the two propagating modes are assumed. For the given parameters in Figure 7.11, the complex propagation constant $\beta_0 - j\alpha_0$ for the fundamental leaky mode of $n'=0$ along the guided direction (i.e., $z$-axis) is found to be 302.519 [rad/m]-$j$5.866 [Np/ m], from which radiation pattern from the viewpoint of the leaky wave antenna can be obtained by use of the above method. The primed number is used for the leaky wave mode to differentiate from the Floquet mode for the scattering (reflection grating) problem. Figure 7.11(b) shows the radiation pattern, where the two maximum beam angles are $-30.81°$ and $24.19°$ which corresponds to $n'=-2$ and $n'=-1$ space harmonics for the leaky wave case respectively.

Note that the two incident angles at which the relative power of $n = 0$ and $n = -1$ modes become maximum and minimum in Figure 7.11(a) correspond to 24.19° and 30.81' respectively. Note that these two incidence angles have been obtained by making two radiation angles ($-30.81°$ and 24.19°) of the leaky wave same sign. At the same time, the maximum power conversion into $n = -1$ spectral order is observed to take place between these two angles under the Bragg condition, in which case the maximum power conversion efficiency is $\sim$99.999%. For this reason, the Bragg blazing of case C is thought of as a kind of Bragg blazing phenomenon that can be observed in the leaky wave supportable reflection grating structures. As mentioned above, the Bragg blazing of case C belongs to the Lorentzian profile type which is symmetrical for the dip position. On the other hand, for the cases of B and D, the shapes of dips in Figure 7.9 are not symmetrical and at the same time go through an abrupt dip to a maximum curve on one half side. This graphical feature belongs to the Fano profile type. As discussed in the dispersion characteristics of Bragg and off-Bragg blazing phenomena in Figure 7.14 appearing later, such types occur under the condition of coupling point [18,19]. In other words, for the B and D cases, the coupling point condition of the standing wave formation along the periodic structure is additionally imposed to the simple transverse resonance condition for the C case of Bragg blazing. So the Bragg blazing phenomena for B and D cases are much more frequency sensitive and narrower in appearance than that of the case C.

As mentioned above the off-Bragg blazing has been observed only if the guided leaky wave is supportable by the present geometry in the TM polarization case. This is also the case with the TE polarization case.

That is, the conditions for the off-Bragg blazing phenomena to occur are the same for both the previous TM and this TE polarization case. The off-Bragg blazing may also occur in the fast-wave region as well as the slow-wave region as in the Bragg blazing case. So the off-Bragg blazing phenomena in the TE polarization case are essentially the same as that in the TM polarization in Figure 7.4. For this reason, we do not repeat the off-Bragg blazing phenomena here. For interested readers see [27,28]. Next, we continue to discuss the two types of Bragg blazing phenomena, the Lorentzian profile-like type and the Fano profile-like type in the TE polarization case.

### 7.2.4   Consecutive occurrence of Bragg blazing of Lorentzian and Fano profile-like type; characteristic mode current distribution

Here we proceed to discuss the two types of Bragg blazing phenomena, (inverted) Lorentzian profile and Fano profile type. To this end, theoretical results for the variation of the specularly reflected power due to the variation of the normalized dielectric thickness ($h/d$) for the incidence (Bragg) angle $\theta_i = \sin^{-1}\left(\frac{\lambda}{2d}\right) = 25°$ ($\lambda$: free space wavelength) for TE polarization are given in Figure 7.12. Here six cases of Bragg blazing phenomena corresponding to points A–F are observed. The reason for the choice of the low dielectric constant in Figure 7.12 is that consecutive occurrences of Bragg blazing of Lorentzian and Fano profile types can be clearly seen.

Comparison between Figures 7.9 and 7.12 shows that consecutive occurrences of the Lorentzian and Fano profile-like type are clearly seen for the low dielectric case, which is also the case of the comparison between Figures 7.2 and 7.5. It is worth noting that the two types of profiles alternate as the dielectric substrate inside the guide increases. This may be a general property of the Bragg blazing phenomenon in this situation.

Similar to the TM polarization case as in Figure 7.5, there are two types, one is the Lorentzian profile type which is almost symmetrical, and the other one is unsymmetrical and goes through an abrupt dip to maximum curve on one half side. To understand the difference of the Bragg blazing between the Lorentz profile and the Fano profile further, we have investigated the induced current distribution over the conducting strip for the two cases. As a representative example among B, D, and F cases for which all the current distributions are almost the same as each other, the current distribution corresponding to case B in Figure 7.12 is illustrated in Figure 7.13(a).

The current distribution over the conducting strip in case B closely resembles the characteristic mode [15–17]. This similarity is evident as the current distribution maintains a constant phase across the strip, even though the magnitude of the current varies significantly.

On the other hand, the current distributions for A, C, and E cases have been observed to be quite different from those for B, D, and F cases. As a sampled example, the induced current distribution for case A in Figure 7.12 is given in Figure 7.13(b) for comparison. It is interesting to note that even under the condition that the leaky wave is supportable, the Bragg blazing phenomena are divided into two categories, depending on the induced current distribution over the conducting strip like the previous TM polarization case. One is the Bragg blazing type whose current distribution qualifies as the characteristic mode current since the current has a constant phase across the strip although the magnitude of the current significantly varies across the strip as shown in Figure 7.13(a). The other one is the Bragg blazing type whose current distribution is quite different from the former type and

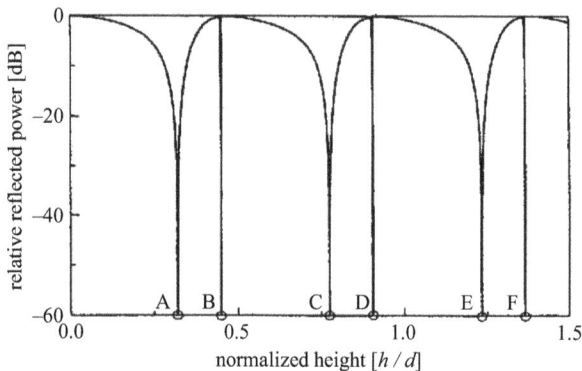

*Figure 7.12    Relative reflected power against normalized dielectric height (h/d).* $(d = 1.1831\lambda, a = 0.5d, \varepsilon_r = 1.03, \theta_i = 25°)$

whose current magnitude is much smaller than the former type as shown in
Figure 7.13(b). Comparison between the current distributions in Figures 7.6(a) and
7.13(a) clearly shows that the two distributions satisfy the appropriate edge conditions according to the TM and TE polarizations, respectively.

## 7.2.5 Dispersion characteristics of Bragg and off-Bragg blazing phenomena

To examine the relationship between the Bragg blazing phenomena and the
leaky waves, the dispersion diagram, which has been obtained for $d = 1.1831\lambda$,
$a = 0.5d$ and $\varepsilon_r = 1.03$ is given in Figure 7.14. In the dispersion curve, six locations

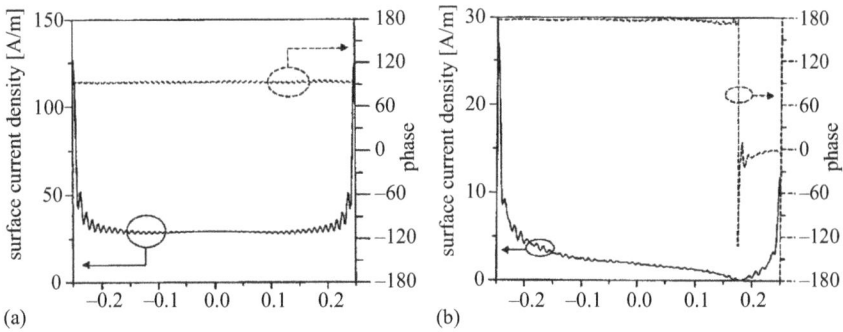

*Figure 7.13* *Induced current distribution on the strip. (a) Corresponding to point
B in Figures 7.12 and 7.14. (b) Corresponding to point A in
Figures 7.12 and 7.14.*

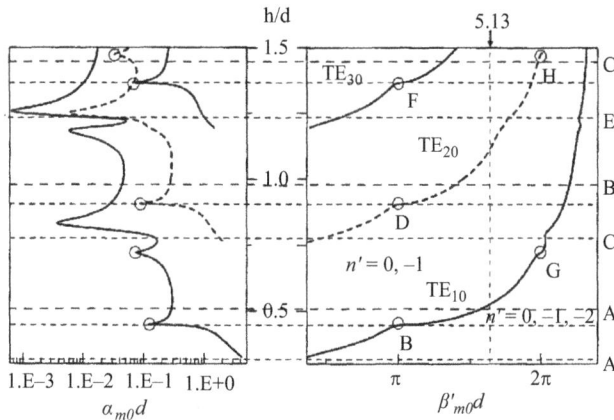

*Figure 7.14* *Dispersion diagrams for $d = 1.1831\lambda, a = 0.5d, \varepsilon_r = 1.03$, where
$\beta'_{m0} - j\alpha_{m0}$; complex root of mth order leaky wave. (a) $\alpha_{m0}d - h/d$
diagram. (b) $\beta_{m0}d - h/d$ diagram.*

(A–F) where Bragg blazing phenomena occur are marked, along with three locations where off-Bragg blazing phenomena occur. The off-Bragg blazing phenomena are observed for the cases A′, B′, and C′ near the coupling points as expected from the previous explanation that the plane waves in the scattering problem possess real wave numbers while the wave number of the leaky wave is complex as mentioned already in Section 7.1.3.

Note that all the Bragg and off-Bragg blazing phenomena were obtained under the assumption of two propagating spectral order modes except the G case where three propagating spectral order modes are assumed.

As pointed out above, the Bragg blazing phenomena are divided into two groups, one group composed of A, C, and E cases, and the other group composed of B, D, and F cases. For the former group, the Bragg blazing phenomena are observed only when the complex transverse resonance equation corresponding to the homogeneous equation is satisfied. That is to say, the Bragg blazing phenomena corresponding to A, C, and E cases can occur only when the complex leaky roots are located on the dispersion curve irrespective of coupling point condition [18] as seen in Figure 7.14(a) and (b). On the other hand, for the latter group, the Bragg blazing phenomena are observed when not only the complex transverse resonance condition but also the coupling point condition is satisfied as discussed already in the TM case.

Under the coupling point condition, also in this TE polarization case, the induced current distribution over the conducting strip justifies as characteristic mode current since the current has a constant phase across the conducting strip although the magnitude of the current significantly varies across the strip. What is more, this current distribution, as a strong resonant current, is much larger than that for the former case. The above discussions on the two kinds of Bragg blazing phenomena are essentially the same as those for the TM polarization case. Recall that B, D, and F cases are located in the region where only the two modes, $n = -1$ and $n = 0$ modes are propagating. According to the terminology in Sections 7.1.3 and 7.1.4, the Bragg blazing of the former group is of non-resonant type and belongs to Lorentzian profile type, on the other hand, the Bragg blazing of the latter group is of resonant type, and belongs to Fano profile type.

As discussed above the Lorentzian profile type takes a symmetrical curve whereas the Fano profile type takes an unsymmetrical curve and at the same time goes through an abrupt dip to a maximum curve on one-half side as seen in Figure 7.12. It should be recalled that also for the Fano profile type in the TM polarization case, the Bragg blazing phenomena are observed only when not only the complex transverse resonance method but also the coupling point condition is at the same time satisfied.

Under the coupling point condition, the standing wave pattern along the guided $z$-direction is formed, as a result of which a strong resonant current distribution like the characteristic mode current results. In summary, for the Fano profile type of Bragg blazing phenomena, if the two propagating modes are assumed for the scattering problem, the coupling points conditions that $\beta_{m0} = \frac{\pi}{d}$ and $\alpha_{m0} = $ minimum should be met as seen in Figure 7.14. If we consider the case that three propagating modes are assumed, the coupling point conditions are given as $\beta_{m0} = \frac{2\pi}{d}$ and $\alpha_{m0} = $ minimum for G and H cases as seen in Figure 7.14. This means that the longitudinal resonant condition—standing wave formation along the guided direction—corresponding to the coupling points should be additionally imposed for the Fano profile type. Recall that for the Lorentzian profile type only complex transverse resonance conditions should be met. It can be concluded that the abrupt transition from the dip to the maximum occurs only in the case of the Fano profile type, where both the complex transverse resonance condition and the longitudinal resonance condition are met.

It is worthwhile to note again that the Lorentzian profile type of Bragg blazing occurs only when the complex transverse resonance condition for the presence of the leaky root is met. In contrast, the Fano profile type occurs when the longitudinal resonance (corresponding to coupled point condition) as well as the complex transverse condition is satisfied. The above consecutive appearance of Lorentzian and Fano profile type is similar to the normal incidence resonance phenomena in weakly modulated gratings as a guided mode resonant grating [14].

So far we have discussed scattering resonance phenomena occurring in the reflection grating structure with the main interest centering on the Bragg and off-Bragg blazing phenomena as a kind of Wood anomalies. Recall that the periodic conducting strip array over the grounded dielectric has been chosen to be the structure under consideration. The reason for this choice is that the structure is expected to give a full explanation for the scattering resonance phenomena associated with a leaky wave in comparison with the rectangular groove reflection grating [29] which has been studied extensively in the application to Bragg and off-Bragg blazing phenomena. Another reason is that the structure is easy to transform into a transmission grating to be considered here.

## 7.3 Retroreflection in the reflection grating and its analogous transmission in the transmission grating and the guided mode resonance filter structures

Here we are to deal with the Bragg blazing phenomena in the reflection grating from the viewpoint of retroreflection and to consider further the refraction analogous to the above retroreflection in the transmission grating structure whose complex propagation constant is chosen to be the same as that for the reflection grating. The relation between leaky waves and the guided mode resonance filter is also briefly touched upon.

### 7.3.1 Retroreflection in the reflection grating and its relation to the transmission in the transmission grating when the two gratings share the same leaky root

Historically speaking, the Wood anomalies of optical diffraction gratings, meaning in a broad sense the rapid change of reflection or transmission characteristics within very small variation of the frequency or incidence angle of the incidence wave, have been of interest since they were discovered by Wood in 1902 [1,30]. But as mentioned above, it was not until Hessel and Oliner reported their work on that topic in 1965 [2] that significant progress in physical understanding was made. They pointed out that there are basically two types of anomalous effects, the Rayleigh type due to one of the spectral orders appearing at the grazing angle and the resonance type due to possible guided complex leaky wave supportable by a grating. They modeled the reflection grating as a periodic reactive surface called the periodically modulated reactance surface (PMRS) model and presented calculated results for the scattering resonances corresponding to the Wood anomalies based on the surface model for the reflection gratings.

Between the two types of anomalies, the Rayleigh type has been well understood, whereas the resonance type, which is associated with the leaky wave, has been investigated in connection with ongoing research topics such as metasurface [31–33], guided mode resonance filter (GMRF) [34,35] in the integrated optics area, and extraordinary optical transmission (EOT) problem [6,7] in a nanostructured slab. Recently various promising metasurface structures [31–33] such as a bipartite Huygen's surface and plasmonic interface (consisting of an array of V-antennas) have been proposed. These types belong to the discretized metasurface which does not involve a transfer of power through surface or leaky waves. In contrast, the Bragg blazing phenomena that have been dealt with here can be used as a retroreflecting surface as a continuous type that can support guided leaky waves [28].

Here we are to investigate the retroreflection [36,37] (Bragg angle reflection) phenomena in the reflection grating and the transmission phenomena in the transmission grating structure whose leaky root (complex propagation constant) is chosen to be the same as that for the reflection grating structure. As a reflection grating for retroreflection, we have chosen the periodic strip grating on a grounded dielectric as shown in Figure 7.15(a), where the frequency $f$ of the incident plane wave is 10 GHz, the period $D$ of the conducting strip array is 30 mm, the strip width $2W$ is 15 mm, and the height $h$ of the dielectric substrate whose relative dielectric constant $\varepsilon_r = 2.57$ is 9.53617 mm.

Figure 7.15(b) shows the transmission grating which comprises double layers of periodic conducting strips with dielectric slabs interposed between them. Note that the thickness $h$ in the reflection grating is chosen to be 1/2 of that in the transmission grating. In the case of the TE polarization under consideration where the incident electric field is parallel to the axis of the conducting strips, if we assume that antisymmetric leaky mode (whose electric field distribution is an odd function for the $x = 0$ plane) be set up in the transmission grating, the upper bisected structure obtained by putting the shorting (conducting) plane at $x = 0$ becomes the same structure to the reflection grating structure. So in this case the complex roots for the homogeneous leaky problem are expected to be the same as each other for both the reflection and transmission gratings as confirmed later in numerical results.

Figure 7.16 shows an example of Bragg blazing phenomena corresponding to the retroreflection.

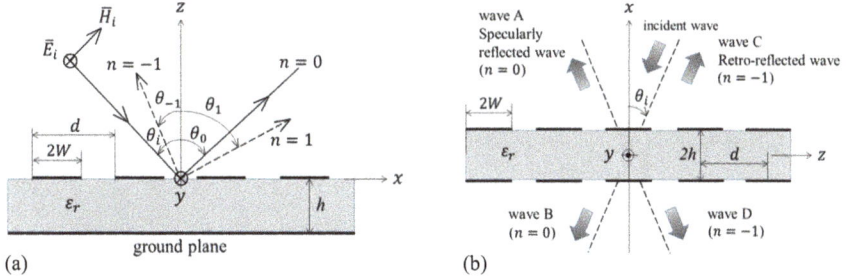

*Figure 7.15   Reflection and Transmission grating having the same leaky wave root complex propagation constant. (a) Reflection grating. (b) Transmission grating.*

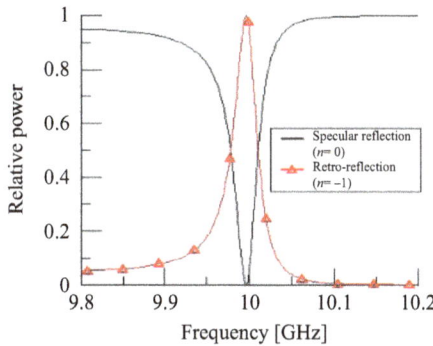

*Figure 7.16   Retroreflection (Bragg blazing) characteristics at Bragg angle (30°) in the reflection grating*

This Bragg blazing (retroreflection) belongs to the Wood anomaly for the case that leaky wave is supportable along the reflection grating structure. A comparison of the frequency characteristics between this leaky wave-related retroreflection and the recently proposed metasurface such as an aggressively discretized metasurface [32] shows that the bandwidth of the present type is significantly narrower than that for the above recently proposed metasurface. Figure 7.17 shows the transmission characteristics of the transmission grating whose complex propagation constant is chosen to be the same as that for the reflection grating.

The maximum transmission angle corresponding to the wave D through the transmission grating is seen to be symmetrical to the Bragg angle (corresponding to

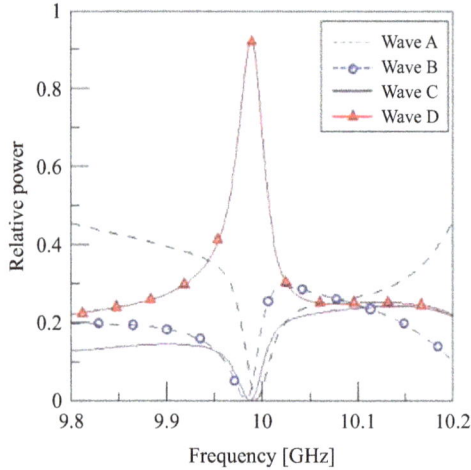

*Figure 7.17    Refraction characteristics of the transmission grating whose transmission angle is symmetrical to the retroreflection (Bragg) angle for the grating plane*

30°) in the reflection grating case. Note that we deal with the TE polarization case in which the electric field is parallel to the strip axis and that the thickness of the dielectric substrate in the transmission grating is chosen to be double that of the reflection grating.

The frequency for both the Bragg blazing (retroreflection) in the reflection grating and the maximum transmission in the transmission grating is exactly the same to each other. From this, the leaky mode is seen to be anti-symmetrical for $x = 0$ in the transmission grating, as a result of which the leaky electric field in the reflection grating becomes identical to that of the upper half region $0 \leq x \leq h$ in the transmission grating as expected. So even if we put a conducting plane at $x = 0$ in the transmission grating structure, the complex leaky propagation constant does not change from the original complex propagation constant of the transmission grating. For this reason, the Bragg blazing angle in the reflection grating and the maximum transmission angle in the transmission grating are found to be symmetrical for the plane at $x = 0$. There are two types of Bragg blazing (retroreflection) phenomena, a broad and a narrow type in appearance. The difference between the former and latter types is that the former types are observed only when the transverse resonance condition for the leaky wave propagation is met. On the other hand, the latter type is observed to occur under the coupling point condition that the longitudinal standing wave along the periodicity is set up in addition to the above transverse resonance condition. For convenience, the former is called the non-resonant type and on the other hand, the latter is called the resonant type for distinction. This means that all Bragg blazing phenomena that are observed in the leaky wave supportable grating do not necessarily belong to the resonant type having a narrow type in appearance. Recall that the coupling point condition is

$\beta d = \frac{\pi}{2}$ for two propagating diffraction mode problems and $\beta d = \pi$ for three propagating mode problems.

As mentioned above, by use of the leaky waveguiding structures, it is possible for the reflection and transmission gratings to be so implemented that the Bragg blazing (retroreflection) angle of the reflection grating and the transmission angle of the transmission grating may be symmetrical for the horizontal plane at the middle height of the transmission grating.

For convenience, let us consider the case that only the two Floquet modes are assumed to be propagating in the half-space region over the grating surface, zeroth mode (specularly reflected mode) and $m = -1$ mode. In this case for the calculation of diffraction efficiency, we have only to investigate only the specularly reflected zeroth order mode because the two mode powers are summed to be unity (one hundred percent) from the power conservation relation under the assumption of no loss system. As mentioned above, to implement the reflection and transmission gratings whose Bragg (retroreflection) angle and transmission angle may be symmetrical for the grating plane, we may use one between two leaky solutions which gives the broad Bragg blazing of the non-resonant type or narrow Bragg blazing of the resonant type. The leaky solution which gives the desired broad or narrow Bragg blazing can be found by varying the grating thickness. It is worthwhile to mention that the broad Bragg blazing of the non-resonant type and the narrow Bragg blazing of the resonant type are related to the Lorentz type and Fano type resonance respectively as discussed above.

## 7.3.2   Bandpass filter design

Here we consider the bandpass filter of a transmission grating type and investigate the reflection (or transmission) characteristic for the normally incident TE wave under the assumption that only the zeroth order be propagating. Figure 7.18 shows

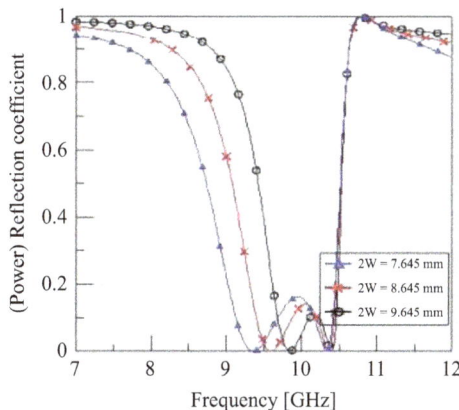

*Figure 7.18   Reflection characteristic of the transmission grating as a bandpass filter*

the reflection characteristics of the transmission grating with the conducting strip width 2W as a parameter, when the period D = 20 mm, 2h = 16 mm, and $\varepsilon_r$ = 2.57.

The present transmission grating is thought of as a kind of metal-based guided mode resonance filter (GMRF) [35] since this structure corresponds to the combined structure of the grating of the double layer and the dielectric substrate between the two layers.

It is interesting to note from Figure 7.18 that the bandpass filter characteristics such as the right side high-frequency limit are very insensitive to the variation of the conducting strip width 2W. This means that the high-frequency passband characteristics are mainly determined by longitudinal periodicity. What is more, the cut-off slope (roll-off rate) of the right side seems to be much stiffer than the conventional filters such as Fabry–Perot [34] type and multilayered type of dielectric slabs. This corresponds to the Fano [4,5,7] type curve which constitutes the abrupt curve from zero to maximum associated with the standing wave formation along the longitudinal periodicity due to the coupling point condition [18,19].

On the other hand, the cutoff characteristics of the left side low-frequency limit are seen to change sensitively according to the variation of the conducting strip width 2W from Figure 7.18. A careful look at the figure reveals that as the conducting strip width 2W increases, the pass band width is seen to decrease. This observation is compatible with the fundamental theory of the leaky wave that as the aperture portion $\left(\frac{D-2W}{D}\right)$ decreases the attenuation constant $\alpha$ decreases and so the bandwidth decreases. So it can be said that the cutoff characteristics of the left side low frequency are determined by the complex root of the transverse resonance condition whereas the abrupt curve of the right side cutoff characteristics like Fano type is determined by the coupling point condition relevant to the standing wave formation along the periodicity.

### 7.3.2.1  *k–β* **Diagram and some Fano-like curve feature**

To assess the Fano-type resonance curve occurring in the bandpass filter under consideration, we have investigated the *k–β* diagram and the resonant field configurations. Some numerical results are illustrated in Figure 7.19. Figure 7.19(a) shows the Brillouin diagram where the numbers in the bracket mean the propagating Froquet mode numbers.

As seen in Figure 7.19(a), the transmission resonance whose curve is composed of the right stiff side and less stiff left side is observed near the 10 GHz where the group velocity becomes minimum. This seems to be compatible with the observation that the resonance type of Wood's anomaly occurs when the group velocity becomes almost zero, which corresponds to the condition that the standing wave is set up along the periodic structure. Figure 7.20(b) illustrates the electric field distribution for two frequencies of the leaky wave guiding structure along the *x*-axis at $= -\frac{h}{2}$.

As seen from the figure, at the frequency of 10.4178 GHz corresponding to the transmission zero of the stiffer side of the transmission curve, a stronger resonance is observed in comparison with the resonance at 9.945 GHz corresponding to the

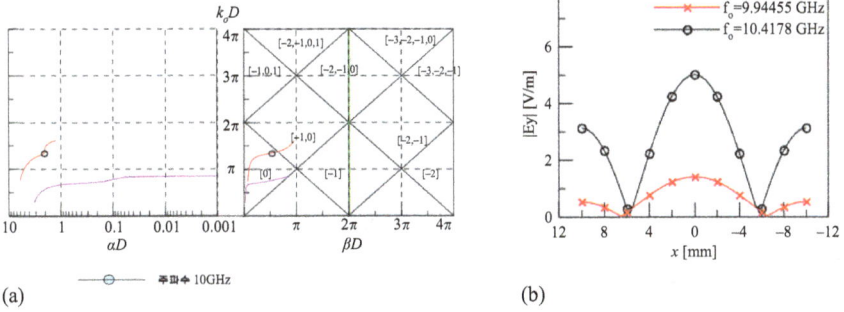

*Figure 7.19    The Brillioun diagram of the transmission filter and the electric field distribution. (a) The Brillioun diagram for the case that $\varepsilon_r = 2.57$, $D = 20$ mm, $2W = 10$ mm, and $h = 16$ mm. (b) y-component electric field distribution for two frequencies of 9.945 GHz and 10.418 GHz for incidence angle $\theta = 0°$.*

transmission zero of the less stiff side of the transmission curve. This suggests that the Fano type of the transmission curve which is featured by abrupt change from transmission zero to peak is related to the formation of the standing wave along the periodic structure.

## 7.4    Further comments

As discussed so far we have seen that the leaky wave plays a key role in the scattering resonance phenomena which can be used in various engineering areas such as Bragg and off-Bragg blazing in the reflection grating area, retroreflecting surface in the metasurface area, and band rejection and bandpass filter in the diffractive optics area. There are some more important areas that deserve further discussions on other application areas relevant to leaky wave radiation as discussed below.

### 7.4.1    Blindspot phenomena in the planar array structure

In the moderate-size array [38], for example, seven collinear elements in an E-plane row by nine parallel elements in an H-plane row, the input impedance of the center element does vary considerably with scan angle and the variation depends on the plane of scan. As the array is scanned in the E-plane to 90° corresponding to endfire, the real part of the input impedance is tending toward zero, which in turn means the element is tending not to radiate. Although all the elements in the array will not have exactly the same behavior, most of them except the edge element will behave similarly and the entire array will tend not to radiate. This phenomenon is known as the blindspot phenomenon in the 2-dimensional dipole array [38].

Let us consider the blindspot phenomenon in the structure under consideration. If we assume that each conducting strip in the periodic strip array structure over the grounded dielectric under consideration be fed by probes, simulated via sheet currents with uniform density and progressive phases, it can be taken as a simplified two-dimensional microstrip-patch element phased array structure. So through the investigation into E-plane scan performance, we can get some physics on the relation between the scan blindness of the phased array as a forced resonance problem and the leaky wave radiation as a free resonance problem.

Analysis [39,40] revealed that the scan angle where the resonant peaks of the real part $R_a$ of the active impedance occur correspond to the main beam angles of the leaky wave from the viewpoint of leaky wave antenna [40]. In particular the resonant peak value of $R_a$ for the certain normalized thickness of the dielectric substrate $\frac{h}{\lambda_0} = 0.30$ for example becomes larger than 10,000 [$\Omega$], in which case the magnitude of the reflection coefficient $|\Gamma|$ looking into the radiating part from the feeding circuit becomes nearly unity. It is worthwhile to mention that this scan blindness which corresponds to total reflection ($|\Gamma(\theta)| = 1$) is seen to be closely related to the leaky wave radiation of the given phased array geometry, which is consistent with a large number of previous discussions on the basic cause of element pattern nulls of certain phased array antennas [41]. This investigation constitutes a good example showing that the scan blindness under discussion is closely related to the Wood anomaly associated with the leaky wave supportable by the phased array structure under consideration.

Note that for both the former dipole array and the latter microstrip phased array cases, the scan blindness phenomena are associated with guided wave structures formed along the E-plane, surface wave, and leaky wave, respectively.

## 7.4.2    Smith–Purcell radiation problem

There is another interesting problem, a forced resonance problem which is associated with the leaky wave radiation as a free resonance problem. Representative is the phenomenon of Smith–Purcell (SP) radiation which is a kind of radiation problem extracted from a moving electric charge over a periodic grating structures. Since this phenomenon [42] was observed in 1953, many researchers have contributed to understanding the nature of this radiation as listed in [42]. Among them, Palocz and Oliner [43] pointed out a strong similarity between SP radiation and leaky wave antennas by use of which they solved the problem of SP radiation for a sheet electron beam of finite thickness passing between two parallel planar strip transmission grating, and Hessel [44] dealt with the problem of radiation from an electric current sheath above an SMRS (sinusoidally modulated reactance surface) which was used as a model of an optical reflection grating. For more details on other prior works, refer to Hessel's work [44]. However, because the SMRS is a kind of an over-simplified model as the author [44] mentioned, the study of such a model is not expected to provide much practically useful information (although it sheds light on the properties common to a large class of grating to some extent).

So some study has been done on the radiation characteristics of the SP type in connection with that of the leaky wave antenna problem in the present periodic strip grating over a grounded dielectric in expectation of the practical results than those from the above oversimplified SMRS model.

Through the study, it has been found that, in the case of the SP-type radiation under consideration, when a particular relation between the velocity $v_0$ of the moving electron beam and the real part $\beta$ of the complex propagation constant ( $\beta - j\alpha$ ) of the leaky wave supportable by a grating structure underneath the electron beam is satisfied, the $n = -1$ space harmonic radiates strongly and peaks sharply to the maximum. In more detail, when $\beta_0 \left( = \frac{k_0 c}{v_0} \right) = \beta$ where $k_0 = \omega \sqrt{\mu_0 \varepsilon_0} = \frac{2\pi}{\lambda}$ and $c$ is the velocity of light in free space, the sharp maximum radiation of the $n = -1$ space harmonic is observed to occur. This feature is in contrast to the case of the rectangular groove (reflection) grating [45], where the sharp fluctuations and so maximum radiation of the radiation intensity of $n = -1$ space harmonic occur at values of $\theta_{-1}$ at which an evanescent wave of neighboring space harmonic changes into a propagating wave, and those at which an evanescent mode in the grooves changes into a propagating mode, and vice versa.

In connection with $k_0 - \beta$ diagram [46] it is also seen that the magnitude of the radiating $n = -1$ increases as the radiation angle directs more toward the grazing direction corresponding to the opposite direction of the current vector. This means that the maximum radiation of the $n = -1$ space harmonic occurs at the operating point just after crossing the boundary where the $n = -1$ space harmonic goes through its cutoff (grazing) point from evanescent to propagating. Some sharp fluctuations of magnitudes of the space harmonics are also observed to occur near the coupling points [18,19], which play an important role in the Fano type of reflection and transmission. The physics of the SP radiation can be applied to various areas [47] such as tunable electromagnetic sources, particle accelerators, free electron lasers, etc.

### 7.4.3   *Bragg/off-Bragg blazing grating, dichroic surface, and grating coupler*

All these three structures use the interference between the reradiated wave above the structures through the excitement of the leaky wave and the reflected wave generated directly at the top surface of the structures. In the Bragg/off-Bragg blazing structure where only the two modes, $n = 0$ and $n = -1$, are assumed to be propagating over the reflection grating, if the reradiated and the reflected wave destructively interferes to be canceled, only $n = -1$ mode remains and so the Bragg or off-Bragg blazing phenomena occur. On the other hand in the case of the dichroic surface in which only one space harmonics is assumed to be propagating in the transmission grating structure, according as the reradiated wave and reflected wave at the top surface of the grating are in phase or out of phase, strong reflection or transmission occurs. Because the phases are frequency-dependent, the overall reflection can have the desired frequency-selective behavior. For this reason dichroic surface [48] feature results. Usually for the (blazed) reflection grating applications such as polarizer, (de)multiplexer, and RCS reduction techniques, the dimensions of the problem are chosen such that only the two space harmonics will propagate independent of infinite or finite [21,25] periodic case.

On the other hand for the grating coupler application [49] which is possible only for the finite periodic case, only the one space harmonic is assumed to propagate, which is also the case with the usual leaky wave application. A finite array of the conducting strip over a grounded dielectric is an example of a grating coupler. In this case, the wave source of finite width such as the Gaussian beam is assumed to be incident upon the finite array and the main concern is how the coupling efficiency of the Gaussian beam into the surface wave is supportable in the grounded dielectric region. As expected from the fact that the leaky wave antenna structure and optical periodic grating coupler are working on the same physical principle, the maximum coupling occurs when the incidence angle of the Gaussian beam corresponds to the maximum radiation angle from the viewpoint of the leaky wave radiator. In general, in the case of the grating coupler, the obliquely incident Gaussian beam is coupled into the finite periodic strip array over the grounded dielectric and reradiated above the grating. This reradiated wave above the grating adds to the reflected plane wave generated directly at the top surface of the grating to give the total reflected field. When the two components are out of phase, the reflected field is small and the maximum coupling into the surface wave guiding structure of the grounded dielectric occurs. On the other hand, in the case of a dichroic surface, the obliquely incident plane wave is coupled into the leaky wave structure, for example, which is composed of an infinite periodic array of dielectric strips [48] and reradiated into the air regions above and below the layer through the same space harmonic, thereby acting as a leaky wave. When the reradiated wave above the grating and the reflected plane wave generated at the top surface of the grating are out of phase, the reflected field is small like the case of the grating coupler. Instead, the strong transmission occurs at a somewhat different frequency from that of the strong reflection. The common feature for the above three structures is to control the interference between the reradiated leaky wave and reflected wave at the top surface of the grating. In other words, the working principles of the above three structures as well as the bandpass filter structure are related to interference of excited leaky mode with the direct generated mode without intervention of the leaky wave structure, which is an important attribute of the Fano resonance.

### 7.4.4 Common features among Wood anomaly, guided mode resonance, and transmission resonance through Bethe's small hole array

Since the relevancy of the leaky wave to the resonant type of Wood anomaly was pointed out [3], the Bragg blazing and off-blazing phenomena associated with the leaky wave have been clarified [20,37]. These topics redrew the research interest in connection with the guided mode resonance (GMR) [34] type of filter in the diffractive optics area and so has been being investigated in that direction for several decades. What is more, it has been lately extended to metasurface research including the retroreflection phenomenon [36].

So, it would be interesting to compare the working principles among the above phenomena. As is well known, the study on the Wood's anomalies of the resonant type was mainly restricted to the reflection grating type made of perfect conducting material under the assumption of two propagating modes, $-1$ order mode and specular reflection mode. In this case, the coupling point is formed as an operating point where the group velocity becomes zero so that the leaky wave stop band is formed outside the guided region. Under this condition, the coupling point is observed where the Bragg condition is met such that the real part of the leaky wave propagation constant $\beta$ corresponds to $\beta = \frac{\pi}{d}, \frac{2\pi}{d}, \ldots$, etc. As a result, the standing wave is set up along the lateral direction and the incident wave is coupled to the leaky wave structure such that the power conversion of the incident wave into the $-1$ order mode corresponding to the resonant Wood's anomaly of the Bragg blazing may occur.

In the guided mode resonant filter (GMRF) case including planar periodic waveguides or photonic crystal slab, the incident fields interact with the filter structure when the stop bands form at the Bragg condition mainly at the second leaky stop band corresponding to $\beta = \frac{n\pi}{d} (n = 1, 2, 3, \ldots)$. Unlike the above Wood anomaly where the two propagating modes are assumed, in the GMRF problem, only one propagating mode is assumed. With the exception of this distinction, in both of the aforementioned problems, the lateral standing wave should be established within the stop band for the interaction between the incident wave and the structure to take place. It deserves further comment that also in the transmission resonance condition for the square array of Bethe's small hole array appearing in the next Chapter 8, the single linear array which is identified as a single linear EOT unit meets the Bragg condition of $\beta = \frac{n\pi}{d}$ for the second leaky stop band along the E-plane direction.

Based upon the above discussions, it can be said that the above three phenomena, Wood anomaly of resonant type, GMRF, and the transmission resonance in Bethe's small hole array structure can be understood in connection with the Bragg condition.

## References

[1]  Wood R.W. 'On a remarkable case of uneven distribution of light in a diffraction grating spectrum.' *Philos. Mag.* 1902

[2]  Ri R. C. M., Oliner A. A. 'Scattering resonances on a fast-wave structure.' *IEEE Trans. Antennas Propagat.* 1965, vol. 13(6), pp. 948–959

[3]  Hessel A., Oliner A. A. 'A new theory of Wood's anomalies on optical gratings.' *Appl. Optics.* 1965; 4(10), pp. 1275–1297

[4]    Hessel A., Oliner A. A. 'Wood's anomalies and leaky waves.' *Proceedings of Symp. on Electromagnetic Theory and Antennas*; Copenhagen, Denmark, 1962. pp. 887–890

[5]    Hägglund J., Sellberg F. 'Reflection, absorption and emission of light by opaque optical grating.' *J. Opt. Soc. A.* 1966, vol. 56(8), pp. 1031–1040

[6]    Sarrazin M., Vigneron J-P. 'Bounded modes to the rescue of optical transmission.' *Europhys. News*, 2007, vol. 38(6), pp. 27–31

[7]    Sarrazin M., Vigneron J-P., Vigoureux J-M. 'Role of Wood anomalies in optical properties of thin metallic films with a bidimensional array of subwavelength holes.' *Phys. Rev. B.* 2003, vol. 67, pp. 085415-1–085415-8

[8]    Cho Y. K., Cho U. H., Ko J. H. 'TM-polarized electromagnetic scattering from a periodic strip array on a grounded dielectric.' *Microwave Opt. Tech. Lett.* 1996, vol. 11(1), pp. 41–45

[9]    Lee C. W., Cho Y. K, 'Periodically slotted dielectrically filled parallel-plate waveguide as a leaky wave antenna for infinite and finite periodic structure.' Proceedings of the URSI International Symposium on Electromagnetic Theory; St. Petersburg, Russia, May 1995, pp. 314–316

[10]   Jose K. A., Nair K. G., 'Reflector-backed perfectly blazed strip gratings simulate corrugated reflector effects.' *Electron. Lett.* 1987, vol. 23(2), pp. 86–87

[11]   Yun L. H., Lee J. I., Cho U. H., Cho Y. K. 'Off-Bragg TM blazing of a periodic strip grating over a grounded dielectric slab.' *Microwave Opt. Tech. Lett.* 1998, vol. 17(6), pp. 375–377

[12]   Lee C. W., Cho Y. K. 'Analysis method for the leaky wave from a periodically slotted parallel-plate waveguide for the TEM incident case.' *Proceedings of the Asia-Pacific Microwave Conference 94*; Tokyo, Japan, 1994, pp. 1117–1120

[13]   Lee C. W., Lee J. I., Cho Y. K. 'Analysis of leaky-waves from a periodically slotted parallel-plate waveguide for finite number of slots.' *Electron. Lett.* 1994, vol. 30(20), pp. 1633–1634

[14]   Brundrett D. L., Glytsis E. N., Gaylord T. K., Bendickson J. M. 'Effects of modulation strength in guided-mode resonant subwavelength gratings at normal incidence.' *J. Opt. Soc. Am. A.* 2000, vol. 17(7), pp. 1221–1230

[15]   Richmond J. H. 'On the edge mode in the theory of TM scattering by a strip or strip grating.' *IEEE Trans. Antennas Propagat.* 1980, vol. 28(6), pp. 883–887

[16]   Garbacz R. J. 'Modal expansion for resonance scattering phenomena.' *Proc. IEEE.* 1965, vol. 53, pp. 856–864

[17]   Harrington R. F., Mautz J. R. 'Theory of characteristic modes for conducting bodies.' *IEEE Trans. Antennas Propagat.* 1971, vol. 19, pp. 622–628

[18]   Jacobsen J. 'Analytical, numerical, and experimental investigation of guided waves on a periodically strip-loaded dielectric slab.' *IEEE Trans. Antennas Propagat.* 1970, vol. 18(3), pp. 379–388

[19]   Lee J. I., Cho U. H., Cho Y. K. 'Analysis for a dielectrically filled parallel-plate waveguide with finite number of periodic slots in its upper wall as a

leaky wave antenna.' *IEEE Trans. Antennas Propagat.* 1999, vol. 47(4), pp. 701–706

[20] Lee C. H., Cho U. H., Cho Y. K. 'Revisited generalized Wood anomalies.' *Proc. IEEE AP-S*; Florida USA, 1999, pp. 1754–1757

[21] Cho Y. K., Cho U. H., Ko J. H. 'Blazing of periodic strip grating on a grounded dielectric: TE polarization case.' *Electron. Lett.* 1995, vol. 31(23), pp. 2035–2037

[22] Butler C. M. 'General solutions of the narrow strip (or slot) integral equations.' *IEEE Trans. Antennas Propagat.* 1985, vol. 33(10), pp. 1085–1090

[23] Lee C. W., Cho Y. K. 'Analysis of electromagnetic scattering by periodic strips on a grounded dielectric slabs.' *IEEE AP-S Int. Symp. Dig.* Newport Beach USA. 1995, pp. 398–401

[24] Tseng D. Y., Hessel A., Oliner A. A. 'Scattering by a multimode corrugated structure with application to P type Wood anomalies.' *Alta Freq.*, 1969, vol. 38, special issue, pp. 82–88

[25] Lee J. I., Cho Y. K. 'Electromagnetic scattering of a Gaussian beam wave from finite periodic slots in a parallel-plate waveguide.' *Proc. Int. Symp. Antennas Propagat.* Chiba Japan, 1996, pp. 29–32

[26] Chen W., Michelson D. G., Jull E. V. 'Off-Bragg TM blazing of rectangular groove grating.' *Proc. URSI Int. Symp. Electromagnetic Theory*, St. Petersburg, Russia, 1995, pp. 311–313

[27] Breidne M., Maysttre D. 'Perfect blaze in non-Littrow mountings: a systematic numerical study.' *Opt. Acta*, 1981, vol. 28(10), pp. 1321–1327

[28] Cho Y. K., Ra J. W., Cho U. H., Lee J. I. 'Off-Bragg TE blazing of periodic strip grating on a grounded dielectric.' *Electron. Lett.* 1997, vol. 33(17), pp. 1446–1447

[29] Hessel A. Shmoys J., Tseng D. Y. 'Bragg-angle blazing of diffraction grating.' *J. Opt. Soc. Am.*, 1975, vol. 65, pp. 380–384

[30] Wood R. W. 'Anomalous diffraction grating.' *Phys. Rev.*, 1935, vol. 48, pp. 928–937

[31] Yu N., Genevet P., Kats M. A., Aieta F., Tetienne J-P., Capasso F., Gaburro Z. 'Light propagation with phase discontinuities: generalized laws of reflection and refraction.' *Science*, 2011, vol. 334(21), pp. 333–337

[32] Wong Alex M. H., Eleftheriades G. V. 'Perfect anomalous reflection with a bipartite Huygens' metasurface.' *Phys. Rev. X.*, 2018, vol. 18(1), pp. 011036 (1-8)

[33] Sun S., Yang K-Y., Wang C-M., Juan T-K., Chen W-T., Liao C. Y., He Q., Xiao S., Kung W-T., Guo G-T., Zhou L., Tsai D.P. 'High efficiency broadband anomalous reflection by gradient metasurfaces.' *Nano Lett.*, 2012, vol. 12, pp. 6223–6229

[34] Tibuleac S., Magnusson R. 'Reflection and transmission guided-mode resonance filters.' *J. Opt. Soc. Am. A.*, 1997 vol. 14(7), pp. 1617–1626

[35] Sharon A., Glasberg S., Rosenblatt D., Friesem A. A. 'Metal-based resonant grating waveguide structure.' *J. Opt. Soc. Am. A.*, 1997, vol. 14, pp. 588–595

[36]  Park S. W., Kim H. T., Cho Y. K., Ko J. W. 'Negative reflection and refraction and filter characteristics in the leaky wave-supportable grating – TE polarization case.' *Proc. EuCAP*, Copenhagen, Denmark 2020

[37]  Memarian M., Li X., Morimoto Y., Itoh T. 'Wide-band/angle blazed surfaces using multiple coupled blazing resonances.' *Sci. Rep.* 2017, pp. 1–12

[38]  Stutzman W. L., Thiele G. A. *Antenna Theory and Design.* 1st edn. New York: Wiley; 1981, pp. 354–356

[39]  Liu C-C., Shmoys J., Hessel A. 'E-plane performance trade-offs in two-dimensional microstrip-patch element phased arrays.' *IEEE Trans. Antennas Propagat.* 1982, vol. 30(6), pp. 1201–1206

[40]  Lee J. I., Cho Y. K. 'An analysis method for two-dimensional microstrip phased array antenna.' *KITE J. Electron. Eng.* 1886, vol. 7(2), pp. 21–25

[41]  Knittel G. H., Hessel A., Oliner A. A. 'Element pattern nulls in phased arrays and their relation to guided waves.' *Proc. IEEE.* 1968, vol. 56(11), pp. 1822–1836

[42]  Smith S. J., Purcell E. M. 'Visible light from localized surface changes moving across a grating.' *Phys. Rev.*, 1953, vol. 92, p. 1069

[43]  Palocz I., Oliner A. A. 'Leaky space-charge waves II: Smith–Purcell radiation.' *Proc. IEEE.* 1967, vol. 55, pp. 46–56

[44]  Hessel A. 'Resonances in the Smith–Purcell effect.' *Can. J. Phys.* 1964, vol. 42, pp. 1195–1211

[45]  Van den Berg P. M., Tan T. H. 'Smith–Purcell radiation from a line-charge moving parallel to a reflection grating with rectangular profile.' *J. Opt. Soc. Am.* 1974, vol. 64, pp. 325–328

[46]  Lee C. H., Lee J. I., Cho U. H., Cho Y. K. 'Radiation of Smith–Purcell type by a periodic strip grating over a grounded dielectric slab.' *Microw. Opt. Tech. Lett.* 1998, vol. 19(4), pp. 292–296

[47]  Haeberle O., Rullhusen P., Salome J. M. 'Calculations of Smith–Purcell radiation generated by electrons of 1–100 MeV.' *Phys. Rev.* 1994, vol. 49, p. 3340

[48]  Bertoni H. L., Cheo L-H. S., Tamir T. 'Frequency-selective reflection and transmission by a periodic dielectric layer.' *IEEE Trans. Antennas Propagat.* 1989, vol. 37(1), pp. 78–83

[49]  Lee J. I., Cho Y. K. 'Electromagnetic scattering by finite periodic slab.' *Proc. IEEE AP-S*; Montreal, Canada, 1997, pp. 298–301

*Chapter 8*

# Transmission of electromagnetic energy through a finite and infinite periodic array of transmission resonant and non-transmission resonant apertures

## Synopsis

Using the relation between the transmission cross section (TCS) through the arbitrarily arranged apertures and the directivity of the radiation pattern through the apertures when illuminated by the normally incident plane wave, we investigate how the TCS varies according to the arrangement of the apertures with main interest centering on the enhanced transmission. For this purpose, we first deal with the smallest basic unit of the two parallel (E-plane) element arrays and compare the two parallel element arrays with the two collinear (H-plane) arrays from the viewpoint of the TCS. The TCS characteristics of the finite linear array of the apertures are also investigated as a function of $N$ (number of apertures) for both E-plane and H-plane arrays as extensions of the parallel and collinear arrays of two elements, respectively.

Comparison of TCS between the two arrays shows that the TCS for the two parallel element arrays is significantly larger than that for the two collinear element arrays and this seems to support the concept of a basic entity of infinite long chain of apertures along the E-plane which was proposed to explain the underlying physics on the two-dimensional-planar extraordinary optical transmission (EOT) structure.

The effect of the aperture periodicity on the TCS in comparison with a single aperture is also discussed for a 2-D array of both transmission resonant aperture (TRA) and conventional circular aperture as a non-TRA in an infinite perfect conducting plane of zero thickness.

Additionally, enhanced transmission through periodic arrays of subwavelength holes in metallic film is discussed with the main interest centering on the role of the surface plasmon polariton (SPP) in the enhanced transmission phenomena.

## 8.1 Two-element array of TRA

As the smallest array composing the infinitely long linear chain as the basic entity [1] of extraordinary optical transmission (EOT) phenomena in a 2-D planar array, the transmission cross section (TCS) characteristics for both the parallel two-element array and the collinear two-element array are investigated and compared between them. The two apertures are placed along the E-plane direction in the parallel case while, in the collinear case, they are aligned perpendicular to the E-plane direction. The terminologies of "parallel" and "collinear" are used from the viewpoint of the directions of the equivalent magnetic current sources as discussed in more detail at the end of this subsection. Here the transmission resonant aperture (TRA) to be considered is chosen to be a small circular aperture loaded with ridges as discussed in Chapter 3.

### 8.1.1   Parallel two-element array of TRA

We consider first the two-element array problem of electromagnetic transmission-resonant apertures (TRA) of the parallel arrangement in a thin conducting plane. Recall that the TCS of the small TRA is given by $\frac{3\lambda^2}{4\pi}$ regardless of the shape or the size of the aperture under the transmission resonance. In addition, since the radiating source of the small TRA into the half-space opposite to the incident side is represented by a small magnetic dipole and its directivity $G$ corresponds to 1.5, the TCS can be expressed to be $\frac{2G\lambda^2}{4\pi}$ [m²]. Here this expression is assumed to hold also for the more complex distribution of the aperture source such as arbitrary array and the validity is checked for the discrete array arrangement of a finite number of apertures later.

The effects of the distance between the resonant apertures on the TCS curve as the transmission characteristics are numerically studied using the Method of Moments (MoM) [2] and the Finite Difference Time Domain (FDTD) method [3].

To begin with, the numerical results for the variations of the TCS on the parallel case where the two TRA are aligned along the E-plane are compared with those in the single aperture case. Figure 8.1(a) shows the geometry of two identical resonant apertures allocated in an infinite perfect electric conductor (PEC) plane of zero thickness with distances of $X_d$ in the x-direction and $Y_d$ in the y-direction. As discussed in Chapter 4, the non-resonant small circular aperture can be resonated by loading double ridges across the aperture.

In Figure 8.1(a), $D$ is the diameter of the aperture, $W$ is the width of the ridge, and $G$ is the gap between the ridge edges. The thickness $t$ of the PEC plane is assumed to be zero. All media except for the PEC plane are assumed to be free space with constitutive parameters $(\varepsilon_0, \mu_0)$. A plane wave polarized parallel to the ridge axis (y-axis direction) and propagating in the z-direction is assumed to be normally incident upon the aperture on the plane of $z = 0$. To solve the problem of electromagnetic coupling problem through two identical apertures as shown in Figure 8.1, both the MoM and the FDTD methods were used [2]. The MoM of the Rao–Wilton–Glisson [3] (RWG) type for both the single resonant aperture problem and the two aperture coupling problem is used. For the FDTD, the conformal finite-difference time-domain (CFDTD) [4] code is used with the perfectly matched layer (PML) type of absorbing boundaries [5]. The analysis procedures for the problem are described in detail, respectively, in [3,5] for the MoM and in [4,5] for the CFDTD and so are not given here.

(a)

(b)                    (c)

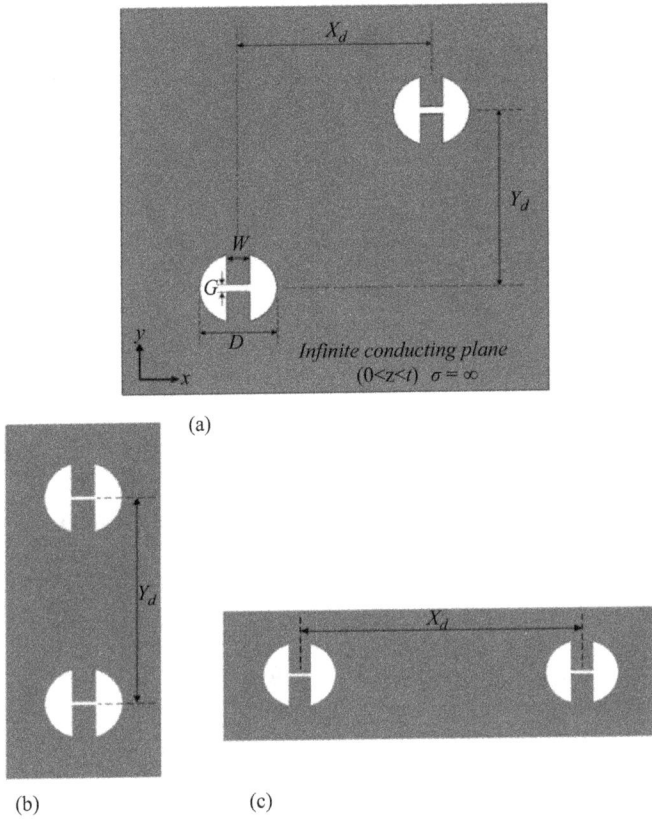

*Figure 8.1   Geometry of a pair of ridge-loaded circular apertures in an infinite
conducting plane. (a) Aperture pair in an arbitrary position in the
plane. (b) Parallel case ($X_d = 0$). (c) Collinear case ($Y_d = 0$).*

To investigate the effects of the incident field polarization and the distance
between the two resonant apertures, two different configurations have been considered, parallel and collinear arrangements. As mentioned above, the two apertures are placed along the ridge direction or E-plane direction for the parallel case
($X_d = 0$) while they are allocated perpendicular to the ridge direction for the
collinear case ($Y_d = 0$) as shown in Figure 8.1(b) and 8.1(c), respectively.

According to the Poynting theorem, the transmitted power $P_t$ to the region
$z > 0$ through the two resonant apertures can be obtained via the surface integral over
the aperture area at $z = 0^+$. The transmission cross section (TCS) is then calculated by
normalizing $P_t$ with the power density $P_o$ of the incident plane wave as

$$\mathrm{TCS} = \frac{P_t}{P_o} \left[\mathrm{m}^2\right] \tag{8.1}$$

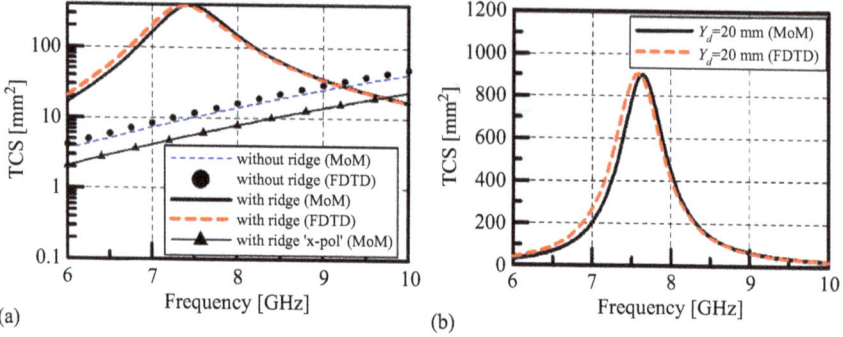

*Figure 8.2*    TCS variations. (a) Single aperture cases. (b) Dual apertures
($X_d$ = 0 mm, $Y_d$ = 20 mm). t = 0 mm, D = 10 mm, W = 3 mm, and
G = 0.5 mm.

in which $P_o = \frac{E_o^2}{2\eta_o} \left[\frac{W}{m^2}\right]$. Here $E_o$ is the magnitude of the incident electric field and $\eta_0 = \sqrt{\mu_0/\varepsilon_0}$. Figures 8.2(a) and (b) show the variations of the TCS of a single transmission resonant aperture and dual apertures in a parallel arrangement of Figure 8.1(b) with geometrical parameters $t$ = 0 mm, $D$ = 10 mm, $W$ = 3 mm, and $G$ = 0.5 mm. As shown in Figure 8.2(a), the TCS obtained by the MoM becomes a maximum of 396 mm$^2$ at the resonant frequency $f$ = 7.44 GHz, which is about a 40-fold (16 dB) increase from the value of 9.7 mm$^2$ for the aperture without the ridge [5]. The maximum value of 396 mm$^2$ agrees well with the TCS value (= $\frac{3\lambda^2}{4\pi}$ = 388 mm$^2$) of a small transmission resonant aperture at the frequency of 7.44 GHz. We note that the TCS value are plotted in log scale to show the differences clearly.

In Figure 8.2(a), the "x-pol." case denotes that the electric field vector of the incident plane wave has been changed to the x-direction, which is perpendicular to the ridge axis. The TCS for the "x-pol." case is decreased significantly from that for the circular aperture without the ridges since the loaded ridge blocks the aperture, making the effective aperture area and so the TCS even less.

Figure 8.2(b) shows the TCS of the dual resonant apertures for the parallel case with $X_d$ = 0 and $Y_d$ = 20 mm. In Figure 8.2(b), when it is compared to the single aperture case, the resonant frequency, obtained by the MoM, is found to be shifted to $f$ = 7.64 GHz, and the maximum TCS obtained amounts to 901 mm$^2$, which is significantly larger than twice the TCS of 390 mm$^2$ of the single resonant aperture. It may be due to the mutual coupling between two adjacent apertures that the transmission characteristics such as TCS, the resonant frequency of the transmission maximum change, and the transmitted field pattern may change in either the parallel case for $X_d$ = 0 or in the collinear case for $Y_d$ = 0.

In Figures 8.2(a) and 8.2(b), the results obtained by the FDTD are also plotted to check the validity of the data calculated by the MoM. The agreements between the data obtained by the two methods verify the results for the TCS. To examine the effects of the distance $Y_d$ on the transmission characteristics for the parallel arrangement case, some data for the TCS are given for sampled distances of $Y_d$ in Figure 8.3(a).

The data for the resonant frequency and the maximum TCS are also presented as a function of the distance between the apertures for the parallel arrangements in Figure 8.3(b).

From Figure 8.3, it is observed that the resonant frequency and the maximum TCS change according to the variation of distance between two identical resonant apertures. They change more sensitively to the distance $Y_d$ in the parallel case, rather than $X_d$ in the collinear case appearing next.

Note also that the transmission characteristics are not significantly affected by the distance $X_d$ when it is larger than some value (roughly 70 mm as seen in Figure 8.4(b)). These phenomena seem natural, considering that the mutual impedance between two identical resonant dipoles collinearly arranged varies much less

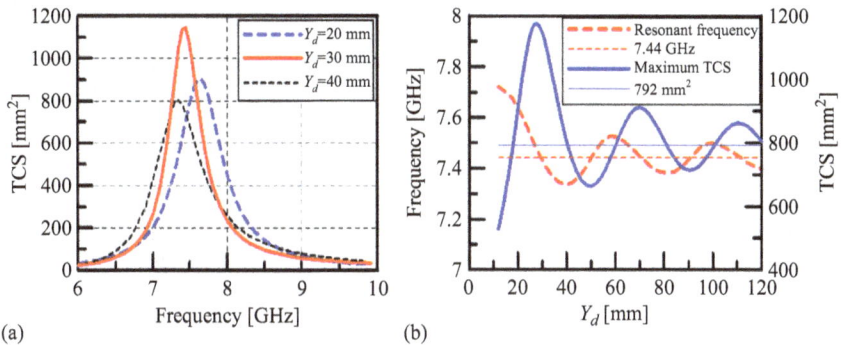

*Figure 8.3* *Transmission characteristics for the parallel arrangement. (a) TCS characteristics. (b) Resonant frequency and maximum TCS obtained by MoM.*

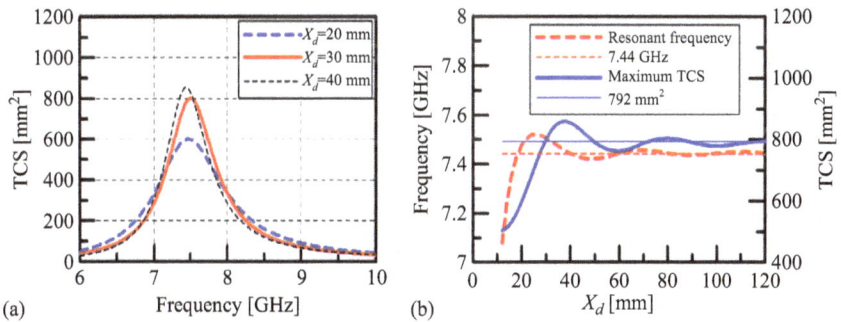

*Figure 8.4* *Transmission characteristics for the collinear arrangement. (a) TCS characteristics. (b) Resonant frequency and maximum TCS obtained by MoM.*

sensitively to the distance than the parallel arranged case [6,7], because the mutual coupling between two resonant dipoles is the dual problem to this coupling between the two resonant aperture. In a study on the gain enhancement of two parallel dipoles fed by uniform current sources [6], it was observed that the gain of an array of the two identical resonant dipoles in a parallel arrangement is maximized to about three times the gain of a single dipole when the spacing between the dipoles approaches to about $0.7\lambda$ at the resonant frequency (see Figure 14 in [6]) where the gain becomes maximum.

From the duality consideration, if two resonant slots, which is the dual structure of the two resonant dipoles in the parallel arrangement are fed by uniform voltage sources, the gain of the array of two slots would be maximized to about three times the gain of a single resonant slot radiation when the spacing between the slots is near to $0.7\lambda$.

As expected from the above discussion, the TCS of the array of two resonant slots in parallel arrangement would be maximized to about three times the TCS of a single resonant aperture when the plane wave of the electric field component polarized perpendicular to the slot axis is incident on the slots from the broadside direction which is the maximum radiation direction when the slots are fed by uniform voltage sources.

Similarly in this study on the TCS of two identical small resonant aperture loaded by symmetrical ridges, the maximum TCS of $1.174$ mm$^2$ in parallel arrangement case at the resonant frequency of $7.44$ GHz is found to correspond to about three times compared to the single aperture case. In addition, the spacing $Y_d$ $=29$ mm for the maximum TCS corresponding to $0.72\lambda$ at $f = 7.44$ GHz agrees well with the spacing of $0.7\lambda$ between the two dipoles for the maximum gain.

The two-element aperture array of the parallel arrangement under consideration is thought to be more useful and fundamental than that of the collinear arrangement as the smallest radiation unit of two elements from the viewpoint of the directivity (and so the TCS). Recall that the directivity is closely related to the transmission efficiency, i.e., TCS.

The two-slot model [8] of a rectangular patch antenna which has been widely used for determining radiation patterns constitutes a good example of the parallel arrangement. The two equivalent magnetic dipole sources on the two slot positions correspond to the dual sources of the two electric dipoles in the parallel arrangement. This is the reason for calling the arrangement in Figure 8.1(b) the parallel one. In contrast, note that for the collinear arrangement in Figure 8.1(c), the equivalent magnetic current sources between the ridges are collinearly arranged along the direction.

It is also found that the variation in the directivity of the two dipole arrays in parallel arrangement against the separation between the electric dipoles has nearly the same shaped curve as the TCS of two resonant apertures in parallel arrangement [6] as expected from the foregoing discussion on the duality between electric and magnetic dipole arrays.

## 8.1.2 Collinear two-element array of TRA

Here we consider the transmission characteristics of the two elements of TRA in a collinear arrangement to compare with those in the foregoing parallel arrangement.

To examine the effects of the distance $X_d$ on the transmission characteristics, we obtained the TCS as a function of frequency, and some plots for sampled distances are shown in Figure 8.4(a). The data for the resonant frequency and the maximum TCS are searched in terms of varying the distance between the collinear cases and the results are given in Figure 8.4(b).

In both cases of Figure 8.3(b) and 8.4(b), as the distances $Y_d$ and $X_d$ are increased, the resonant frequency converges to $f = 7.44$ GHz in the single aperture case, and the maximum TCS at resonant frequency approaches 792 mm$^2$, which is double the maximum TCS, 396 mm$^2$, of the single aperture as expected.

From a comparison between Figure 8.3 and Figure 8.4, it is observed that the TCS of the two elements for the parallel arrangement is significantly larger than that for the collinear arrangement near the separation between the two elements where the TCS becomes maximum and that the sensitivity of the TCS to the separation is also larger for the parallel arrangement as discussed above.

It is worthwhile to note that the parallel arrangement of the two elements of apertures constitutes a fundamental two-slot model of the rectangular microstrip antenna structure as well as the simplest two-element array for the basic single linear EOT structure.

Based upon the discussions so far, as Huygens' source, the two aperture array of parallel arrangement is seen to produce significantly higher directivity than that of collinear arrangement does. As a result, the aperture source of the parallel arrangement has a larger TCS than that of the collinear arrangement.

## 8.2 Finite periodic array of apertures

We deal with two kinds of linear chains of a finite number of apertures. One is a finite periodic linear array of apertures extended along the above two-element array axis of the parallel arrangement. The other is a finite periodic linear array of apertures extended along the two-element array axis of the collinear arrangement.

Two types of apertures are considered. One is a conventional rectangular or circular aperture. The other is a type of transmission resonant aperture (TRA), which has a much larger transmission cross section at a significantly lower operating frequency as discussed in Chapter 4.

Here, the main concern is focused on investigating the TCS characteristics of the above two kinds of linear finite periodic aperture arrays, i.e., TCS as a function of the number of apertures for both the conventional aperture and the type of TRA numerically and comparing the numerical results for the TCS with those obtained by use of the analytical expression of TCS, $\frac{2G\lambda^2}{4\pi}$ to the finite number of aperture arrays of the above two kinds.

A comparison reveals that the larger the TCS of the elementary apertures, the better the agreement between the two results for TCS obtained by the use of the two approaches.

Two additional comments may be in order. First, to implement the transmission enhancement in the aperture array structure, we can employ the transmission resonant aperture as a single aperture element, rather than the usual aperture such as circular or rectangular apertures. Second, as another method for transmission enhancement, an array scheme can be also employed. Between the two representative arrays, the parallel and collinear arrangements, the parallel arrangement is much more favored in that this arrangement gives significantly larger directivity [9] and larger TCS than those of the collinear arrangement for any number of finite linear arrays. Note that these two types of physical mechanisms seem to be analogous to the site and periodicity resonance, respectively, which have been proposed as the main causes for the EOT phenomena [10].

It deserves mentioning that the coaxial aperture which has been widely investigated in connection with super-enhanced light transmission phenomena [11–13] belongs to the TRA type in the sense that the TCS of this aperture can be made to be significantly increased in comparison with the physical aperture area as discussed in Chapter 4.

## 8.2.1   Finite array of rectangular aperture as a non-TRA

Here we are to consider two kinds of finite arrays of square aperture arranged along the array axis of the parallel configuration case and the array axis of the collinear configuration case from the viewpoint of the TCS and compare between the two cases. First, let us look into the TCS characteristics curve versus frequency for the single square aperture as shown in Figure 8.5(a) for reference. The maximum TCS is seen to be 171.6 [mm$^2$] at 12.82 GHz as shown in Figure 8.5(b), whereas the physical area of the aperture is 100 [mm$^2$]. This TCS is significantly smaller than that for the TRA.

Next, we consider the finite array arranged along the axis of the parallel configuration, where the axis corresponds to the direction of the incident electric field as shown in Figure 8.6(a). Figure 8.6(b) shows the TCS of the above type of finite array as a function of the number $N$ of the aperture. The solid line marked with a triangle represents the TCS versus $N$ which has been obtained by use of the numerical Method of Moments (MoM) of Rao–Wilton–Glosson (RWG) type. Here the numerical analysis method is summarized as follows.

An integral or integro-differential equation whose unknowns are equivalent magnetic surface currents over the finite number of apertures in an infinite conducting plane is formulated under the normal incidence of the plane wave, for example, and is solved numerically by use of the MoM. Through the use of the complex Poynting vector theorem [14], the transmitted power through all the apertures into the half-space opposite to the incident side is calculated, from which the TCS is obtained by dividing the total transmitted power by the incident power density of the Poynting vector magnitude.

$f$=12.82 GHz, 171.6248    $\lambda_o$=23.4 mm

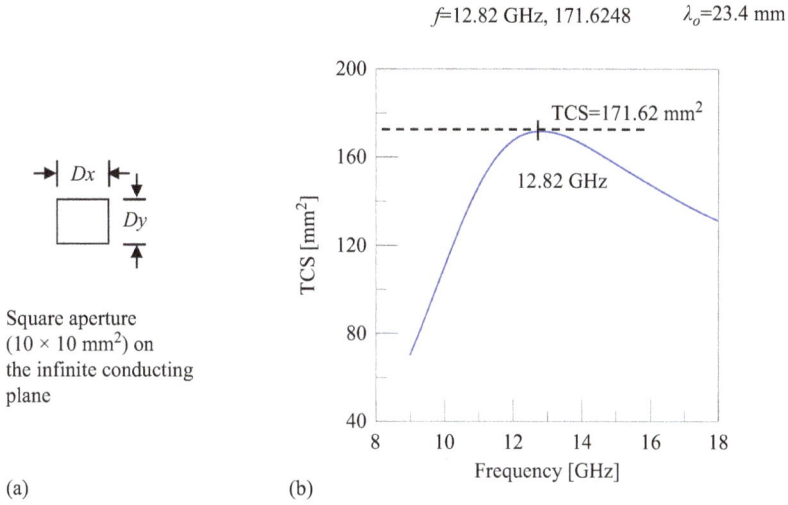

Square aperture
($10 \times 10$ mm$^2$) on
the infinite conducting
plane

(a)                    (b)

*Figure 8.5    Single square aperture and its TCS characteristics. (a) Single square aperture ($10 \times 10$ mm$^2$) and (b) its TCS versus frequencies.*

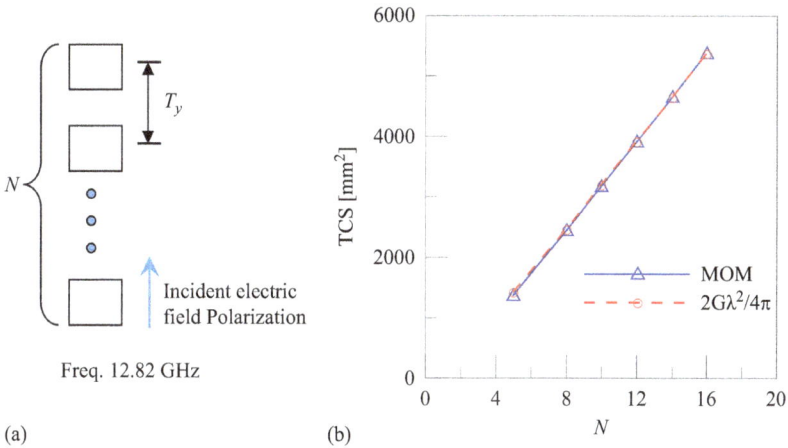

Incident electric
field Polarization

Freq. 12.82 GHz

(a)                    (b)

*Figure 8.6    Finite array of apertures and its TCS. (a) Finite array of the square apertures with the number* N *and the separation between two adjacent apertures* T$_y$. *(b) Comparison of the TCS versus* N *between results through the use of the MoM and those obtained by use of* $\frac{2G\lambda^2}{4\pi}$.

Note that at every step for calculating the TCS for every aperture number $N$, the separation between two adjacent apertures of $N$-1th and $N$th is searched such that the total TCS for $N$ number of the aperture may be maximum. For the separation between the first two adjacent apertures, the separation for the two elements of the parallel configuration is chosen as the desired separation $T_y$ for

$N = 2$. Next, the separation between the second aperture and the third aperture is obtained by searching for the separation for which the total transmitted power through the three apertures becomes maximum while varying the separation.

By repeating the calculation procedure for the increased number (by 1 by 1) $N$ of the apertures, the TCS curve versus the aperture number $N$ is obtained as shown in Figure 8.6. This curve is compared with that obtained by use of the analytic expression of $\frac{2G\lambda^2}{4\pi}$. For each number $N$ of apertures, Table 8.1 provides numerical data for the separation $T_y$ between two adjacent apertures, the TCS obtained using the numerical MoM, the TCS obtained using the analytic expression $\frac{2G\lambda^2}{4\pi}$, and the directivity resulting from the transmission through all the apertures.

From Table 8.1, it is seen that the separation between two adjacent apertures is very slightly increased as the aperture number $N$ is increased. All the separations between adjacent apertures can be approximated as being equal to the separation between the first adjacent two apertures.

The values listed in the fifth column represent the directivity by the transmitted field through all apertures in the half-space opposite to the incident side. Those in the fourth column represent the transmission cross section for all the apertures which have been obtained by using the directivity of the fifth column in the analytic expression $\frac{2G\lambda^2}{4\pi}$ in the fourth column.

Next, we consider the finite array arranged along the axis of the collinear configuration, where the axis corresponds to the direction of the equivalent magnetic current vector as shown in Figure 8.7(a).

The two TCS curves of the collinear configuration versus $N$ as in Figure 8.7(a), were obtained through the use of the MoM method and the expression of $\frac{2G\lambda^2}{4\pi}$, are compared in Figure 8.7(b). First by calculating the directivity $G$ as Huygens' source and inserting the result into the $\frac{2G\lambda^2}{4\pi}$, the approximate result for TCS is obtained, and is compared with the accurate result obtained through the use of the MoM in Figure 8.7(b).

To compare the TCS for the two kinds of finite arrays in Figure 8.6(a) and 8.7 (a), we have investigated the directivity as Huygens' source as a function of aperture number $N$ and listed in Table 8.1 and Table 8.2. Comparison shows that the directivity for the E-plane array corresponding to the parallel arrangement is

*Table 8.1   Variation of $T_y$, TCS (MoM), TCS$(\frac{2G\lambda^2}{4\pi})$, directivity versus aperture number N for E-plane array of parallel configuration*

| Aperture number N | Separation between two adjacent aperture along y-direction, $T_y$ | TCS (MoM) | TCS $\left(\frac{2G\lambda^2}{4\pi}\right)$ | Directivity |
|---|---|---|---|---|
| 5 | 21.0 | 1359.77 | 1420.88 | 16.3 |
| 8 | 21.6 | 2431.45 | 2457.58 | 28.2 |
| 10 | 21.7 | 3164.72 | 3198.21 | 36.7 |
| 12 | 21.9 | 3903.59 | 3914.95 | 44.9 |
| 14 | 22.0 | 4640.53 | 4638.18 | 53.2 |

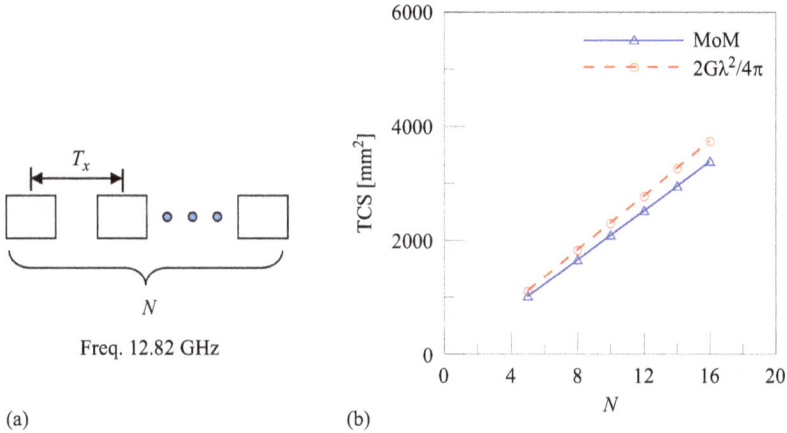

(a)                                    (b)

*Figure 8.7*   *Finite array of apertures and its TCS. (a) Finite array of the square apertures with the number N and the separation between two adjacent apertures T$_x$, (b) comparison of the TCS versus N between results obtained through the use of the MoM and those obtained by use of $\frac{2G\lambda^2}{4\pi}$.*

*Table 8.2*   *Variation of T$_x$, TCS(MoM), TCS $\left(\frac{2G\lambda^2}{4\pi}\right)$, directivity versus aperture number N for H-plane array*

| Aperture number N | Separation between two adjacent aperture along x-direction | TCS (MoM) | TCS $\left(\frac{2G\lambda^2}{4\pi}\right)$ | Directivity |
|---|---|---|---|---|
| 5 | 24.34 | 992.67 | 1089.39 | 12.5 |
| 8 | 24.0 | 1633.87 | 1813.30 | 20.81 |
| 10 | 24.07 | 2064.55 | 2282.24 | 2619 |
| 12 | 24.06 | 2494.20 | 2756.55 | 31.63 |
| 14 | 24.0 | 292493 | 3244.83 | 37.23 |

significantly larger than that for the H-plane array corresponding to the collinear arrangement.

As a result, the TCS obtained through the use of the $\frac{2G\lambda^2}{4\pi}$ for the E-plane array is expected to be larger than that of the H-plane array as seen in Figures 8.6(b) and 8.7(b).

To validate the expression of $\frac{2G\lambda^2}{4\pi}$ for the TCS for finite aperture array, we have calculated the accurate TCS through the use of the MoM and compared it with the above approximate TCS in Figure 8.7(b). A comparison shows that the approximate TCS (expression of $\frac{2G\lambda^2}{4\pi}$) is larger than the accurate TCS (MoM results). This difference, which becomes more clearly seen as the aperture number $N$ increases, is conspicuous in comparison with the foregoing E-plane array case corresponding to the parallel arrangement case where the two results from the accurate MoM and approximate expression of TCS $(\frac{2G\lambda^2}{4\pi})$ agree well to each other.

This suggests that the reduction factor $\gamma$ less than unity in the expression for TCS should be introduced as in $\frac{2G\lambda^2}{4\pi}$ to take into account the reduction of the TCS due to the significant reduction of the directivity as a Huygens' source of finite array along the H-plane. Recall that, in the E-plane array case corresponding to the parallel arrangement, the directivity of the finite array is significantly larger than that for the H-plane array case and the two TCS results from the MoM and the expression of $\frac{2G\lambda^2}{4\pi}$ correspond well with each other unlike the H-plane case corresponding to the collinear arrangement. This observation is compatible with the prior work on the 1-D (dimensional) chain of the holes as the basic unit of the extraordinary transmission phenomenon observed in a two-dimensional hole array [1] in that present discussion and prior discussion on the basic unit of EOT is based upon the E-plane array corresponding to the parallel arrangement.

So far we have discussed the transmission characteristics of the finite array composed of non-TRA types such as square aperture. Next, we deal with the transmission characteristics of the finite array composed of TRA type which shows more improved transmission characteristics.

## 8.2.2   Finite array of transmission resonant aperture (TRA)

As discussed in Chapter 4, the aperture of the TRA type has a significantly larger TCS than the conventional apertures such as rectangular or circular apertures. Here we choose the H-shaped aperture as a TRA and consider finite arrays of both the parallel and collinear configurations and compare two configurations from the viewpoint of the TCS.

Figures 8.8(a) and 8.8(b) show an H-shaped aperture as a TRA element and its TCS characteristics versus frequencies, respectively. Figure 8.9(a) shows a finite array of H-shaped arrays in the parallel configuration. In Figure 8.9(b) the TCS characteristics obtained by use of the MoM for the finite array in the parallel configuration are given as a function of the TRA number $N$ and compared with that obtained by use of the expression of $\frac{2G\lambda^2}{4\pi}$. A comparison shows good agreement.

For comparison with the above parallel configuration case, the TCS data for the collinear configuration case in Figure 8.10(a) which has been obtained by use of MoM is illustrated as a function of the aperture number $N$ as shown in Figure 8.10(b).

As seen from Figure 8.9(b) and Figure 8.10(b), in this TRA case too, the TCS for the parallel configuration is significantly larger than that for the collinear configuration similar to the conventional aperture case. Comparison between the TCS obtained by use of both MoM and the analytic expression of $\frac{2G\lambda^2}{4\pi}$ is also made in Figure 8.10(b) and shows good agreement. This is different from the finite array case of the collinear arrangement of the usual apertures in which the analytic expression related to the directivity $G$ gives the larger value of TCS obtained by the use of the MoM.

In summary, we have seen that the TCS for the finite array of the parallel configuration is significantly larger than that for the finite array of the collinear

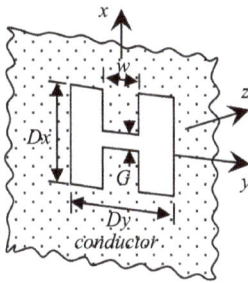

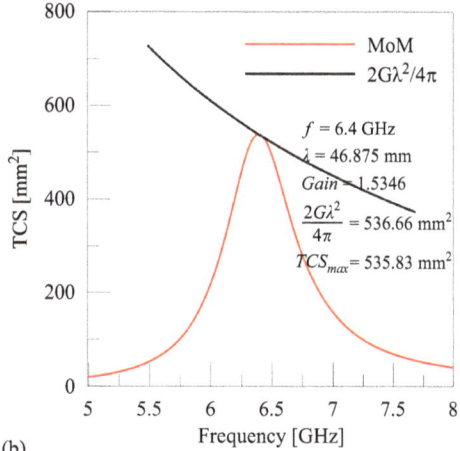

$D_x = D_y = 10$ mm

$W = 3$ mm, $G = 0.5$ mm

(a)                    (b)

*Figure 8.8*  *Single H-shaped aperture and its TCS characteristics, (a) single H-shaped aperture as an example for a TRA and (b) its TCS characteristics*

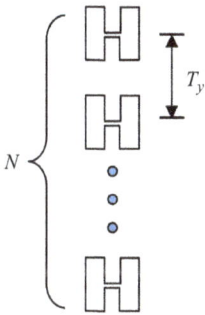

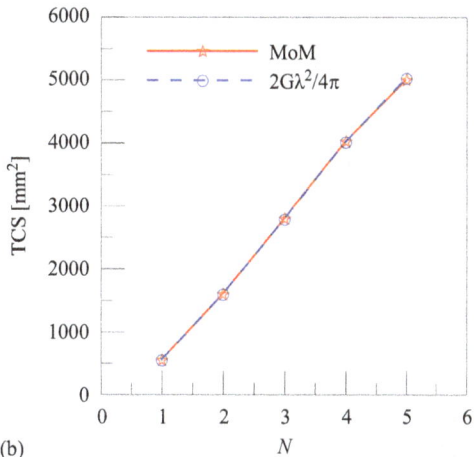

Freq. =6.4 GHz

(a)                    (b)

*Figure 8.9*  *Finite array of H-shaped aperture in the parallel configuration and its TCS. (a) Finite array in the parallel configuration and (b) its TCS characteristics.*

configuration under the same condition of aperture number and overall finite array length. This observation is valid irrespective of whether the finite array is composed of the non-TRA type or the TRA type. As already mentioned, the non-TRA type here means the circular or rectangular aperture whose TCS is much smaller than that of the TRA as discussed in Chapter 4.

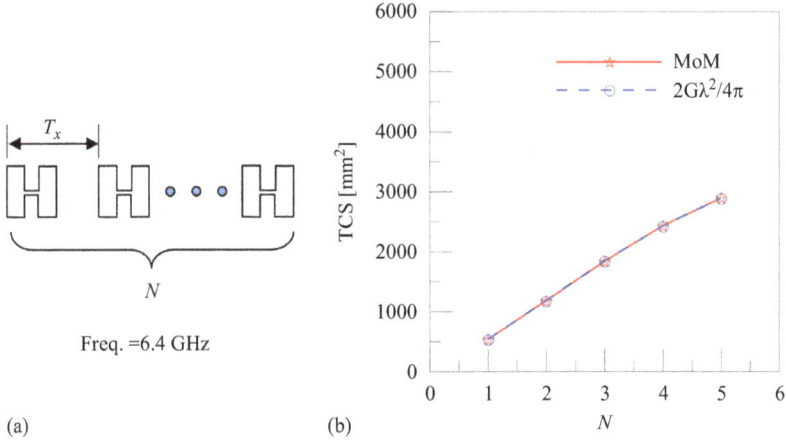

*Figure 8.10    Finite array of H-shaped aperture in collinear configuration and its TCS characteristics. (a) Finite array in collinear configuration and (b) its TCS characteristics.*

Note that in the case of the parallel configuration, the directivity of the overall array is significantly larger than that for the collinear configuration. Note also that the TCS is determined by the directivity as seen from the expression of $\frac{2G\lambda^2}{4\pi}$ if the directivity of the finite array exceeds some critical value between those of the parallel and collinear arrays.

It is interesting to see from Figure 8.6(b) and Figure 8.7(b) that for the usual aperture, i.e., non-TRA case, the two results for TCS obtained through the use of the MoM and the analytic expression ($\frac{2G\lambda^2}{4\pi}$) agree well to each other only in case of the parallel configuration whereas in the collinear case, the TCS values obtained by use of the MoM are smaller than those obtained by use of the analytic expression of $\frac{2G\lambda^2}{4\pi}$ for any number $N$ of apertures.

On the other hand, for the finite array made of the TRA, the TCS values obtained by the above two methods agree well with each other for both parallel and collinear configuration types of arrays though the TCS values for the collinear configuration case are always smaller than for the parallel configuration case. This suggests that the transmission resonance through the aperture source arrangement as well as each TRA plays an important role in the transmission enhancement.

Based on the above discussion, the correspondence between the two results for the TCS obtained by use of the MoM and the analytic expression of $\frac{2G\lambda^2}{4\pi}$ can be used as a simple measure for the transmission enhancement for the finite array of apertures.

In other words as the overall directivity of the finite array system increases [9], the expression for the TCS of $\frac{2G\lambda^2}{4\pi}$ becomes accurate. So when the directivity exceeds a certain critical value, the correspondence between the two results is observed as discussed below.

As a result, the finite array of the collinear arrangement whose element aperture is a usual aperture, i.e., non-TRA type such as circular or rectangular aperture gives the accurate TCS value (obtained by use of the MoM) different from the TCS result obtained by use of the analytic expression because the above finite array arrangement gives significantly smaller directivity than the finite array of the parallel arrangement composed of the TRA type.

It is clearly seen that the correspondence between the two results for TCS by MoM and directivity-related expression ($\frac{2G\lambda^2}{4\pi}$) holds for the three cases of Figures 8.6(b), 8.9(b), and 8.10(b) where the directivity and the TCS are relatively large except Figure 8.7(b) where the directivity and the TCS are low. From this observation, it is found that for significant transmission enhancement to occur at least one resonance type should contribute between the aperture resonance and the discrete source of aperture array distribution.

In more detail, when the TRA type whose TCS is much larger than the usual (non-TRA) type of aperture is used as the aperture element for a finite array of parallel or collinear arrangements, transmission enhancement is seen to occur. In contrast, when the usual (non-TRA) type of aperture is used as the aperture element, the transmission enhancement is seen to occur only for the parallel arrangement. So the finite array composed of the usual (non-TRA) type of aperture of collinear arrangement has been found to give the smallest directivity. Based upon the proportional relation between the directivity and the TCS, the TCS for the same condition as above is also the smallest among the above four cases. In this case, the TCS value obtained by use of the MoM is smaller than that obtained by use of $\frac{2G\lambda^2}{4\pi}$, which is different from the above three cases. Here it is interesting to note that the TCS of the finite array of the TRA of collinear arrangement is smaller than that of non-TRA of parallel arrangement. This emphasizes the role that the finite periodic aperture (source) distribution plays in the contribution to the transmission enhancement, comparable to the transmission resonance of the aperture itself. Interestingly the transmission resonance of the aperture itself and the transmission enhancement due to the discrete aperture source distribution here are analogous to the site resonance and periodicity resonance, respectively, in the previous work [10] on the extraordinary optical transmission phenomena.

Based upon the above discussion, the condition for the transmission enhancement in the finite array of the usual aperture (non-TRA) is observed to be similar to that in the infinite long chain array [1] as a basic EOT unit in the two-dimensional hole array (2DHA) structure in the sense that the two arrays belong to the type of parallel arrangement. It is interesting to note also that the parallel arrangement of the finite aperture array under consideration can be taken as a dual problem of the parasitic array [15,16] of the director structure in the Yagi–Uda array antenna if we focus our concern on the possibility of the guidance of something like a surface wave [15,17,18]. Only the difference is that the finite aperture array here is a kind of broadside array structure whereas the director array in the Yagi–Uda array is an end-fire array. Note that these two structures are dual ones to each other as arrays of parallel configurations.

As discussed above, the transmission characteristics of the finite array can be divided into two cases according to whether the finite array is made of the TRA or non-TRA. In other words, the transmission characteristics of the finite array structure made of TRA are quite different from that made of the non-TRA type in that for the latter case the property of the transmission enhancement is mainly affected by the aperture source array arrangement, i.e., finite periodic arrangement of apertures rather than by the transmission characteristics of the single non-TRA aperture itself. On the other hand, for the former case, the property of the transmission enhancement is mainly affected by the transmission characteristics of each TRA rather than the aperture source arrangement. These two observations provide some clues for dividing the main cause for the transmission enhancement into two kinds, site resonance corresponding to the transmission resonance through the TRA itself and lattice resonance corresponding to the aperture source (parallel) arrangement as discussed in [10].

What is more, the above two observations are similar to the previous works on the various types of array effects on the transmission enhancement depending on whether the aperture element is a TRA or non-TRA (usual aperture such as circular and rectangular aperture). Through the previous study, it was found that when the TRA type is employed as an aperture element, the spectral location and the shape of peaks of enhanced transmission are independent of whether any arrangement is chosen among periodic, quasi-periodic, and random arrays and so the particular spatial arrangement is of minor importance. For reference, the TRA which is employed as an aperture element in the previous work [19] is the narrow slot of the Fabry–Perot type in the thick conducting screen.

On the other hand, if we employ a usual (non-TRA) aperture as an array element, the transmission enhancement characteristics are determined only by the lattice resonance, which is a conspicuous difference from the former TRA case. This difference is due to the main resonance mechanism which primarily evokes the transmission enhancement between localized resonance in individual aperture and the lattice resonance. More clear difference between the two resonance types has been reported through the study on the optical transmission enhancement in coaxial waveguide arrays [11,12]. Here it deserves to emphasize the meaning of the parallel arrangement again in connection with the transmission enhancement phenomena.

If the finite periodic structure of the parallel arrangement is extended to the infinite case, it is expected to be the infinite linear array as the basic entity of EOT [1]. It is worthwhile to mention also that this parallel arrangement is taken as a dual structure of the director array in the Yagi–Uda antenna structure [15,16].

Here it should be also noticed that even in the limit of taking the TCS of each aperture to be zero, the total transmission phenomena due to the periodicity resonance is observed. As explained in [17], the origin of this total transmission can be found in phase accumulation during long-range dipole–dipole interaction among apertures when the wavelength is close to the onset of the first diffraction order under the above limiting condition.

## 8.3   Further comments on the array of the parallel configuration type

As discussed above, in the case of the finite array of the non-TRA type, the TCS for the parallel arrangement is significantly larger than that for the collinear arrangement. This is compatible with the previous study on the basic unit of the extraordinary optical transmission phenomena in two-dimensional hole array (2DHA) in that it corresponds to the linear chain of holes of the parallel configuration.

In addition, it is interesting to point out that the polarization direction of the incident wave under the condition of the parallel configuration corresponds to the electric field direction of the transverse magnetic (TM) mode for the surface plasmon. The previous study [20,21] on the transmission in two-dimensional hole array (2DHA) also revealed that the transmission minimum phenomenon called Wood's anomaly is observed only in the case of an array of the parallel configuration case.

It is worthwhile to mention that this type of Wood's anomaly is observed as a blind spot phenomenon [20,21] in a two-dimensional phased array made of dipole antennas. It is interesting to note that when this array is scanned in the E-plane to 90° the real part of the input impedance tends toward zero and so very poor radiation results in the endfire direction. Important is that the two phenomena—the EOT through a linear chain of holes and the Wood's anomaly—are observed only for the E-plane array case. Recall that the finite array of the parallel configuration under consideration corresponds to the dual problem of the parasitic array of directors in the Yagi–Uda array antenna.

Because the parasitic array of the parallel configuration [15,16] can support a surface wave type of traveling wave, its dual structure, the aperture array of the parallel configuration too can support the surface wave [16,17,22].

On the other hand, the traveling wave such as the surface wave is not supportable along the array of the collinear arrangement. For this reason, no Wood's anomalies appear in the 2DHA composed of collinear arrangement type.

Note that the structure for the Wood anomaly corresponding to the blind spot in the 2-D dipole array and the structure for the aperture array of the parallel configuration belongs to a category of the E-plane array structure. From the duality consideration, the aperture array structure of the parallel configuration is seen to constitute the guiding structure of the surface wave type like the parasitic director array structure of the Yagi–Uda array antenna. This is similar to the TM polarization case for the surface wave type like the surface plasmon [23].

Finally, it is worthwhile to mention that recently reported works on the transmission enhancement through finite aperture arrays of rectangular holes [22] in a conducting layer with finite thickness and of circular holes [24] in a conducting plane with zero thickness correspond, respectively, to the finite array of the TRA and non-TRA types of parallel arrangement. This parallel arrangement means the field polarization under which the surface wave is supportable along the array direction for both the two structures [22,24]. Recall that the transmittance through the finite array of apertures is determined mainly by the overall directivity as seen from $T = \frac{2G\lambda^2}{4\pi}$ [mm$^2$]. This is thought to be compatible with the suggestion that the transmission characteristics through the finite arrays of apertures follow the usual Fraunhofer diffraction as the sum of the electromagnetic waves emitted from the overall apertures [22].

It is worth commenting that in the present work on the transmission enhancement through the finite array of the small numbers of aperture, a linear dependence of the transmittance at the peak frequency on the number of apertures is observed irrespective of the E-plane array of the parallel arrangement or H-plane array of the collinear arrangement for both cases where the array is composed of the TRA or the non-TRA as seen Figures 8.6(b), 8.7(b), 8.9(b), and 8.10(b). This observation is compatible with the previous works on the finite array of apertures [22,24–27].

This observation on the linear dependence in the finite aperture array problem is similar to the linear dependence of the gain of the finite array of the short dipoles [28] from the duality consideration.

As discussed above, the E-plane array structure corresponding to the parallel arrangement which is found to have significantly larger directivity than that of the H-plane array structure corresponding to the collinear arrangement can be a structure where something like a surface wave is supportable. What is more, the inclusion of the gain in the expression for the TCS of $\frac{2G\lambda^2}{4\pi}$ implies that the transmitted waves follow the Fraunhofer diffraction as the sum of the electromagnetic waves emitted from the finite array of apertures, which suggest also that the transmittance shows the linear dependence on the aperture number [22]. These descriptions of the relevance of the transmission enhancement through the finite array of apertures on the surface wave coupling and the Fraunhofer diffraction seem to be compatible with those of previous works on the same topics [22,24]. This tentative explanation, however, needs further experimental and theoretical studies.

## 8.4 Equivalence of transmission cross section between continuous long resonant slit and discrete periodic resonant apertures

The topic in this section is somewhat deviated from the main context of this chapter and so may be skipped. However, this section deals with how the transmission resonance through a long narrow slit which has been widely investigated in connection with the EOT problem is related to the discrete periodic collinear array. So the study on the topic to be discussed may help in understanding the equivalence between the continuous source and the discrete array source from the viewpoint of the diffraction effect such as the TCS.

Among the previous works having a direct bearing on the transmission resonance relevant to the extraordinary optical transmission (EOT) phenomena [29], the first treatment was done for the transmission cross section per unit length ($\frac{\lambda}{\pi}$ [m]) through an infinitely long narrow slit [30] in thick conducting screen when the electric field of the normal incident plane wave lies in the plane of incidence.

The research work at the time of publication dealt with in connection with the applications to the electromagnetic interference (EMI) [30] problem in the AP (Antenna and Propagation) EMC (Electromagnetic Capability) societies. Later, however, since the observation of the enhanced transmission of light through a metallic film perforated by an array of subwavelength holes [29] was reported, the slit coupling structure aroused concern anew as a key ingredient of the simplified two-dimensional structure [31–33] which has been widely investigated to explain the underlying physics on the enhanced transmission problem. Here we are going to consider the problem of whether the above long slit source can be replaced with the periodic discrete transmission resonant aperture (TRA) array source from the viewpoint of the transmission cross section (TCS).

For this purpose, we discuss the equivalence between the TCS for the one-dimensional long slit in a thick conducting screen as a continuous source and that for the periodic TRA array as a discrete source as shown in Figure 8.11.

Recall that the expression of $\frac{2G\lambda^2}{4\pi}$ [m$^2$] for the transmission cross section holds for the finite number of resonant aperture as well as the single resonant aperture. In more detail, the above expression applies to both arrays of the parallel configuration type and the collinear configuration type if the element aperture belongs to the TRA class whereas the expression holds only for the array of the parallel configuration if the element aperture belongs to the non-TRA class as discussed in the previous section.

Assuming that the continuous source is substituted for discrete source distribution [34], we expand the continuous long magnetic current source into the discrete periodic magnetic current source of the collinear array type as shown in Figure 8.11. For calculational convenience, let us first consider the high gain broadside array whose elements are assumed to be isotropic radiators and relative currents $a_n$ are assumed to be constant $a_0$, In this case, the space factor $S(\theta)$ is expressed by

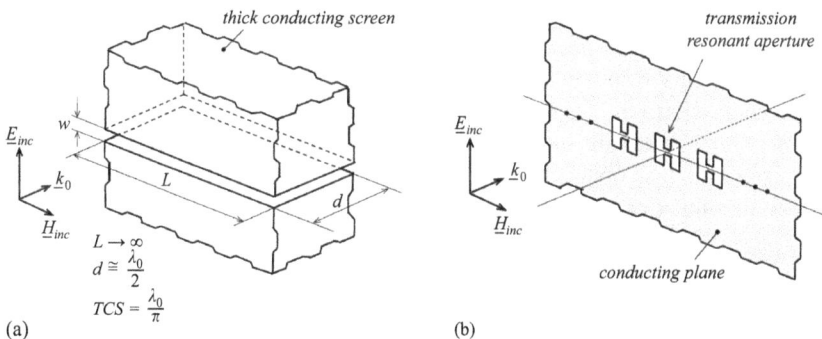

(a)                                                          (b)

*Figure 8.11    Equivalence between continuous long slit and discrete aperture source from the viewpoint of transmission cross section (TCS). (a) The long slit in a thick conducting screen as a continuous source, (b) periodic transmission resonant aperture (TRA) array as a discrete source section (TCS).*

$$S(\theta) = a_0 \sum_{n=0}^{N-1} e^{jnkd\cos\theta} = a_0 \frac{1 - e^{jNkd\cos\theta}}{1 - e^{jkd\cos\theta}}, \tag{8.2}$$

where $N$ is the number of elements, $d$ is element spacing, and $k$ is the wavenumber in the free space. For large $N$, the magnitude of the space factor $|S(\theta)|$ is approximated to be

$$|S(\theta)| \cong Na_0 \frac{\left|\sin\frac{1}{2}(Nkd\cos\theta)\right|}{\left|\frac{1}{2}Nkd\cos\theta\right|} \tag{8.3}$$

The space factor of an array is expressed as though the elements were isotropic radiators. It is of course modified if actual elements having some directivity are employed, but for high-gain arrays, the modification is small. So the expression for the directivity of the long linear array is approximately given by Ramo *et al.* [35]

$$G = \frac{4\pi|S|_{\max}^2}{2\pi\int_0^\pi |S|^2\sin\theta d\theta} \tag{8.4}$$

Since, for the high-gain broadside array, most contribution to the integral in the denominator of the above equation for $G$ comes from the region near $\theta = \frac{\pi}{2}$, the integral is approximated to be

$$\int_0^\pi |S|^2\sin\theta d\theta = \frac{2(Na_0)^2}{Nkd}\int_0^{NKd} \frac{\sin^2\left(\frac{a}{2}\right)}{(a/2)^2}da \tag{8.5}$$

$$\cong \frac{4(Na_0)^2}{Nkd}\cdot\frac{\pi}{2}$$

where $\alpha = Nkd\cos\theta$, $Nkd$ is taken effectively infinite, and $\int_0^\infty \frac{\sin^2 t}{t^2}dt = \frac{\pi}{2}$ is used. Using $|S|_{\max} = Na_0$ and this integration results in the expression for $G$, we get the expression for the directivity as

$$G = \frac{2l}{\lambda} \text{ for large } \frac{l}{\lambda}, \tag{8.6}$$

where $l$ means the effective overall length of the array, i.e., $l \cong Nd$ for large $N$.

Inserting this result for $G$ into the expression of $\frac{2G\lambda^2}{4\pi}$ for TCS leads to the desired expression $\frac{\lambda}{\pi}$ [m] for the TCS per unit length of transmission resonant cavity of a one-dimensional slit in the thick conducting screen. This corresponds to the TCS result which is obtained by use of MoM for the infinitely long narrow slot problem in a thick conducting screen [30].

## 8.5 Electromagnetic scattering by a two-dimensional infinite periodic array of small transmission resonant apertures (TRAs) and conventional circular non-TRA

We consider the electromagnetic scattering problem such as frequency selective surface (FSS) type of characteristics by a two-dimensional infinite array which is composed of small transmission resonant apertures (TRAs) and compare its transmission properties as an FSS structure with that of a two-dimensional infinite array which is composed of the conventional circular apertures. Here the conventional circular aperture means the circular aperture with no additional resonant structure like the ridge as employed in the TRAs.

Before dealing with the FSS characteristics of the infinite periodic structure of the TRAs, let us take a look at the single TRA structure and its TCS characteristics versus frequencies as shown in Figure 8.12.

Figure 8.12 shows the TCS characteristics for a ridged aperture as a representative TRA in which the transmission resonance phenomena occur. Here the transmission resonance means the phenomena in which case the power incident upon the area corresponding to the TCS of $\frac{2G\lambda^2}{4\pi}$ in the incident side much larger than the physical aperture area is maximally transmitted at resonance frequency through the TRA as discussed in Chapter 4. The TCS curve of the TRA is illustrated as a function of frequency in Figure 8.12, where the solid curve marked with small triangles, circles, and crosses represents the TCS for the TRA for the case that $D = 10$ mm, $W = 3$ mm, and $g = 0.2$ (point a case), $0.5$ (point b case), $0.8$ mm (point c case) and the solid line with a very gentle slope to 22 GHz represents the TCS for the circular aperture with no ridge as a conventional aperture. A comparison shows that the TCS for the TRA becomes significantly increased at a much-reduced transmission resonance frequency in comparison with the circular aperture with no ridge. In order words the TCS is seen to be made to be remarkably increased by modifying the aperture in shape as in the above ridged aperture in Figure 8.12.

For reference, the TCS values which have been obtained through the use of the expression of $\frac{2G\lambda^2}{4\pi}$ [m$^2$] is represented by the dashed line.

Comparison of this curve with the TCS curve obtained by use of the MoM reveals that these two curves correspond to each other at the transmission resonance

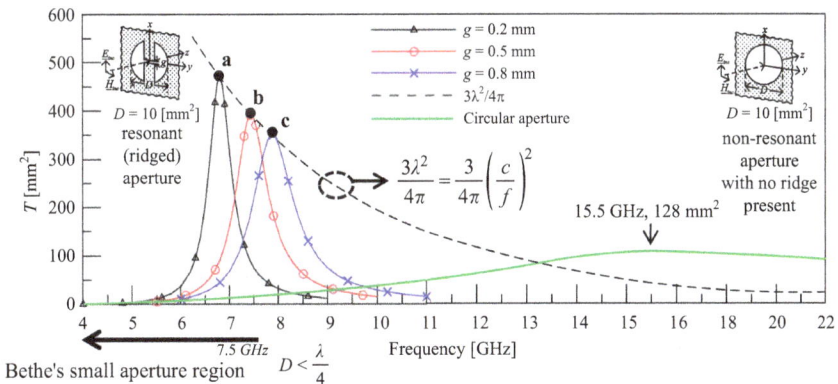

*Figure 8.12* *TCS characteristics versus frequencies for circular ridged aperture in comparison with that for the circular aperture with no ridge. For reference the TCS curve of $\frac{2G\lambda^2}{4\pi}$ is also given.*

frequencies of points a, b, and c. Recall that the TCS is defined as a ratio of the transmitted power [Watts] through the aperture to the incident power density [Watts/m$^2$] when the incident power density is assumed to be 1 Watts/m$^2$.

It seems rare that research has been reported on the two-dimensional frequency selective surface (FSS) structure composed of the above TRA, i.e., the small resonant aperture type of ridged circular aperture. So we investigate the transmission characteristics of small circular apertures with/without a ridge through the calculation of transmission cross section (TCS) and compare the results for the TCS characteristics for the FSS structures in which the resonant circular aperture with a ridge and a conventional circular aperture are arranged in the conducting plane.

### 8.5.1    Transmission characteristics of the frequency selective surface (FSS) composed of TRAs

Here we deal with the transmission characteristics of the FSS composed of the TRA of the ridged circular aperture type as shown in Figure 8.13(a).

The structure to be analyzed here belongs to the FSS problem of transmission type which corresponds to the upper constituent FSS in the absorption FSS problem in Chapter 6. Note that absorptive FSS comprises the present FSS structure backed by the lossy ground plane.

The analysis method for the absorption FSS problem in Chapter 6 is, by eliminating the lossy conducting plane, simplified to be applied to the present transmission FSS problem. So the detailed analysis method is omitted here.

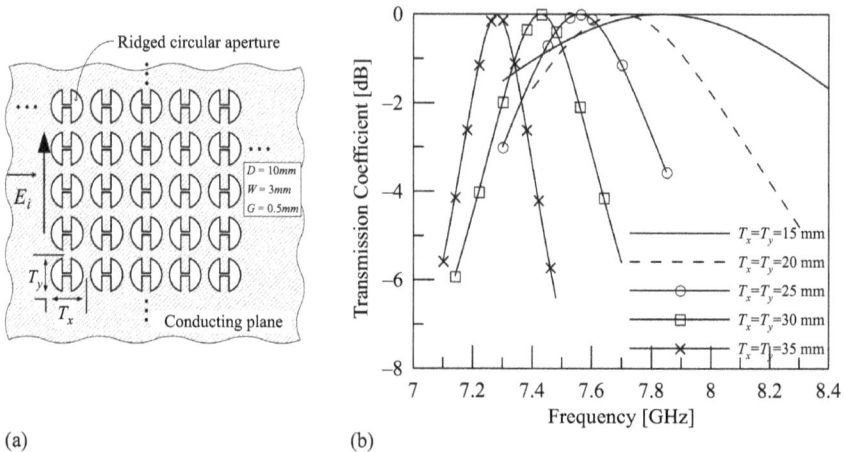

(a)                                            (b)

*Figure 8.13    Effect of periodicity on periodic array composed of ridged circular aperture. (a) Periodic array (FSS) composed of ridged circular aperture, (b) its transmission coefficient for five cases for which $T_x = T_y = 15, 20, 25, 30,$ and 35 mm.*

In the present problem of Figure 8.13(a), it is assumed that the plane wave whose electric field has only the $y$-component is normally incident upon the FSS screen.

To look into the variation of the transmission characteristics according to the variation of the periodicity of the FSS, we have investigated the transmission coefficient for various cases of periodicities. As numerical examples, Figure 8.13 (b) shows the variations of the transmission coefficients as a function of frequencies for five periodicities of a square array in which cases $T_x$ (period along $x$-axis) = $T_y$ (period along $y$ -axis) = $15, 20, 25, 30,$ and $35$mm. The frequencies of the zero reflection for which the transmission coefficient becomes zero [dB], i.e., the total transmission occurs for each case are found to be, respectively, 7.825 GHz, 7.7 GHz, 7.56 GHz, 7.43 GHz, and 7.28 GHz.

As expected, as the periodicity or interelement spacing ($T_x = T_y$ ) increases, the resonance frequency of the array as well as the bandwidth is observed to decrease. This means also that the smaller the ratio of TCS to unit cell area (= $T_x \cdot T_y$ ) is, the larger the bandwidth is.

## 8.5.2    Transmission characteristics of the frequency selective surface (FSS) composed of the conventional circular aperture with no ridge

To look at the difference from the viewpoint of the transmission cross section as a single element between the above TRA and the conventional circular aperture with no ridge, we have investigated the TCS versus frequencies for the conventional circular aperture, i.e., non-TRA type as shown in Figure 8.14(a).

Figure 8.14(b) shows the frequency characteristics of the transmission cross-section for the circular aperture. Comparison between the TCS curve for the ridged circular aperture as a TRA in Figure 8.12 and that for the circular aperture as a non-TRA in Figure 8.14(b) shows that the TCS can be remarkably enhanced for the TRA of the ridged aperture in comparison with that the non-TRA, i.e., the conventional circular aperture with no ridge, whereas the transmission resonance frequency of 15.5 GHz for the non-TRA case is significantly reduced to the 7.44 GHz for the TRA case as mentioned already.

Next, we investigate the transmission characteristics of the frequency selective surface (FSS) structure as shown in Figure 8.15(a), while varying the periodicities. In Figure 8.15(b), the transmission characteristics are shown for four cases of square arrays where $T_x = T_y = 15$ mm, $T_x = T_y = 20$ mm, $T_x = T_y = 25$ mm, $T_x = T_y = 30$ mm. From Figure 8.15(b), it is seen that the total transmission corresponding to the zero reflection occurs at frequencies of 9.88 GHz, 11.65 GHz, 14.08 GHz, and 18.11 GHz.

Note that the larger the TCS of the aperture is than the physical area of the unit cell, the larger the transmission bandwidth is, which is similar to the aforementioned TRA case. It is also observed that the TCS of the non-TRA, i.e., a conventional circular aperture with no ridge is smaller than the smallest unit cell area among the above four unit cell areas ($225, 400, 625, 900$ mm$^2$), whereas the TCS of

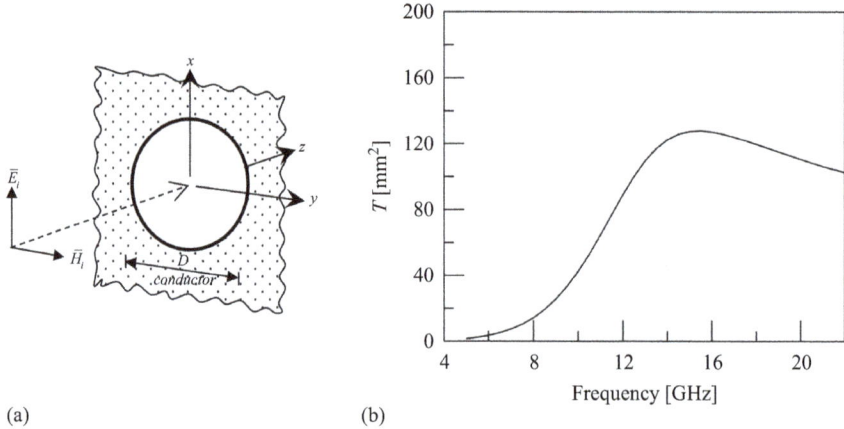

*Figure 8.14    Small circular aperture structure and its TCS, (a) small circular aperture structure and (b) its TCS characteristics versus frequencies*

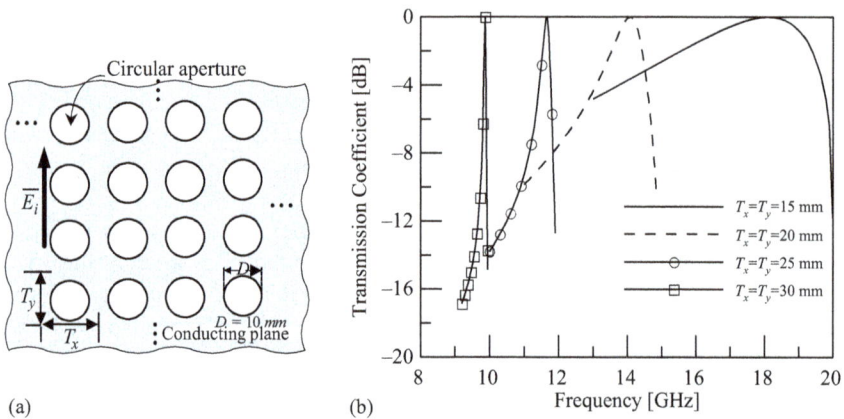

*Figure 8.15    Effect of the periodicity on periodic aperture (FSS) composed of circular aperture with no ridge, (a) periodic aperture array (FSS) composed of circular aperture with no ridge and (b) its transmission characteristics versus frequencies for four cases for which $T_x = T_y = 15$, 20, 25, and 30 mm*

the TRA of the circular ridged type is comparable to roughly the medium size among the above five unit cell sizes. This shows also that the TCS has been significantly enhanced for the FSS structure composed of the TRA.

When the TRA in which the enhanced transmission occurs at each aperture is used as the aperture element for the finite array of apertures, the enhanced transmission for the finite array is also observed. The expression of $\frac{2G\lambda^2}{4\pi}$ hold for both parallel arrangement and collinear arrangement as discussed in Section 8.2.2. This suggests that such an enhanced transmission may occur irrespective of the arrangement of the apertures in a finite array of apertures. Recall that this is analogous to the prior work [13] on the effects of the periodic, quasi-periodic, and random arrangements on enhanced transmission. In the prior work, it was found that when the TRA type is used as an element aperture in various arrangements such as periodic, quasiperiodic, and random arrays, the particular spatial arrangement is of minor importance in transmission enhancement. Note that the annular nano-aperture which was used as a unit cell of the periodic structure in the previous work [11,12] belongs to the TRA in the sense that Fabry–Perot type resonance including the near-cutoff type resonance should occur for the transmission resonance like that inside the slot in the thick conducting screen as discussed in Chapter 3.

On the other hand, when the conventional circular aperture, i.e., non-TRA type is used as an element aperture in particular for the infinite periodic FSS structure, there are two kinds of transmission resonance corresponding to the total transmission which can be called also a kind of impedance matching. One is the total transmission usually occurring in the typical FSS structure composed of the conventional circular aperture.

The other is a total transmission observed even when the TCS of the element aperture approaches zero, i.e., the element aperture area approaches zero. This type of total transmission phenomena is observed near $\lambda = d$ as discussed in the previous work [36] on the extraordinary transmission through an array of electrically small holes from a circuit theory perspective. It is interesting to note that these two types of transmission enhancements are analogous, respectively, to the site and lattice resonances in the recent work [10] on extraordinary optical transmission.

For the case of the total transmission, i.e., the zero reflection, the smallest ratio of the actual perforated circular aperture area to the unit cell area among the data in Figure 8.14(b) is found to be roughly 8.72%, which is larger than that (3.22%) for the previous case composed of the ridged circular aperture. This shows clearly that while the physical aperture area is significantly reduced by employing the TRA instead of the non-TRA, i.e. conventional aperture, the TCS or the transmitted power through the TRA is significantly increased. In addition, the increased transmission bandwidth at resonance frequencies for the TRA case as seen in Figure 8.13(b) in comparison with the non-TRA case in Figure 8.14(b) is thought to be due to the significant increase of TCS for the TRA, in other words, the increased transmitted power through the TRA, which means the lowered aperture Q (Quality factor).

Summarizing, under the same periodicity, if we employ the TRA as an element of the periodic array, the transmission resonance frequency of the FSS along with the aperture Q can be made to be significantly reduced in comparison with those for the conventional circular aperture, i.e., non-TRA.

It is worth commenting that it is easy to adjust the ratio of transmission resonant wavelength to the interelement spacing, $\lambda_0/T$ such that $\lambda_0/T > (1 + \sin\theta_i)$ where $\theta_i$ is an incident angle. Under this condition, even for significant deviation from the normal incidence, the undesired grating lobe problem can be significantly alleviated.

## 8.6    Comparison of enhanced transmission of electromagnetic energy through the aperture in ideal conductor and real metal plane of finite thickness

Since the phenomenon of enhanced transmission (called also extraordinary optical transmission, EOT) through the finite thickness metal film with periodic arrays of subwavelength holes was reported in 1998, a huge amount of research work has been devoted to analyzing the physical mechanism behind it. Though the underlying mechanism for the phenomenon itself is intricate and still merits scientific attention, it can be generally attributed to the interplay between the transmission characteristics of individual holes and the transmission characteristics of the periodicity under the assumption of the role of the surface plasmon. Nonetheless, there is research indicating that enhanced transmission can be achieved in spectral domains and polarizations where surface plasmons are either absent or nonfunctional, as reported in [37].

This means that the extraordinary transmission is then linked with the excitation of other electromagnetic resonances through various transmission resonant structures as discussed in Chapters 4 and 5. Concurrently, basic principles from the theory of frequency-selective surfaces can be applied to explain the extraordinary optical transmission, although not with absolute precision [36].

So it is necessary to look into the phenomena relevant to the enhanced transmission that can be observed in the conventional FSS structure before going into the comparison of the enhanced transmission phenomena between perfect conducting medium and real metal plane cases

### 8.6.1    *Wood's anomaly, periodicity resonance, and Fano profile characteristics in the conventional FSS structure*

First, we discuss the effect of aperture size and periodicity on the frequency shift of transmission resonance in the FSS structure which is composed of conventional non-TRA type square apertures ($D_x = D_y = D$) in the square array ($T_x = T_y = T$) arrangement as shown in Figure 8.16(a), where the incident electric field is assumed to be vertically polarized along the $y$-direction.

As reference data, the reflection characteristics versus frequencies for the geometrical parameters as in Figure 8.16(a) are given for the cases that $D_x = D_y = D = 10, 5,$ and 3 mm in Figure 8.16(b).

From Figure 8.16(b), it is seen that the typical Wood's anomaly (total reflection) occurs very near the frequency of the total transmission, i.e., at a slightly higher frequency. The closer these two phenomena occur, the smaller the element aperture size is. This results in a Fano-like resonance curve of rapid change from the dip (total transmission) to zero (total reflection) within a very narrow frequency range.

Next, to discuss interesting diffraction phenomena in more detail such as Wood's anomaly, periodicity resonance, and Fano profile characteristics, we

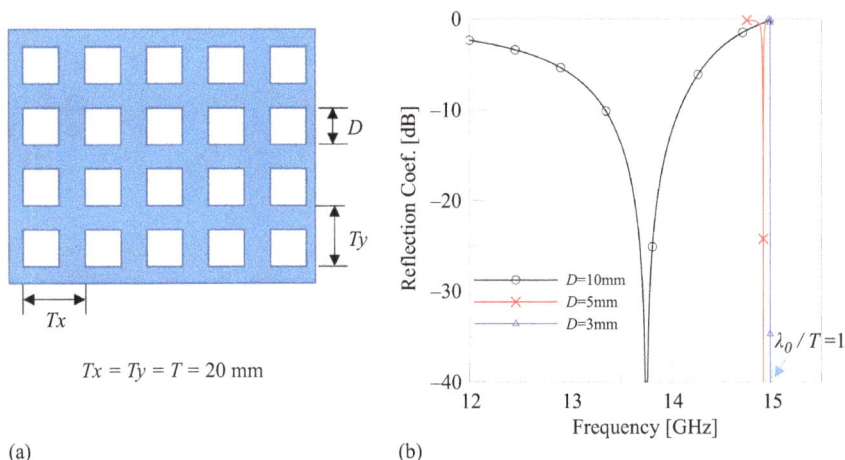

(a)                                        (b)

*Figure 8.16    Periodic square aperture array and its transmission resonance according to the aperture size. (a) Periodic square aperture array structure ($D_x = D_y = D$), (b) its reflection characteristics versus frequencies for the three cases for which $D_x = D_y = D = 10, 5,$ and 3 mm.*

consider a conventional FSS structure which is composed of a two-dimensional periodic array of square apertures as shown in Figure 8.16(a). As a numerical example, we investigate the reflection coefficients of the FSS versus frequencies for the case that $D_x = D_y = 10$ mm with the periodicity $T_y$ along the $y$-direction as a parameter while keeping $T_x = 20$ mm constant. Figure 8.17(a) illustrates reflection characteristics for three cases that $T_y = 19$ mm, 20 mm, and 21 mm. As seen from Figure 8.17(a), the transmission-zero phenomena corresponding to the Wood's anomalies occur at the frequencies of $\frac{c}{T_y} = \frac{300}{21}, \frac{300}{20}$, and $\frac{300}{19}$, respectively. Here $c$ means the light velocity.

Note that the cusp at 15 GHz of the curve for $T_y = 19$ mm in Figure 8.17(a) corresponds to the frequency at which the first higher Floquet mode is at onset as expected.

To confirm the relevancy of Wood's anomaly to the periodicity along the E-plane, we investigate the reflection characteristics versus frequency for the case that $D_x = D_y = 10$ mm with the $T_x$ as a parameter while keeping the $T_y$ constant. Figure 8.17(b) shows that the transmission zero phenomena corresponding to Wood's anomalies are observed at the same frequency for the three cases irrespective of the periodicity $T_x$ along the $x$-axis normal to the incident electric field polarization direction. The cusp at the frequency of $f = \frac{300}{21}$ GHz corresponds to the frequency at which the first higher Floquet mode is at onset.

It is also seen from Figures 8.17(a) and (b) that the frequency shift of the total transmission (zero reflection) according to the variation of the periodicity $T_y$ along

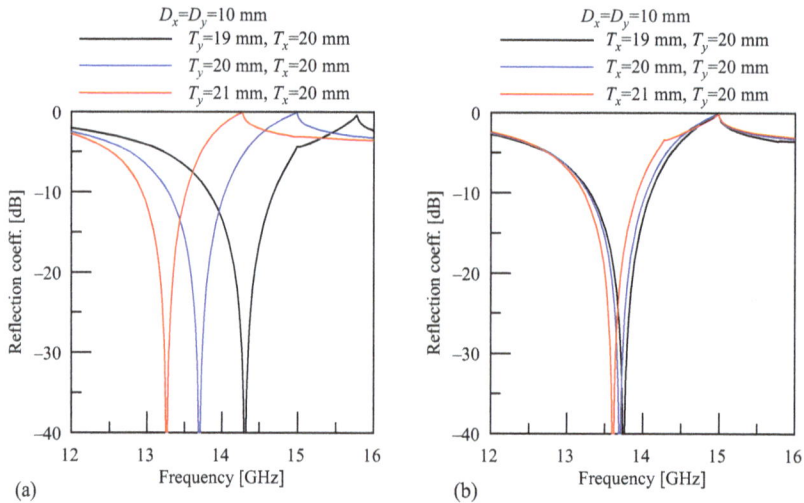

*Figure 8.17    The variations of the reflection characteristics versus frequencies (a) according to the variation of the periodicity $T_y$ along the direction of an incident electric field with $T_x$ fixed constant (b) according to the variation of the $T_x$ with $T_y$ fixed constant*

the electric field direction of the incident electric field is much larger than that according to the variation of the periodicity $T_x$.

It is worthwhile to mention that the dependence of the transmission resonance frequency shift on the variation of the periodicity $T_x$ and $T_y$ can be significantly reduced by making the quality factor of the aperture element larger, i.e., by reducing the aperture height $D_y$ as seen from the comparison between Figures 8.17 and 8.18. As reference data, Figures 8.18(a) and (b) show the variation of the zero reflection (total transmission) frequency depending on the variation of periodicities along the x- and y-axis, respectively, for the case of $D_x = 10$ mm and $D_y = 5$ mm.

In general, the total transmission frequency of the two-dimensional FSS structure is modified from the transmission resonance frequency of the single aperture as in the typical FSS case as discussed above.

In contrast to this, in the limit where the TCS of the aperture element of the FSS decreases to zero, the transmission resonance wavelength of the FSS approaches the periodicity $T$ for the square array, which corresponds to the minimum limit ($\frac{\lambda_0}{T} = 1$) for grating lobeless condition ($\frac{\lambda_0}{T} > 1$) under the normal incidence. This is a kind of transmission resonance that is related to the periodicity of the FSS structure as discussed in the prior work on the extraordinary transmission through arrays of electrically small holes from a circuit theory perspective [36].

To look into the transmission characteristics of such a transmission resonance relevant to the periodicity, we have investigated the reflection characteristics versus frequencies while varying an individual aperture area from $10 \times 10 [\text{mm}^2]$ to

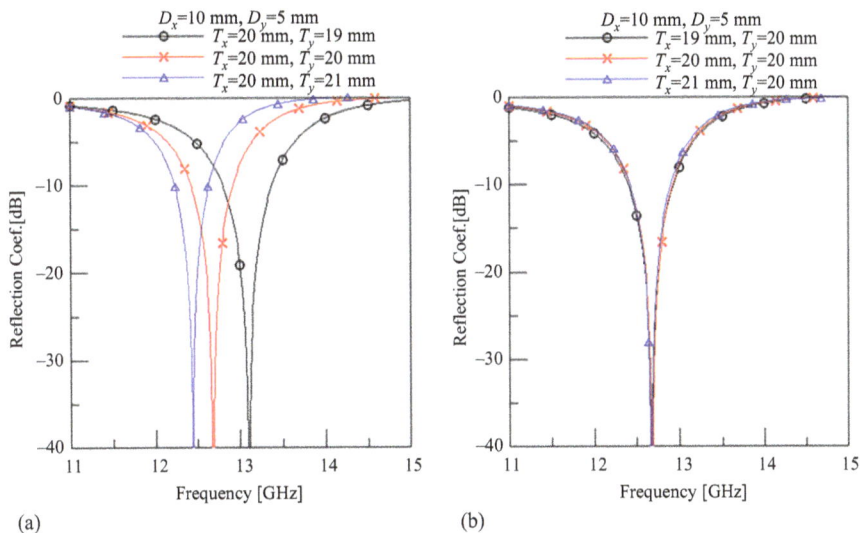

*Figure 8.18* *The variations of the reflection characteristics (a) according to the variation of the periodicity $T_y$ along the direction of an incident electric field with $T_x$ fixed constant (b) according to the variation of the $T_x$ with $T_y$ fixed constant*

$1 \times 1$ [mm$^2$] so that its transmission cross section approaches zero with the periodicity $T_x = T_y = 20$ mm kept constant. Figures 8.19(a), (b), (c), (d), and (e) illustrate, respectively, the gradual change of the reflection (transmission) characteristics as the area $D \times D$ of the aperture element approaches zero with the periodicities $T_x = T_y = 20$ mm fixed.

As mentioned above it is seen from Figure 8.18 that the transmission resonance occurs when the wavelength approaches the periodicity $T$ of the FSS as the TCS of the aperture element reduces to zero.

Based upon the foregoing discussion it can be said that there are two kinds of transmission resonance phenomena in the FSS structure which have been dealt with here. One is a transmission resonance occurring in the typical FSS structure, which is observed near the transmission resonance frequency of the isolated aperture. When the apertures constitute the periodic aperture array, the transmission resonance frequency of the array undergoes some frequency shift from that of the isolated aperture because of the mutual coupling among apertures. In this case the larger the TCS of the isolated aperture is, the less frequency shift the transmission resonance frequency of the periodic array undergoes. So the frequency shift of transmission resonance frequency in the FSS composed of TRAs from that of the isolated TRA is much smaller than the frequency shift of the transmission resonance frequency in the FSS composed of non-TRA, i.e., conventional circular or rectangular aperture.

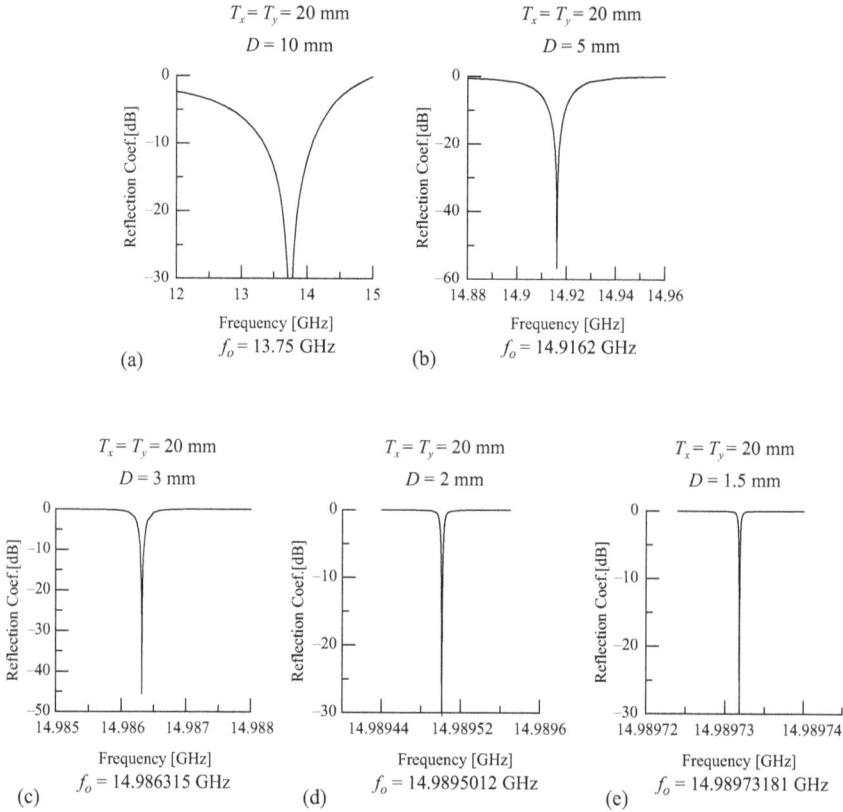

$T_x = T_y = 20$ mm
$D = 10$ mm

$T_x = T_y = 20$ mm
$D = 5$ mm

Frequency [GHz]
$f_o = 13.75$ GHz
(a)

Frequency [GHz]
$f_o = 14.9162$ GHz
(b)

$T_x = T_y = 20$ mm
$D = 3$ mm

$T_x = T_y = 20$ mm
$D = 2$ mm

$T_x = T_y = 20$ mm
$D = 1.5$ mm

Frequency [GHz]
$f_o = 14.986315$ GHz
(c)

Frequency [GHz]
$f_o = 14.9895012$ GHz
(d)

Frequency [GHz]
$f_o = 14.98973181$ GHz
(e)

*Figure 8.19    Change of the reflection characteristic curve for the 2-D periodic aperture structure as the aperture size approaches zero. D means the one-side length of the square aperture and the frequencies in each graph mean those of the zero reflection (total transmission). (a) D = 10 mm, (b) D = 5 mm, (c) D = 3 mm, (d) D = 2 mm, and (e) D = 1 mm.*

The other is the periodicity resonance which is observed as the physical size of the aperture and the TCS of the aperture is gradually reduced to zero. Note that the transmission resonance relevant to the periodicity approaches the frequency which is determined by the periodicity along the polarization direction of the incident electric field, i.e., $c/T_y$, as the TCS of the single aperture reduces to zero.

It is worthwhile to mention that this limit of $c/T_y[f]$ gives the upper bound of the frequency range over which the grating lobeless condition is satisfied.

Recall that the TCS of the aperture array is mainly determined by the directivity of the array system as discussed in [9]. So for the linear uniform aperture array to be most directive, we can see that it has to satisfy the above grating lobeless condition and $T_y \rightarrow \lambda$, i.e., $T_y$ approaches $\lambda$ with $T_y$ kept smaller than $\lambda$

under the assumption of the broadside array. The above transmission resonance is due to the periodicity resonance which is observed at a very slightly lower frequency than $\frac{c}{T_y} = \frac{c}{\lambda}$ is thought to meet such a condition for the transmission resonance.

The combined appearance of this type of transmission peak and the aforementioned transmission zero phenomena result in the Fano profile characteristics [10] associated with the periodicity resonance because the combined appearance shows the rapid variation of transmission characteristics from transmission peak to zero within a very narrow frequency range as discussed in the previous work [10] on the basic theory for extraordinary transmission through the periodic aperture in the conducting screen of zero thickness.

As mentioned above there are two types of transmission resonances. One is the transmission resonance relevant to the transmission resonance frequency of the isolated aperture. The other is due to the periodicity resonance. Note that, for the periodic FSS structure in the PEC plane of zero thickness, only the transmission resonance relevant to the transmission resonance frequency of the isolated aperture is expressed in the transmission characteristics curve versus frequency. That is to say, in the FSS structure made of the perfect conducting plane of zero thickness, the transmission resonance due to the periodicity cannot be observed. This means that the FSS structure made of a PEC plane of zero thickness cannot simulate the EOT phenomena observed in the periodic arrays of subwavelength holes in the real metal of finite thickness. So to simulate the extraordinary optical transmission through arrays of holes in thick real metal in terms of the FSS structure in the PEC plane, it has been turned out [38] that the FSS structure composed of TRA in a thick PEC plane should be used as discussed in the next section.

It goes without saying that the polarization direction of the incident electric field along which Wood's anomaly is observed is the same as that along which the fundamental minimal linear array for EOT is established [1].

### 8.6.2 General enhanced transmission through single aperture and its periodic array in PEC screen of finite thickness

Since Ebbesen's work [29] on the EOT through drilled subwavelength hole array in the finite thickness real metal was reported, tremendous research work has been done to understand the underlying physics of the phenomena including theoretical [38] and experimental [39,40] work.

There are too many reports on the topic to list them all. Among them, not a few works dealt with the enhanced transmission through a single aperture in a thick PEC screen in expectation of a significant correlation between the transmission resonances through a single aperture and its array.

As a general transmission characteristic through a periodic subwavelength hole array in a real metal of finite thickness, the transmission characteristics curve versus frequencies exhibit two types of transmission resonance. One is the aperture resonance, which is related to the transmission characteristics of the isolated aperture itself. The other is periodicity resonance, associated with the periodic structure.

While the original EOT phenomena were observed in the real metal medium case, not a few attempts to gain the underlying physics using the simplified model where a real metal medium is replaced with a PEC medium have been made. The attempt realized the anticipated result though not perfectly as discussed in [36].

As an example of a successful result, we can mention the periodic array model of TRA in the PEC plane of finite thickness [39], which can simulate the optical transmission characteristics of the periodic subwavelength hole array drilled in the real metal plane of finite thickness. Here the optical transmission characteristics in the above real metal medium slab case mean the simultaneous appearance of the aperture resonance and periodicity resonance.

As mentioned above, if the TRA which has a significantly larger TCS and so much lower quality Q than that of the non-TRA is employed as an aperture element for periodic aperture structure in the PEC screen of finite thickness, the simultaneous appearance of both the aperture resonance and the periodicity resonance corresponding to the transmission characteristics in the real metal medium slab case is observed also in the PEC slab case.

For reference, the TRA used in [13] belongs to the narrow slot type [19] in thick PEC screen, whose transmission resonances can be divided into two kinds, i.e., near-cutoff (transverse resonance) type and Fabry–Perot (longitudinal resonance) type as discussed in Chapter 3.

The previous observation [39] of the simultaneous occurrence of the aperture resonance and periodicity resonance is essentially the same as that in the study on the periodic screen with large holes in the prior work [36] based on the circuit theory perspective.

Since the presence of such two kinds of transmission resonances was recognized, various terminologies for the two resonances have been used such as waveguide resonance and resonance of coupled plasmon polaritons (SPPs), site and lattice resonances [10], vertical and horizontal surface resonances [41], hole shape and periodicity resonances [40], and the localized surface plasmon (LSP) modes [42] of individual apertures and the surface plasmon polariton (SPPs) of the periodic structure [43], etc.

To summarize, though enhanced transmission is a complex phenomenon that depends on many parameters [43], two key ingredients as main mechanisms for enhanced transmission can be briefly outlined as above. In addition, the overall enhanced transmission of the periodic hole array is thought to be attributed to an interplay [43] of the above two ingredients. This is essentially the same also to the interplay [39] between the shape of the hole or the scatterer the periodicity in the determination of the enhanced optical properties for photonic crystals, such as an increased width of the photonic bandgap.

The presence of the above two types of transmission resonances was also experimentally confirmed by the recent research [39] work on enhanced transmission through periodic arrays of subwavelength holes for the case of Au films.

Recall that as the physical area of each aperture composing the periodic structure gradually reduces to zero and so does the TCS of each aperture, the transmission resonance frequency, called periodicity resonance frequency approaches the frequency corresponding to $\frac{c}{T_y}$ for the case that the periodic structure is made of the PEC plane of zero thickness. Here $c$ is the velocity of light and $T_y$ is the periodicity along the polarization direction of the incident electric field. Note that the simultaneous occurrence of the aperture resonance and the periodicity resonance cannot be observed in the case of the PEC plane of zero thickness [36], because of the vanishment of the aperture resonance.

As discussed above, for both aperture resonance and the periodicity resonance to be observed in the transmission characteristics curve versus frequencies, the PEC plane of finite thickness should be used as a periodic aperture plane.

Recent studies revealed [39] that periodic aperture structures composed of TRA in a PEC plane of finite thickness can simulate well the EOT phenomena occurring in periodic aperture arrays drilled in a real metal slab. Here the narrow slot type of aperture was chosen as an elementary TRA. As mentioned above, the TRA has a significantly larger TCS than the physical aperture area of the TRA. The prior work based on the circuit theory perspective [36] also supports the validity of the above simulation result on the EOT phenomena occurring in the periodic aperture array drilled in the PEC conducting slab. These two studies explain the important transmission characteristics, i.e., the simultaneous occurrence of the aperture resonance and periodicity resonance observed in the periodic apertures in real metal slabs. We have already pointed out that a periodic aperture structure composed of TRA in a PEC plane of finite thickness can simulate well the EOT phenomena occurring in periodic subwavelength hole arrays drilled in real metal slabs. Here the TRA has a significantly larger TCS than the physical aperture area of the TRA and such a TRA can be taken as a large aperture from the viewpoint of the previous work [36] on the extraordinary transmission through arrays of electrically small holes from a circuit theory perspective. For this reason present model by use of the TRA is compatible with the large aperture model in previous work [36] and the two models' descriptions explain the transmission characteristics, i.e., the simultaneous occurrence of the aperture resonance and periodicity resonance observed in the periodic apertures in a real metal slab.

The transmission resonance problem in the periodic small aperture in the PEC plane of finite thickness still remains to be discussed.

Before going into the transmission resonance problem through the periodic small apertures drilled in thick PEC slab, we recall that, in the transmission resonance problem through the periodic small aperture in an infinitely thin PEC plane, the simultaneous occurrence of the aperture resonance and the periodicity resonance cannot be observed, i.e., only one of either the aperture resonance or the periodicity resonance is observed. What is more, as the aperture size and its TCS gradually reduce to zero, the transmission resonance frequency called periodicity resonance frequency approaches Wood's anomaly frequency ($\frac{c}{T_y}$).

On the other hand, for the case of the transmission resonance through periodic small apertures in a thick PEC slab, two perfect transmission peaks approach

Wood's anomaly frequency as the aperture size approaches zero. In addition for a progressive increase in thickness, the two peaks collapse into a single one whose peak diminishes to zero for very thick screens as discussed in [36]. This is thought of as a difference between transmission resonance through the periodic hole array in the real metal slab and that in the PEC slab.

Note that the cylindrical hole structure inside the real metal medium can support a guided mode, however small the diameter of the hole is. This along with the foregoing discussion is helpful to understanding the above difference in transmission resonance characteristics relevant to the EOT phenomena between the PEC medium slab case and real metal medium slab case.

### 8.6.3 Enhanced transmission through a small hole and its periodic hole array in the real metal slab

As discussed above there are two kinds of transmission resonance phenomena through a small slot in the PEC slab. One is the transmission resonance type whose phase velocity along the longitudinal direction inside the guided structure becomes almost zero at near-cutoff frequency and the other is the longitudinal resonance type corresponding to the Fabry–Perot resonance [19,41,44,45]. Here the narrow rectangular resonant slot structure in the thick conducting screen has been chosen to be a small hole structure as a representative TRA because this structure has been widely used to explain the localized waveguide resonant mode as well as to keep the computations as easy as possible.

The investigation was also extended to the enhanced transmission phenomena through a small hole problem in a real metal [46] film to understand the underlying physics of the EOT phenomena associated with the surface plasmon polariton (SPP).

To this end, through the studies on the transmission resonance through the small slots in real metal slabs, it has been revealed that the transmission resonance, in this case, is very similar to those previously found for the single rectangular slot perforated on perfect conductors, except for the redshift in the resonance peak resulting from the change in the cutoff wavelength. That is to say, for both PEC and real metal cases, the transverse resonance type whose phase velocity becomes zero and the longitudinal type corresponding to the Fabry–Perot resonance are observed [45–47], except for the redshift of the transmission resonance peaks.

In more detail, the TM propagation and reflection of light in a subwavelength slit in a real metal differ significantly from the perfect electric conductor (PEC) due to the existence of surface plasmon [48]. As the slit becomes narrower, i.e., as the slit width is reduced, there is more penetration of the field into the metal as a result of coupling between surface plasmon on the long edges of the slot, so that the cutoff wavelength is found to be much larger than that for PEC case.

The redshift effect on the transmission resonance through the narrow slot in the real metal slab was experimentally confirmed for the transverse resonant type occurring in the case of zero phase velocity corresponding to the Fabry–Perot resonance of fundamental (zeroth) order.

Such a redshift effect for the non-zeroth order of Fabry–Perot resonance was also experimentally confirmed in the previous work [43] on the role of shape and localized resonance in EOT through periodic arrays of subwavelength holes.

If we constitute a two-dimensional periodic structure composed of the above TRA like the narrow resonant slot in the thick PEC slab, then the two types of transmission resonances are observed as discussed in this section: the aperture resonance called localized waveguide resonance and periodicity resonance. To verify that the transmission resonance frequency of the localized waveguide type is mainly determined by the aperture geometry and almost independent of the periodicity, one usually investigates the transmission characteristics through PEC film where the same size aperture is randomly distributed. Through the use of this method, it is seen that the random distribution of small TRAs maintains a peak transmission resonance through an aperture but removes the higher frequency peak due to the periodicity resonance as shown in [13]. This confirms electromagnetic field localized in each aperture is almost independent of the periodicity resonance.

In contrast to the above PEC slab case, the transmission through the two-dimensional periodic apertures in the real metal slab case is different because of the presence of the surface plasmon polariton (SPP). Though similar behavior of the two types of resonance to the PEC slab cases can also be expected for the real metal case, the aperture resonance type undergoes significant redshift. On the other hand, the position of the other transmission peak due to the surface plasmon depends on the periodicity, which can be given approximately by the surface plasmon dispersion for the smooth film [47,48],

$$\lambda_{max} = \frac{d}{\sqrt{i^2 + j^2}} \sqrt{\frac{\varepsilon_1 \varepsilon_2}{\varepsilon_1 + \varepsilon_2}} \tag{8.7}$$

where the integer index $\{(i,j)|, i^2 + j^2 > 1\}$ is corresponding to the different sets of peaks, $\varepsilon_1$ and $\varepsilon_2$ are dielectric constants of the substrate and the metal, respectively, and $d$ is the periodicity.

It is worthwhile to mention that there is one more structure that is essentially the same as the narrow resonant slot structure in a thick real metal slab as a TRA. The annular aperture (or coaxial aperture) also can show the same transmission resonance characteristics as that of the narrow resonant rectangular slot in the real metal slab. In more detail, if the $TE_{11}$ mode is used for coupling of the linear polarized incident wave, the two kinds of transmission resonances are observed like the narrow rectangular slot case: Fabry–Perot type occurring at cutoff frequency or propagating frequencies, respectively, corresponding to transverse or longitudinal resonance types. So under this condition, the annular aperture structure in the real metal slab can be taken as a Fabry–Perot resonator. In fact, this annular aperture of the Fabry–Perot resonator type has been widely investigated in connection with a super-enhanced transmission through a 2-D periodic array of the metallic annular aperture as a study of photonic bands [12] as discussed above.

This Fabry–Perot type resonance type in the annular (coaxial) aperture is observed both in the PEC and real metal slab just as the Fabry–Perot resonance type in the narrow rectangular slot is observed both in the PEC and real metal slab.

Summarizing the Fabry–Perot type resonances including the zeroth order (corresponding to the transverse resonance type) and the higher order (corresponding to the longitudinal resonance type) are observed in the narrow slot types such as a rectangular or annular (coaxial) slot for both the PEC and the real metal slab.

In general, the use of the real metal instead of PEC leads to a shift in the transmission peak toward a larger value of wavelength. So it can be said that the physical origin of the transmission resonances appearing in both real metals and perfect conductors is the same, except for the redshift in the resonance peak resulting from the change in the cutoff wavelength [46].

## 8.7    Relation between the transmission cross section (TCS) and the directivity and its implication

In the case of the TRA in Chapter 4 and the TRC in Chapter 5 as a single transmission element, under the transmission resonance condition where the maximum transmission occurs through the single element, the linear relation between the directivity and the TCS has been observed to hold. This may suggest that for such a relation to hold the single aperture source should be an efficient radiating source where nearly reactive energy is almost negligible in comparison with the radiating energy. This means that the aperture source satisfying the relation should be the efficient radiating Huygens' source. Borrowing the impedance matching concept, such radiations through an aperture mean those occurring under the impedance matching as discussed in Chapters 4 and 5.

It is worthwhile to mention also that the meaning of the relation between the transmission cross section (TCS) and the directivity is equivalent to saying that the TCS is determined by the far field (Fraunhofer) diffraction property [24] because the directivity is entirely determined by the far field pattern from the aperture source where the reactive energy becomes negligible similar to the above single transmission element case.

There is an additional remark which is worthwhile to mention. Through the previous study on the TCS of the finite linear array composed of non-resonant usual rectangular aperture in Section 8.2, we made the following two observations: the first one is that the TCS of the finite array is approximately proportional to the number $N$ of aperture for both parallel and collinear configurations and the second is that the slope of the TCS curve versus the aperture number $N$ for the parallel configuration is seen to be somewhat smaller than the double times that for the collinear configuration. The first observation is compatible with the nearly linear dependence of the observed transmission on the total aperture number $N$ [24] in the recent work on the finite array size effect on the transmission property.

Recall that as a Huygens' source, the finite array of aperture sources of the parallel arrangement shapes a much more directive field than that of the collinear arrangement under the same condition of the same number of apertures along the same overall length. So if we extend the finite aperture array source of the parallel arrangement to the infinite-size array, it is expected to constitute the single chain of small apertures in which the EOT phenomena are already present [1]. As a result, the above second observation is thought to provide the basis for the formation of the concept of a basic unit of the EOT phenomena as an infinite single chain type as discussed in [1].

In the finite aperture array of a parallel arrangement, a quasi-bound-state (QBS) mode similar to a surface wave is suggested to explain the wave generated close to the onset of the first propagating channel, i.e., close to a Rayleigh frequency due to the interaction of the scattered field with diffraction modes. The coupling between the incident field and the QBS mode field leads to the transmission resonance. This way of explanation for the transmission resonance is compatible with that of previous work on the extraordinary transmission through small numbers of apertures where transmission characteristics are discussed in connection with the conversion of the incident wave to the surface mode and the Fraunhofer diffraction relevant to the directivity.

It is to be noted that the surface wave is well known to be supportable along the structure of the directors in the Yagi–Uda array [15,16]. This structure corresponds to the dipole array of parallel arrangement. From the duality consideration, it is seen that the surface wave type can be supportable along the aperture array of the parallel arrangement. This is also compatible with the polarization direction for the propagation of the surface plasmon as expected.

## 8.8   Single aperture and its array structure for enhanced transmission

Since the extraordinary optical transmission (EOT) was reported, as the topic has been actively researched, the enhanced transmission has been also used in parallel with the EOT in the sense that the two words mean the case that incident power on the significantly larger area than the total physical perforated area is transmitted through the perforated aperture area irrespective of single aperture or the aperture array type.

Based upon the discussion on the transmission resonant aperture (TRA) in Chapter 3, transmission resonant cavity (TRC), and transmitted power through the finite array of apertures in Chapter 8, it can be said that when the plane wave is incident on the infinite perfect electric conducting( PEC) plane having a single aperture or the aperture array, there are two types of the aperture distribution which can give the above-enhanced transmission phenomena. One is the type of single transmission resonant structures such as TRA and TRC. The other is a finite array type which is arranged in a parallel configuration [6]. This type of array as a Huygens' source is found to give a more directive far field in comparison with that arrayed in the collinear configuration [6].

The above discussion means that the extraordinary optical transmission phenomena, before the development of 2-D EOT phenomena, is already present in a single finite chain of the small hole array in the parallel configuration irrespective of whether its element is a type of TRA or non-TRA [22] as well as a single transmission resonant structure [22]. This is thought to be compatible with the previous description [1] that the single chain can be considered as a basic entity of EOT.

In summary, the TRA and TRC exemplify the structure for the enhanced transmission which can be implemented in terms of a single aperture structure. On the other hand, the finite or infinite array type in parallel configuration exemplifies the structure that can be implemented in terms of array type for enhanced transmission. Here the array type in parallel configuration means the case that the in-phase component of the incident electric field points along the direction of the array. This discussion is compatible with the previous work on the transmission of electromagnetic energy through a finite array of holes including a single hole [22,24].

## 8.9   Transmission resonance through a Bethe's small hole in the waveguide iris and Wood's anomaly phenomenon

We have already seen that the small hole limits the transmission resonance through the square array [49,50] of the small holes can occur at wavelength $\lambda$ immediately above the lattice period $d$ in Section 8.6. It is also well known that in the square hole array, the array arranged along the incident electric field (pointing the array axis) constitutes the aperture array source which can produce the more directive far field in the half-space opposite to the incident side in comparison with the aperture array source arranged along the normal direction to the above incident electric field. Note that an equivalent magnetic current source array producing a more directive far field corresponds to the array in parallel configuration [6]. The type of array arranged along the incident electric field has been found to constitute the minimal system that shows EOT as a basic unit of a single linear chain [1].

Here we investigate the transmission resonance characteristics through a Bethe's small hole [51] in the waveguide iris with the main concern on the comparison of the transmission resonance between the present structure and the 2-D planar square array structure of small holes.

For this purpose, we consider the transmission resonance for the case of the square waveguide with an iris having a small square hole at the center. In more detail, the one-side length of the square waveguide is chosen to be $10\sqrt{5}$ [mm] such that the frequency at which Wood's anomaly for the present case may be almost the same as that for the 2-D square array case in Section 8.6. For reference, the frequency for Wood's anomaly in the waveguide case whose cross section is $a \times b$ [mm] is given by the expression of $f_{TM/2}^{c} = \frac{c}{2\pi}\sqrt{\left(\frac{\pi}{a}\right)^2 + \left(\frac{2\pi}{b}\right)^2}$ [49,50] for the cutoff frequency of the $TM_{12}$ mode. In order for this cutoff frequency to be the same as the frequency ($\cong$ 15GHz) of the Wood's anomaly in case of 2-D planar square array in Section 8.6, the one side of the square waveguide is calculated to be $a = b = 10\sqrt{5}$ [mm].

The numerical result for the transmission resonance through the small hole at the center of the iris in the waveguide in Figure 8.20(a) is illustrated in Figure 8.20 (b). As seen in Figure 8.20(b), the frequency (14.9903 GHz) of the transmission zero (obtained by use of the cutoff frequency of the $TM_{12}$ mode) is observed to be very near the transmission peak frequency as an example of the Wood's anomaly.

It is also seen that the transmission resonance frequency (14.9509 GHz) for the waveguide case is almost the same as that (14.986 GHz) for the 2-D planar array case. This observation is compatible with the previous comment that the transmission resonance frequencies are mainly determined by the array effect arranged along the direction of the incident electric field for both two cases.

The cutoff frequency of the $TM_{12}$ mode corresponds to the onset frequency of the first grating lobe by an image array of the equivalent magnetic current source along the direction of the incident electric field. This seems to be compatible with the previous observation that the image along the direction of the incident electric field constitutes a basic single linear array source of EOT present before the development of 2-D EOT phenomena. So for both the waveguide case and the 2-D planar square array cases, the frequencies for Wood's anomalies are mainly determined by the onset frequency of the lowest grating lobe [36] along the array of the equivalent magnetic current arranged along the direction of the incident electric field, though the Wood's anomalies for the wide band including more high frequency band are rather complicated to describe in a word.

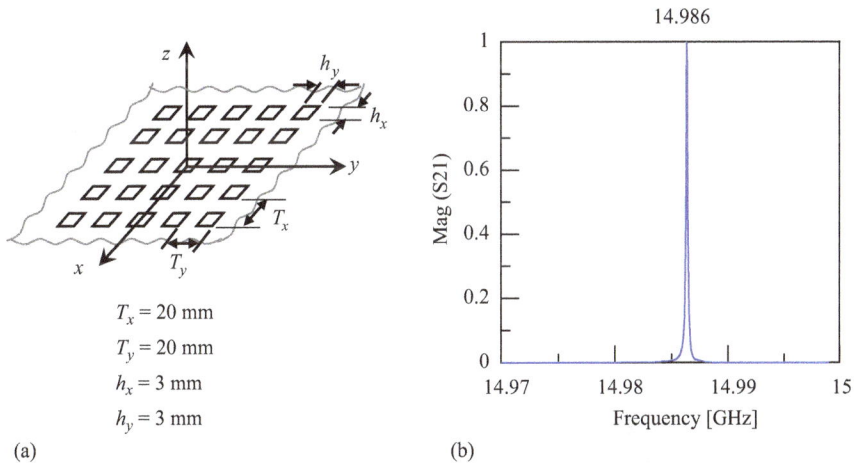

*Figure 8.20    Small hole in a waveguide screen and transmission resonance through the small hole. (a) Small hole in a waveguide screen. (b) Transmission peak (resonance) and zero.*

Here it should be pointed out that for the transmission resonance condition to be satisfied, the required capacitive component should be compensated from the reactive stored energy due to higher evanescent TM mode, not from the capacitive component of the circuitry.

In more detail, the desired reactance for the transmission resonance can be obtained in the form of the reactive energy and so even when the impedance matching condition is met under the transmission resonance, the reactive energy due to the evanescent $TM_{12}$ mode remains, unlike the usual impedance matching situation where reactive energy is canceled. Even under the impedance matching, the reactive stored energy due to the evanescent TM mode exists on the holed iris surface while some power is leaked through the hole. This stored energy distribution due to the evanescent mode seems to be similar to the surface wave mode but different in that the power leak is going on at the same time. In this sense, it is called "Quasi bound" mode [17] to distinguish from the surface wave with no power leak or the leaky wave mode with no stored energy.

It has been also observed that in the case of the transmission resonance through the small hole in the waveguide iris, the Wood's anomaly occurs at the Rayleigh wavelength where the lowest grating lobe is onset. In this situation, the QBS mode is set up. This is in contrast with the Wood's anomaly associated with a leaky wave which is observed in connection with the Bragg condition as discussed in Chapter 7. These two types of Wood's anomalies, present Rayleigh wavelength type and leaky wave-related type are compatible with the previous work on the new theory on the Wood's anomaly by Hessel and Oliner.

## 8.10 Correspondence of the guided channel between small hole and narrow slit in the real metal medium

At the time when the research work by Ebbesen *et al.* on the extraordinary optical transmission (EOT) phenomena in 2-D arrays of small holes was reported, the most existing theories which aim to explain the EOT dealt with a 1-D transmission grating with infinite slits [32,33,54].

Through the studies on 1-D grating, two important types of transmission resonance mechanisms were predicted: slit-guided modes [40] and coupled surface plasmon polariton (SPP) resonances. It was also found that enhanced transmission associated with the slit waveguide mode resonance is also present in single narrow slits [52,53].

However, hole arrays and slit arrays are two different geometries in the sense that transverse electromagnetic (TEM) modes without cutoff frequency are possible in the slit structure but not in the cylindrical hole structure in the PEC medium case. So the clue for the correspondence between both two structures seems to be lacking.

During an investigation on such correspondence, it was found that using simple concepts from the conventional theory of FSSs, waveguides, and transmission lines allows us to explain extraordinary transmission for both thin and thick periodically perforated perfect conducting screens including most of the details of the observed transmission spectra [36].

The prior work [36] is very helpful to understand the basic physics of the extraordinary transmission through the periodic hole structure in the PEC medium slab but the fundamental correspondence between the slit and cylindrical hole geometries in the real metal slab remains to be investigated further.

There seem to be several observations that may support the correspondence between cylindrical aperture (or hole) structures and slit structures in the real metal slab case from the viewpoint of underlying physics on the transmission resonance phenomena.

As discussed in [31], the positions of the transmission peaks in both structures of 2-D aperture array and 1-D infinitely long slit array are determined mainly by the periodicity of the array systems and are almost independent of the diameter of the circular aperture or the slit width and of the particular metal used.

What is more, the dispersion of the transmission peaks in a 2-D small aperture array as a function of incident angle [31] is quite similar to the one obtained for coupled SPPs in transmission gratings (see Figure 2(a) in [31]).

But note that in the perfect conducting medium case, there exists a cutoff frequency in the small hole structure, unlike the infinitely long slit structure where there does not exist a cutoff frequency of the lowest TEM mode, which is a conspicuous difference between the two structures.

Despite this seeming difference between the two structures in the PEC medium case, it has been found that the small aperture such as a narrow slot of rectangular shape or the small hole of circular shape in the real metal slab supports an efficient channel for the transmission of electromagnetic energy. In more detail in the narrow slot of the rectangular shape, the cutoff wavelength for the fundamental mode whose electric field is normal to the long edges of the slot is significantly increased as the slot width is gradually reduced to zero. Similar to this trend, in the cylindrical small hole case too, the guided mode is supportable, however small the diameter of the cylindrical hole is. So the transmission cavity with its input and output apertures open-ended (i.e., terminated with a lossy magnetic wall) is expected to be formed, which is a key ingredient to the underlying physics on the vertical resonance called also as the site resonance [10] or the localized waveguide resonance [54]. Here the formation of the transmission cavity as the resonant transmission path presupposes the presence of the guided mode irrespective of the infinite long slit or hole structure.

Such a channel is specific to single-hole structures and their two-dimensional arrays with finite conductivity and explains the extraordinary optical transmission in them [38]. Here note that the hole structure is defined to have simply connected aperture boundary such that the small narrow rectangular slot as well as the small circular aperture belongs to this category.

Recall that the two kinds of Fabry–Perot transmission resonance, transverse resonance (corresponding to the zeroth order) and longitudinal resonance (corresponding to the higher order) were observed in the narrow rectangular slot structure perforated in the conductor slab of finite thickness irrespective of the perfect or real metal slab. It is interesting to note that such Fabry–Perot type of transmission resonances are observed also in the annular (coaxial) aperture which has been widely investigated in connection with a super-enhanced transmission through a 2-D metallic annular aperture array as a study of photonic bands [11].

It is to be noted here that, though the fundamental lowest mode of the annular (coaxial) guide is TEM mode, the $TE_{11}$ mode is most easily excited inside the coaxial aperture if the linear polarized illuminating field from outside the aperture is incident upon and coupled into the annular guide perforated in the conducting metal slab [12]. So when the linear polarized field is normally incident upon the annular aperture input, the strong transmission resonant field as a standing wave type of the $TE_{11}$ mode is established inside the annular (coaxial) cavity.

Consequently, the transmission through the annular aperture is resonantly enhanced at two wavelengths depending on the Fabry–Perot resonance of the zeroth order and the next higher order, which is essentially the same as the transmission resonance phenomena observed in the narrow rectangular slot in the thick conducting slab.

In the sense that the aperture boundary on the transverse plane of the guiding structure inside the conducting slab is simply connected, the above rectangular slot belongs to the hole type of aperture structure. In contrast, the aperture boundary of the annular slot as well as one dimensional infinitely long slit belongs to the doubly connected curve. So for the present discussion's convenience, the aperture structures can be divided into two categories, hole and infinitely long slit structures. It is important to note that, for both types of the narrow rectangular slot belonging to the hole structure type and the infinitely long slit type, the transmission resonance cavity of the Fabry–Perot type can be formed in the real metal slab case.

As a result, the flatness of the dispersion curves at near-normal incidence angle, i.e., large bandwidth in incidence is strong evidence for the highly localized nature of the modes due to the presence of the transmission cavity structure that mediates the enhanced transmission.

Summarizing, the enhanced transmission is observed for both 1-D long slit [36,51] array structures and 2-D hole [55,56] (like the narrow rectangular slot) array structures because the guiding structure as above is set up so that the two types of Fabry–Perot resonances can be observed.

This may be taken as a correspondence between the 1-D long slit and 2-D narrow rectangular slot type of hole array structures from the viewpoint of the transmission resonance through the single 1-D long slit and 2-D narrow rectangular slot type of hole in the real metal slab due to the presence of the above guiding structure despite the seeming difference between the two structures.

There is one more thing to be supplemented concerning the small circular cylindrical hole in the real metal slab. Though the narrow rectangular slot type seems to be quite different from the small circular cylindrical hole shape in appearance, it is categorized into the hole type because the boundary of the narrow slot is singly connected like the cylindrical hole in contrast to the infinite long slit geometry as discussed above and because even in the limit where the physical area of the circular cylindrical aperture approaches zero, the guided mode is supportable [57].

Here note that the aperture boundary of the annular (coaxial) type is doubly connected, which is fundamentally different from the singly connected curve. Remember, however, that the easily excited mode inside the annular aperture is not TEM mode but $TE_{11}$ mode when a linearly polarized uniform plane wave is incident upon and coupled into the aperture. Because the cutoff wavelength of the annular aperture in metal slab becomes significantly increased as the difference between the inner and outer radius decreases, for the small outer radius under the assumption that the inner radius approaches zero, the transmission properties of the single annular aperture and its array as well are transformed into those of the single circular aperture and its array. So the dominant resonant mode of the $TE_{11}$ mode in the coaxial waveguide morphs into that of $TE_{11}$ mode in the circular waveguide. This suggests that even the very small cylindrical hole may support the guided mode and constitute the transmission cavity. As a result, the enhanced transmission through the cylindrical hole inside the real metal slab occurs.

For comparison even at a lower frequency than the cutoff frequency inside the cylindrical hole in the PEC medium slab, the enhanced transmission through the cylindrical hole can be made to occur by filling the hole with a high index of refraction material to constitute the transmission cavity resonance inside the cylindrical hole [55,56].

In summary, in the case of the circular cylindrical hole in the real metal slab, however small the diameter is, the guided mode is supportable [56,58] and so the transmission resonance cavity is expected to be formed along the cylindrical hole as in the narrow rectangular slot.

For these reasons, the apertures whose boundary is singly connected such as circular aperture as well as the rectangular aperture have been categorized into the hole type of aperture as discussed above.

Based upon the foregoing discussion, it seems reasonable to explain the enhanced transmission phenomena through the aperture array in the real metal slab in terms of the transmission resonance through the single cylindrical hole relevant to the transmission cavity inside the hole as well as the contribution due to the periodicity resonance.

In other words, as an origin of the enhanced transmission (called also the EOT) it seems reasonable to think of, for both the PEC and real metal medium cases, the channel for the electromagnetic wave transmission not only for the small hole array but also for the one-dimensional long slit array structure [38].

Finally, it is worthwhile to mention also that Fano-like transmission resonance evidenced by wavelengths of vanishing transmission close to transmission maxima is observed for both 1-D long slit array structure and 2-D hole array structure [28,58,59] in the real metal slab.

As is well known, the Fano-like transmission resonance presupposes the existence of discrete states localized at the hole cavity. As a result, coupling between the discrete states and the continuum of states outside the slab gives rise to Fano-like transmission resonance. So the appearance of the Fano-like transmission curve for both hole and slit array structures in the real metal slab provides a clue that the transmission cavities in the real metal slab are established for both structures. This constitutes the correspondence between the hole and slit structures from the viewpoint of the transmission resonance structure.

## 8.11    Comparison between extraordinary optical transmission (EOT) and guided mode resonance (GMR) problems

The guided mode resonances are spectrally or angularly abrupt variations in the amplitudes of the specular (zeroth order) reflected and transmitted waves [60] as a result of the coupling of the incident wave to a leaky waveguide mode in the grating structure by phase-matching with the leaky wave stopband [61].

As is well known, in the case of the EOT problem, the transmission characteristics are determined by an interplay between localized resonances in individual holes and lattice resonances originating in the array periodicity, called periodicity resonance.

These two structures, i.e., EOT and guided mode resonance (GMR) structures can be taken mainly as those of the bandpass filter even though they can be designed as other types of a filter such as a band rejection filter. It is interesting to compare the two structures from the viewpoint of the filter structures.

As a common feature, the above two structures of the EOT and GMR need particular constituents for the interaction of the incident wave with the structures. For the former, the 2-D small hole array structure in a thick conducting slab is used as a typical example of the EOT structure as in [62]. In more detail, in the EOT problem case, the localized resonant mode inside the hole plays an important role in increasing the coupling efficiency of surface plasmons between the input (incident) side and output side which connect through periodically drilled small hole array, resulting in electromagnetic energy squeezing (or funneling) through holes and a higher transmissivity.

On the other hand, in the GMR problem case, the above periodically localized resonant mode is replaced by the standing wave pattern produced along the longitudinal direction over the leaky wave structure at the second order stopband [63] called leaky wave stopband [64] for example of the dielectric waveguide grating structures.

Though these two structures are different from each other, the two structures show a typical common transmission characteristic. For example, let us consider the two simple structures. One is the transmission metallic grating [44] as an EOT structure. The other is a transmission grating structure [65] which comprises double layers of periodic conducting strips with dielectric slabs interposed between them as a GMR structure [66].

As discussed above, the main difference between the EOT and the GMR problem is that, in the former problem, the enhanced transmission resonance occurs as the result of the interplay [10] between the vertical plasmon mode (or waveguide mode or localized resonant mode) inside the hole and the horizontal plasmon mode, whereas, in the latter problem, the guided mode resonance occurs when the incident wave is coupled to a leaky waveguide mode by phase matching with a waveguide grating structure.

In more detail, in the EOT problem, the localized resonant mode is set up along the hole perforated through the thick metal slab. On the other hand, in the GMR problem, the standing wave is built up along the longitudinal direction over the leaky wave structure at the stopband region. A closer look at the transmission characteristic curve [54,59] reveals that the Fano profiles of rapid change from the transmission zero to the transmission peak within a very small wavelength or frequency range are observed for both problems.

The existence of the localized resonant mode at the hole cavity for the EOT problem and the existence of the standing wave mode along the leaky wave structure at the stopband region for the GMR problem is thought to give rise to Fano-like transmission resonances by coupling the incident plane wave outside the two structures under consideration.

Comparison of the equivalent circuit representation between the EOT problem [18] and the GMR problem [67] shows that the two problems can be represented by the lossy resonator circuit of a parallel RLC circuit, whose complex root corresponds to the leaky wave solution of the fundamental mode. Note that the leaky wave solutions for the two problems are found from the transverse resonance condition in the equivalent RLC circuit representation though the above two problems deal with different polarization cases. In terms of their relevance to the leaky wave, these two problems can be regarded as essentially similar to each other.

The appearance of the Fano-like curve in both EOT and GMR problems may also support the above argument on the essential similarity between the two phenomena. As is well known, the Fano type of resonant curve results from a superposition of resonant and non-resonant contribution to the zero diffraction order. In the EOT problem, the transmission resonance through the hole contributes to the resonant transmission. On the other hand, in the GMR problem, the main contribution is the incident wave coupling to a standing wave formed along the leaky wave guiding structure at the stopbands such as the first leaky stopband in Section 7.3.2 for metal-based resonant bandpass filter and the second leaky stopband for usual dielectric waveguide filter.

The main difference between the two can be clearly understood if we look at the viewpoint of application. In the case of the EOT problem, the main application is in the area of the subwavelength optics where electromagnetic energy passing through the single subwavelength aperture structure is required mainly for overcoming the diffraction limit. Recall that the larger the TCS of the single aperture, the less the change in the transmission characteristics of the single aperture in the 2-D array situation undergoes. On the other hand, the GMR is used mainly in the design area of the filter in the diffractive optics.

A usual FSS structure with a circular or rectangular aperture as an element aperture can be taken as a kind of EOT structure if the concern is restricted to the periodicity resonance. That is, even for the case of the 2-D array of the very small aperture with the TCS of each aperture approaching zero, the total transmission can be achieved when $D = \lambda$. Here $D$ corresponds to the periodicity along the E-plane array direction.

In general, there are two types of transmission resonance in the EOT phenomena, site resonance due to the single aperture resonance and lattice resonance due to the periodicity. On the other hand, in the case of the GMR, the transmission resonance occurs under the Bragg condition where the standing wave along the leaky wave structure is set up. Associated with this resonant standing wave mode in the GMR phenomena and the resonant mode inside the hole cavity in the EOT phenomena, the Fano-type transmission resonance curves are observed to appear for both the EOT and GMR phenomena.

## 8.12    Transmission resonance through an infinitely long narrow slit and a small hole array in the parallel-plate waveguide screen

It is interesting to investigate the transmission resonance characteristics through an infinite long narrow slit in the parallel-plate waveguide (PPW) screen in the narrow slit limit and to compare with that through the small hole array in the PPW screen in the small hole limit.

First let us consider the latter type of the transmission resonance problem, i.e., the infinite periodic structure of small holes as shown in Figure 8.21(a).

In this problem, the TEM mode having a $y$-component electric field is assumed to be incident upon the periodic hole structure in the screen from $+z$. Figure 8.21(b) shows some numerical data on the transmission resonance characteristics for three cases of square aperture sizes. From the figure, it is seen that, as the small aperture

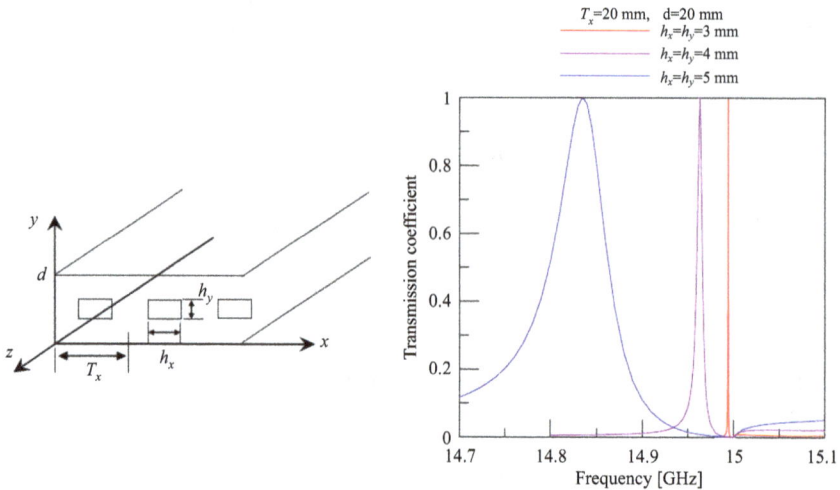

*Figure 8.21    Periodic structure of small holes in the parallel-plate waveguide screen and its transmission resonance characteristics. (a) Small hole array structure in the parallel-plate waveguide screen. (b) Transmission resonance characteristics for three cases of the square apertures whose sides are 3, 4, and 5 [mm] with $T_x = d = 20$ [mm] fixed.*

size decreases, the transmission resonance frequency approaches the particular frequency of the transmission zero very near 15 GHz.

As is well known, this frequency of the transmission zero corresponds to the cutoff frequency of the $TM_{02}$ mode in the PPW [36]. Remind that this frequency corresponds to the onset frequency of the lowest grating mode in the array which is constituted by an array of the equivalent magnetic current source over the narrow slit along the $y$-axis. Furthermore, if we put the two magnetic walls along the $x$-axis at appropriate positions, the present structure is reduced to the unit cell structure for the 2-D extraordinary transmission problem [49] through arrays of small holes. A previous study [51] on the transmission resonance through Bethe's small hole in a square waveguide screen also supports the important role of the TM mode in the transmission resonance process through Bethe's small hole in such a physical situation.

In contrast to the periodic hole array case, the transmission resonance characteristics for the infinite long narrow slit case in Figure 8.22(a) are quite different.

As shown in Figure 8.22(b), the transmission resonance occurs at the far extreme end to the zero frequency (or infinite wavelength limit) from the transmission zero at 15 GHz corresponding to the lowest grating lobe onset frequency. This is quite different from the case of the above periodic hole array where the transmission peak (resonance) and zero occur very near to each other called Wood anomaly. It is interesting to note that the transmission zeros at 30 and 45 GHz correspond, respectively, to the onset frequency of the second and third grating lobes irrespective of the aperture size.

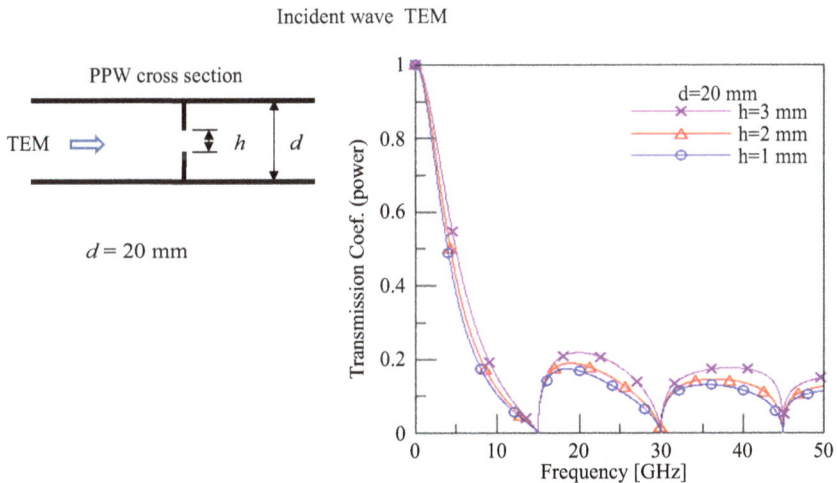

*Figure 8.22* *Infinitely long narrow structure in the parallel-plate waveguide screen and its transmission resonance characteristics. (a) Narrow slit structure in the parallel plate waveguide screen. (b) Transmission resonance characteristics for three cases of the slit width $h = 1, 2,$ and $3$ [mm].*

The reason for the occurrence of the transmission peak (resonance) at the zero frequency limit can be explained by pointing out that in the limit where the guide height $d$ is much smaller than the wavelength, i.e., $d \ll \lambda_0$, the transverse slit structure does not significantly perturb the characteristic impedance of the original PPW without the capacitive loading.

# References

[1]   Bravo-Abad, J., Garcia-Vidal, F.J., Martin-Moreno, L. 'Resonant transmission of light through finite chains of subwavelength holes in a metallic films'. *Phys. Rev. Lett.* 2004, vol. 93, p. 227401

[2]   Lee, J. I., Cho, Y. K., Ko, J. H., Yeo, J. 'Resonant transmission through a pair of ridge-loaded circular sub-wavelength apertures'. *Prog. Electromagn. Res.* 2012, vol. 24, pp. 113–126

[3]   Rao, S. M., Wilton, D. R., Glisson, A. W. 'Electromagnetic scattering by surfaces of arbitrary shape'. *IEEE Trans. Antennas Propagat.* 1982, vol. 30 (3), pp. 409–418

[4]   Yu, W., Mittra, R. 'CFDTD: Conformal finite difference time domain Maxwell's equations solver'. *Software and User's Guide*: Artech House, London; 2003

[5]   Park, J. E., Yeo, J., Lee, J. I., Ko, J. H., Cho, Y. K. 'Resonant transmission of an electrically small aperture with a ridge'. *J. Electromagn. Waves Appl.* 2009, vol. 23(14–15), pp. 1981–1990

[6]   P. S. Carter, 'Circuit relations in radiating systems and applications to antenna problems'. *Proc. IRE.* 1932, vol. 20(6), pp. 1004–1041

[7]   C. A. Balanis, *Antenna Theory: Analysis and Design.* Wiley, New York; 1997

[8]   J. R. James P. S. Hall, *Handbook of Microstrip Antennas*, 'Chapter 10: Transmission-line model for rectangular microstrip antenna'. London: Peter Peregrinus, 1989, vol. 1, pp. 527–578

[9]   L. Verslegers, Z. Yu, P. B. Catrysse, S. Fan. 'Temporal coupled-mode theory for resonant apertures'. *J. Opt. Soc. An. B.* 2010, vol. 27(10), pp. 1947–1956

[10]  F. S. Garcia de Abajo, J. J. Saenz, L. Campillo, J. S. Dolado. 'Site and lattice resonances in metallic hole arrays'. *Opt. Express.* 2006, vol. 14(1), p. 9769

[11]  S. M. Orbons, A. Roberts, D. N. Jamieson, M. I. Haftel, C. Schlockermann, D. Freeman, B. Luther-Davies. 'Extraordinary optical transmission with coaxial apertures'. *Appl. Phys. Lett.* 2007, vol. 90, pp. 251107-1–251107-3

[12]  F. I. Baida, D. Vanlabeke, G. Granet, A. Moreau, A. Belkhir. 'Origin of the super-enhanced light transmission through a 2-D metallic annular aperture array: a study of photonic bands'. *Applied Physics B.* 2004, 79, pp. 1–8

[13]  C. Rockstuhl, F. Lederer, T. Zentgraf, H. Giessen. 'Enhanced transmission of periodic, quasiperiodic and random nanoaperture arrays'. *Appl. Phys. Lett.* 2007, vol. 91, pp. 151109-1–151109-3

[14] R. H. Harrington, *Time-Harmonic Electromagnetic Fields*. McGraw-Hill Book Company, Inc.; 1961. pp. 16–26

[15] R. J. Mailloux. 'Excitation of a surface wave along an infinite Yagi–Uda array'. *IEEE Trans. Antennas Propagat.* 1965, vol. 13, pp. 719–724

[16] R. J. Mailloux. 'Antenna and wave theories of infinite Yagi–Uda arrays'. *IEEE Trans. Antennas Propagat.* 1965, vol. 13, pp. 499–506

[17] F. J. Garcia de Abajo, R. Gomez-Medina J. J. Saenz. 'Full transmission through perfect-conductor subwavelength hole arrays'. *Phys. Rev.* 2005, E72, p. 016608

[18] R. Marques, F. Mesa, L. Jelinek, F. Medina. 'Analytical theory of extra-ordinary transmission through metallic diffraction screens perforated by small holes'. *Opt. Express.* 2009, vol. 17(7), pp. 5571–5579

[19] Y. K. Cho, K. W. Kim, J. H. Ko, J. I. Lee. 'Transmission through a narrow slot in a thick conducting screen'. *IEEE Trans. Antennas Propagat.* 2009, vol. 57(3), pp. 813–816

[20] W. L. Stutzman G. A. Thiele, *Antenna Theory and Design*. Wiley, New York; 1981

[21] C. Liu, J. Shmoy, A. Hessel. 'E-plane performance trade-offs in two dimensional microstrip-patch element phased array'. *IEEE Trans. Antennas Propagat.* 1982, vol. 30(6), pp. 1201–1206

[22] J. M. Brok H. P. Urbach. 'Extraordinary transmission through 1,2, and 3 holes in a perfect conductor, modelled by a mode expansion technique'. *Opt. Express.* 2006, vol. 14(7), pp. 2552–2572

[23] Ishimaru A. *Electromagnetic Wave Propagation, Radiation, and Scattering*, Chapter III. Prentice-Hall International Edition, Englewood Cliffs, NJ; 1991

[24] F. Miyamaru M. Hangyo. 'Finite size effect of transmission property for metal hole arrays in subterahertz region'. *Appl. Phys. Lett.* 2004, vol. 84(15), pp. 2742–2744

[25] M. Beruete, M. Sorolla, I. Campillo, J. S. Dolado, L. Martin-Moreno, J. Brovo-Abad, F. J. Garcia-Vidal. 'Enhanced millimeter wave transmission through quasioptical subwavelength perforated plate'. *IEEE Trans. Antennas Propagat.* 2005, vol. 53(6), 1897–1903

[26] M. Beruete, M. Sorolla, I. Campillo, J. S. Dolado. 'Increase of the transmission in cut-off metallic hole arrays'. *IEEE Microw. Wirel. Compon. Lett.* 2005, vol. 15(2), pp. 116–118

[27] J. Bravo-Abad, L. Martin-Moreno, F. J. Garcia-Vidal. 'Resonant transmission of light through subwavelength holes in thick metal films'. *IEEE J. Sel. Top. Quantum Electron.* 2006, vol. 12(6), pp. 1221–1227

[28] C. T. Tai. 'The nominal directivity of uniformly spaced array of dipoles'. *Microw. J.* 1964, vol. 7, pp. 51–55

[29] T. W. Ebbesen, H. J. Lezec, H. F. Ghaemi, T. Thio, P. A. Wolff. 'Extraordinary optical transmission through subwavelength hole arrays'. *Nature (London)*, 1998, vol. 391(12), pp. 667–669

[30] R. F. Harrington, D. T. Auckland. 'Electromagnetic transmission through narrow slots in thick conducting screen'. *IEEE Trans. Antennas Propagat.* 1980, vol. 28(5), pp. 616–622

[31] J. A. Porto, F. J. Garcia-Vidal, J. B. Pendry. 'Transmission resonances on metallic gratings with very narrow slit'. *Phys. Rev. Lett.* 1999, vol. 83(14), pp. 2789–2791

[32] S. Collin, F. Pardo, R. Teissier, J. L. Pelouard. 'Strong discontinuities in the complex photonic band structure of transmission metallic gratings'. *Phys. Rev. Lett.* 2001, vol. 63(3), p. 033107

[33] A. Barbara, P. Quemerais, E. Bustarret, T. Lopez-Rios, T. Fournier. 'Electromagnetic resonances of subwavelength rectangular metallic gratings'. *Eur. Phys. J. D.* 2003, vol. 23, pp. 143–154

[34] A. Ksienski. 'Equivalence between continuous and discrete radiating arrays'. *Can. J. Phys.* 1961, vol. 39, pp. 335–349

[35] S. Ramo, J. R. Whinnery, T. Van duzer. *Fields and Waves in Communication Electronics*, 3rd edn. Wiley, New York; 1993. Ch. 12

[36] F. Medina, F. Mesa, R. Marques. 'Extraordinary transmission through arrays of electrically small holes from a circuit theory perspective'. *IEEE Trans. Microwave Theory Tech.* 2008, vol. 56(12), pp. 3108–3120

[37] E. Popov, S. Enoch, G. Tayeb, M. Neviere, B. Gralak, N. Bonod. 'Enhanced transmission due to nonplasmon resonances in one- and two-dimensional grating'. *Appl. Opt.* 2004, vol. 43(5), pp. 999–1008

[38] E. Popov, M. Neviere, S. Enoch, R. Reinisch. 'Theory of light transmission through subwavelength periodic hole arrays'. *Phys. Rev. B.* 2000, vol. 62 (23), pp. 16100–16108

[39] Z. Ruan M. Qiu. 'Enhanced transmission through periodic arrays of subwavelength holes: The role of localized waveguide resonance'. *Phys. Rev. Lett.* 2006, 96, p. 233901

[40] K. J. Klein Koerkamp, S. Enoch, F. B. Segerink, N. F. van Hulst, L. Kuipers. 'Strong influence of hole shape on extraordinary transmission through periodic arrays of subwavelength holes'. *Phys. Rev. Lett.*, vol. 92(18), p. 183901

[41] S. Collin, F. Pardo, R. Teissier J. L. Pelouard. 'Horizontal and vertical surface resonances in transmission metallic gratings'. *J. Opt. A: Pure Appl. Opt.* 2002, vol. 4, pp. S154–S160

[42] A. Degiron T. W. Ebbesen. 'The role of localized surface plasmon modes in the enhanced transmission of periodic subwavelength apertures'. *J. Opt. A: Pure Appl. Opt.* 2005, vol. 7, pp. S90–S96

[43] K. L. Van der Molen, K. J. Klein Koerkamp, S. Enoch, N. F. van Hulst, L. Kuiper. 'Role of shape and localized resonances in extraordinary transmission through periodic arrays of subwavelength holes: Experiment and theory'. *Phys. Rev. B.* 2005, 72, 4, p. 045421

[44] H. J. Garcia-Vidal, E. Moreno, J. A. Porto, L. Martin-Moreno. 'Transmission of light through a single rectangular hole'. *Phys. Rev. Lett.* 2005, vol. 95(10), p. 103901

[45] F. J. Garcia-Vidal, L. Martin-Moreno, Esteban Moreno, L. K. S. Kumar, R. Gordon. 'Transmission of light through a single rectangular hole in a real metal'. *Phys. Rev. B.* 2006, 74(15), p. 153411

[46] R. Gordon, L. K. S. Kumar, A. G. Brolo. 'Resonant light transmission through a nanohole in a metal film'. *IEEE Trans. Nonotechnol.* 2006, vol. 5 (3), pp. 291–294

[47] R. Gordon. 'Light in a subwavelength slit in a metal: Propagation and reflection'. *Phys. Rev. B.* 2006, vol. 73(15), p. 153405

[48] H. F. Ghaemi, T. Thio, D. E. Grupp, T. W. Ebbesen, H. J. Lezec. 'Surface plasmons enhance optical transmission through subwavelength holes.' *Phys. Rev. B.* 1998, 58(11), pp. 6779–6782

[49] R. Gordon. 'Bethe's aperture theory for arrays'. *Phys. Rev. A.* 2007, vol. 76, p. 053806

[50] F. Medina, F. Mesa, R. Marques. 'Equivalent circuit model to explain extraordinary transmission.' *IEEE MTT-S International Microwave Symposium.* Digest. Atlanta, GA, July 2008, pp. 213–216

[51] Y. Pang, A. N. Hone, P. P. M. So, R. Gordon. 'Total optical transmission through a small hole in a metal waveguide screen'. *Opt. Express.* 2009, vol. 17(6), pp. 4433–4441

[52] Y. Takakura. 'Optical resonance in a narrow slit in a thick metallic screen'. *Phys. Rev. Lett.* 2001, vol. 86(24), pp. 5601–5603

[53] F. Yang, J. R. Sambles. 'Resonant transmission of microwave through a narrow metallic slit'. *Phys. Rev. Lett.* 2002, vol. 89(6), p. 063901

[54] S. Collin, G. Vincent, R. Hardar, N. Bardou, S. Rommeluere, J-L Pelouard. 'Nearly perfect Fano transmission resonances through nanoslits drilled in a metallic membrane'. *Phys. Rev. Lett.* 2010, vol. 104(2), p. 027401

[55] P. B. Catrysse, S. Fan. 'Propagating plasmonic mode in nanoscale apertures and its implications for extraordinary transmission'. *J. Nauophotonics.* 2008, vol. 2, p. 021790

[56] H. Shin, P. B. Catrysse, S. Fan. 'Effect of the plasmonic dispersion relation on the transmission properties of subwavelength cylindrical holes'. *Phys. Rev. B.* 2005, vol. 72, p. 085436

[57] K. Y. Kim, Y. K. Cho, H. S. Tae, J. H. Lee. 'Optical guided dispersion and subwavelength transmission in dispersive plasmonic circular holes'. *Opto-Electron. Rev.* 2006, vol. 14, pp. 233–241

[58] C. Genet, M. P. Van-Exter, J. P. Woerdman. 'Fano-type interpretation of red shifts and red tails in hole array transmission spectra'. *Opt. Commun.* 2003, vol. 225, pp. 331–336

[59] M. Sarrazin, J. P. Vigneson, J. M. Vigoureux. 'Role of Wood anomalies in optical properties of thin metallic films with a bidimensional array of sub-wavelength holes'. *Phys. Rev. B.* 2003, vol. 67, p. 085415

[60] D. L. Brundrett, E. N. Glytsis, T. K. Gaylord, J. M. Beudickson. 'Effects of modulation strength in guided-mode resonant subwavelength gratings at normal incidence'. *J. Opt. Soc. Am.* 2000, vol. 17(7), pp. 1221–1230

[61]    Y. Ding R. Magnusson. 'Resonant leaky-mode spectral-band engineering and device application'. *Opt. Express.* 2004, vol. 12(23), pp. 5661–5674

[62]    L. Martin-Moreno F. J. Garcia-Vidal. 'Optical transmission through circular hole arrays in optically thick metal films'. *Opt. Express.* 2004, vol. 12(16), pp. 3619–3628

[63]    Y. Ding R. Magnusson. 'Use of nondegenerated resonant leaky modes to fashion diverse optical spectra'. *Opt. Express.* 2004, pp. 1885–1891

[64]    A. Hessel. 'General characteristics of travelling-wave antennas' in *Antenna Theory Part 2*, R. E. Collin F. J. Zucker, (eds.), McGraw-Hill, New York, 1969, Chap. 19, pp. 151–258

[65]    S. Park, H. Kim, Y. K. Cho, J. H. Ko. 'Negative reflection and refraction and filter characteristics in the leaky wave-supportable gratings – TE polarization case'. *14th EUCAP European Conference on Antenna and Propagation*; 2020

[66]    S. S. Wang R. Magnusson. 'Theory and applications of guided-mode resonance filter'. *Appl. Opt.* 1993, vol. 32(14), pp. 2606–2613

[67]    M. C. M. Li A. A. Oliner. 'Scattering resonance on a fast-wave structure'. *IEEE Trans. Antennas Propagat.* 1965, vol. 13(6), pp. 948–959

# Index

www.ingramcontent.com/pod-product-compliance
Lightning Source LLC
Chambersburg PA
CBHW050512190326
41458CB00005B/1508